I0867131

Electronic Instrumentation for Distributed Generation and Power Processes

Electronic Instrumentation for Distributed Generation and Power Processes

Felix Alberto Farret, Marcelo Godoy Simões, and
Danilo Iglesias Brandão

CRC Press
Taylor & Francis Group
Boca Raton London New York

CRC Press is an imprint of the
Taylor & Francis Group, an **informa** business

MATLAB® is a trademark of The MathWorks, Inc. and is used with permission. The MathWorks does not warrant the accuracy of the text or exercises in this book. This book's use or discussion of MATLAB® software or related products does not constitute endorsement or sponsorship by The MathWorks of a particular pedagogical approach or particular use of the MATLAB® software.

CRC Press
Taylor & Francis Group
6000 Broken Sound Parkway NW, Suite 300
Boca Raton, FL 33487-2742

© 2018 by Taylor & Francis Group, LLC
CRC Press is an imprint of Taylor & Francis Group, an Informa business

No claim to original U.S. Government works

Printed on acid-free paper

International Standard Book Number-13: 978-1-138-74613-8 (Paperback)
International Standard Book Number-13: 978-1-4987-8241-8 (Hardback)

This book contains information obtained from authentic and highly regarded sources. Reasonable efforts have been made to publish reliable data and information, but the author and publisher cannot assume responsibility for the validity of all materials or the consequences of their use. The authors and publishers have attempted to trace the copyright holders of all material reproduced in this publication and apologize to copyright holders if permission to publish in this form has not been obtained. If any copyright material has not been acknowledged please write and let us know so we may rectify in any future reprint.

Except as permitted under U.S. Copyright Law, no part of this book may be reprinted, reproduced, transmitted, or utilized in any form by any electronic, mechanical, or other means, now known or hereafter invented, including photocopying, microfilming, and recording, or in any information storage or retrieval system, without written permission from the publishers.

For permission to photocopy or use material electronically from this work, please access www.copyright.com (http://www.copyright.com/) or contact the Copyright Clearance Center, Inc. (CCC), 222 Rosewood Drive, Danvers, MA 01923, 978-750-8400. CCC is a not-for-profit organization that provides licenses and registration for a variety of users. For organizations that have been granted a photocopy license by the CCC, a separate system of payment has been arranged.

Trademark Notice: Product or corporate names may be trademarks or registered trademarks, and are used only for identification and explanation without intent to infringe.

Visit the Taylor & Francis Web site at
http://www.taylorandfrancis.com

and the CRC Press Web site at
http://www.crcpress.com

Contents

Foreword xiii
Preface xv
Acknowledgments xvii
Authors xix

Chapter 1 Computer Interface and Instrumentation Electronics 1

1.1 Introduction 1
1.1.1 Operational Amplifiers 1
1.1.2 Gains of Differential Amplifiers 2
1.1.3 Calculation of A_d 4
1.1.4 Calculation of A_c 5
1.1.5 Calculation of Common-Mode Rejection Ratio (*CMRR*) 6
1.1.6 Current Mirror Circuit 6
1.1.7 Parameters of Differential Amplifiers 7
1.2 Processing of Analog Signals 9
1.2.1 Inverting Amplifier 9
1.2.2 Summing Amplifier for Inverting Variables 10
1.2.3 Integration Amplifier 11
1.2.4 Derivative Amplifier 12
1.2.5 Noninverting Amplifier 13
1.2.6 Buffer Amplifier 14
1.2.7 Level Detector or Comparator 14
1.2.8 Precision Diode, Precision Rectifier, and Absolute-Value Amplifier 15
1.2.9 High-Gain Amplifier with Low-Value Resistors 16
1.2.9.1 Example of a High-Gain Amplifier with Low-Value Resistors 17
1.3 The Error Equation 17
1.4 Large-Signal Amplifiers 20
Exercises 21
References 23

Chapter 2 Analog-Based Instrumentation Systems 25

2.1 Introduction 25
2.1.1 Voltage Feedback for Differential Amplifiers 25
2.1.2 Standard Differential Amplifier for Instrumentation 27
2.1.3 Programmable Gain Amplifier 29

2.2 Customized Instrumentation Amplifier 30
2.2.1 Isolated Instrumentation Amplifier 30
2.2.2 Magnetic Isolation 30
2.2.3 Optical Isolation 31
2.2.4 Capacitive Isolation 32
2.3 Voltage-to-Current Conversion (V-to-I) 34
2.3.1 Single-Input V-to-I Converter with Ungrounded Load 34
2.3.2 Single-Input V-to-I Converter Load with Grounded Load 34
2.3.3 V-to-I Converter with Differential Input and Ungrounded Load 36
2.3.4 V-to-I Converter with Differential Input and Grounded Load 37
2.4 Analog Signal Processing 38
2.4.1 Signal Linearization 38
2.4.2 Phase-Shift Amplifier 39
2.4.3 Schmitt Trigger Comparator 41
2.4.4 Oscillators for Signal Generators 42
2.4.4.1 Wien Oscillator 43
2.4.4.2 VCO Oscillator 45
2.4.4.3 Relaxation Oscillator 45
2.4.4.4 Crystal Oscillators 47
2.4.4.5 Types of Crystal Oscillators 48
2.4.5 Automatic Gain Control (AGC) 49
2.5 Signal Conditioning 49
2.5.1 Digital-to-Analog Conversion 51
2.5.2 Analog-to-Digital Conversion 53
2.6 Working With Circuit Boards for Signal Conditioning 57
Exercises 58
References 59

Chapter 3 Sensors and Transducers 61

3.1 Introduction 61
3.2 Passive Electric Sensors 62
3.2.1 Resistive Sensors 63
3.2.2 Resistive Sensors and Effects of Temperature on Measurements 63
3.2.3 Compensation of External Influences on Resistive Sensors 66
3.2.4 Capacitive Sensors 69
3.2.5 Inductive Sensors 72

3.3 Active Electric Sensors 75
3.3.1 Thermocouples 75
3.3.2 Piezoelectric Sensors 78
3.3.3 Photovoltaic Sensors 78
3.3.4 Hall Effect Device 79
3.3.5 Photomultipliers and Image Intensifiers 81
3.3.6 Geiger-Müller Tube and Proportional Counter 83
3.3.7 Probes for Biological and Chemical Voltages 84
3.3.8 Alcohol Sensors Used in Blood Measurements 86
3.3.9 Other Magnetic Sensors 87
3.4 Mechanical Sensors 87
3.4.1 Elastic Sensors 87
3.4.2 Pneumatic Sensors 87
3.4.3 Differential Pressure Sensors 88
3.4.4 Turbine Sensors 89
3.4.5 Rotation Sensors 89
3.4.6 Shaft Encoders (Rotation or Position) 90
Exercises 95
References 96

Chapter 4 Electronic Instruments for Electrical Engineering 97

4.1 Introduction 97
4.2 Instrument in Direct Current with Amplifiers 97
4.2.1 Standard DC Electronic Voltmeter 97
4.2.2 Electronic Voltmeter-Ammeter 98
4.3 Common Circuits Used in Instrumentation 99
4.3.1 Operating Principle of the Chopper Amplifier 99
4.3.2 Alternating Current Voltmeters with Rectifier 100
4.3.3 Voltmeters of True Effective Value ($V_{true-rms}$) 102
4.4 Electronic Multimeters 103
4.4.1 General Characteristics of the Analog Voltmeters 103
4.5 General Classification of DVMs 105
4.5.1 Standard Single-Slope or Ramp Voltmeter 105
4.5.2 Dual-Slope Voltmeter 106
4.5.3 Continuous Balance Voltmeter 108
4.5.4 Voltmeter of Successive Approximations 109
4.5.5 General Characteristics of a DVM 110
4.6 Ohmmeters and Megometers 110
4.7 Digital Network Analyzers 113
4.7.1 Analog-to-Digital Network Analyzers 113
4.8 Power Sources and Adapters 115
Exercises 116
References 117
Bibliography 117

Chapter 5 Signal Simulators and Emulators ... 119

5.1 Introduction ... 119
5.1.1 The Attenuator ... 119
5.1.2 Types of Attenuators ... 120
5.2 Waveforms for Electronic Instruments ... 122
5.3 Signal Generators and Simulators with Frequency Synthesizer ... 123
5.3.1 Indirect Method of Frequency Synthesis ... 123
5.3.2 Phase-Locked Loop ... 124
5.3.3 Direct Method of Frequency Synthesis ... 127
5.4 Signal Generators by Frequency Division ... 128
5.5 Signal Generator with Modulation ... 129
5.6 Frequency Sweeping Generator ... 129
5.7 Pulse Generators and Rectangular Waves ... 131
5.8 Function Generators, Simulators, and Audio Generators ... 131
Exercises ... 131
References ... 132

Chapter 6 Advanced Harmonic Analysis for Power Systems ... 135

6.1 Introduction ... 135
6.2 Harmonic Analysis ... 136
6.2.1 Effects of Harmonic Distortions on Network Parameters ... 139
6.2.1.1 Instantaneous Active Power, p_a ... 142
6.2.1.2 Instantaneous Nonactive Power, p_{na} ... 143
6.2.1.3 Active or Average Power, P ... 143
6.2.2 Effects of Harmonic Distortion on the Power Factor ... 144
6.3 Harmonic Distortion Analyzers ... 147
6.3.1 Harmonic Analyzer with a Tuned Filter ... 147
6.3.2 Heterodyne Harmonic Analyzer or Wave Distortion Meter ... 148
6.3.3 Determination of THD ... 149
6.4 Spectrum Analyzers ... 149
6.4.1 Spectrum Analyzers by Frequency Scanning ... 150
6.4.2 Mathematical Analysis of Power Signals ... 153
Exercises ... 155
References ... 156
Bibliography ... 156

Chapter 7 Instrumentation and Monitoring for Distributed Generation Systems ... 157

7.1 Introduction ... 157
7.2 General Control Scheme for Distributed Generation ... 159

7.2.1 Current-Based DG 160
7.2.2 Voltage-Based DG 160
7.3 Signal Reference Generator 161
7.3.1 Fundamental Frequency Terms 161
7.3.1.1 Generator of the In-Phase Reference Term 163
7.3.1.2 Quadrature Reference Term Generator 163
7.3.2 Nonfundamental Frequency Terms 164
7.4 Power Quality Standards Applied to DG 166
7.4.1 Harmonic Circulation 166
7.4.2 Power Factor and THD Concerns in DG 167
7.5 Distributed Generator Based on Instrumentation: Case Studies 169
7.5.1 Instrumentation of DG-Based PV 169
7.5.2 Instrumentation of DG-Based Smart Inverter: Voltage-Based DG 169
7.5.3 Instrumentation of DG-Based Automonitoring Inverter: Current-Based DG 170
Exercises 174
References 177

Chapter 8 Fuzzy Logic and Neural Networks for Distributed Generation Instrumentation 179

8.1 Introduction 179
8.1.1 Fuzzy Logic Systems 180
8.1.2 Neural Network Systems 181
8.2 Applications of Artificial Neural Network in Industrial Systems, Energy Conversion, and Power Systems 181
8.2.1 Applications of Fuzzy Logic and Neural Networks in Distributed Generation Systems 182
8.3 Fuzzy Logic and Neural Network Controller Design 183
8.4 Fuzzy Logic and Neural Network Function Optimization 189
8.5 Fuzzy Logic and Neural Network Function Approximation 197
References 200
Bibliography 200

Chapter 9 Instruments for Data Acquisition 203

9.1 Data Acquisition by Computers 203
9.1.1 Fundamentals of Digital Signal Processing 203
9.1.2 The Data Acquisition Board 206
9.1.3 Microcontrollers 207
9.1.4 Example of Data Acquisition 208
9.2 Signal Processors for Instrumentation 210
9.2.1 Digital Signal Processors 210
9.2.2 Parallel Processing 211

9.2.3 Transputers 212
9.2.4 Design of Instruments with ASIC 213
9.2.5 FPGA (Field-Programmable Gate Arrays) in Instrumentation 213
9.2.6 PLDs (Programmable Logic Devices) in Instrumentation 214
9.3 Computer-Based Systems Instrumentation 214
9.3.1 Essential Components 214
9.3.2 Tests Operated by Computer 214
9.3.3 The Parallel Interface IEEE488 215
9.3.4 Example 217
9.3.4.1 Frequency Counter (Hardware) 217
9.3.5 Input and Output Drivers as Open Collectors (Transceiver) 217
9.3.6 Output Drivers 218
9.3.7 Serial Interface IEEE485 219
9.3.8 USB Connectors 221
9.3.8.1 USB Hub 222
9.3.8.2 USB Process 223
9.3.8.3 USB Features 224
9.3.9 Signals of Data Transmission 225
9.3.10 RS485 Network Biasing 226
9.3.11 Calculation of Biasing Resistors 227
9.4 Digital Control 227
9.4.1 Instrumentation Controlled by Computer 228
9.4.2 Programmable Logic Controller 228
Exercises 233
References 234

Chapter 10 Software for Electric Power Instrumentation 235

10.1 Introduction 235
10.2 LabVIEW Development System 235
10.2.1 Programming with LabVIEW 238
10.2.2 Virtual Instrument for Power Quality Analysis 239
10.2.3 Electric Power Instrumentation for Distributed Generation 242
10.3 Arduino Development System 247
10.4 Mathworks MATLAB®/Simulink® Development System 248
10.4.1 Interface with LabVIEW 249
10.4.2 Interface with Arduino 250
10.4.2.1 MATLAB® Support Package for Arduino 251
10.4.2.2 Simulink® Support Package for Arduino 251

10.4.3 Interface with PSIM ... 252
10.4.4 Programming DSP through MATLAB®/Simulink® ... 253
Exercises ... 255
References ... 256

Chapter 11 Introduction to Smart Grid Systems ... 257

Luciane N. Canha, Alzenira R. Abaide, and Daniel P. Bernardon

11.1 Introduction ... 257
11.2 Distribution System Automation ... 259
11.3 Advanced Metering Infrastructure ... 260
11.4 Smart Home ... 262
11.5 Information and Communication Technologies ... 262
11.5.1 Standard 61850 ... 263
11.5.2 Standard 62351 ... 263
11.5.3 Attacks on SCADA Systems ... 264
11.6 Electric Vehicles (EV) in Smart Grid Systems ... 265
11.7 A Smart Operation Example ... 266
11.7.1 Methodology for Automatic Restoration of a Power Supply ... 266
11.7.2 Objective Functions and Constraints ... 268
11.7.2.1 Objective Functions ... 268
11.7.2.2 Constraints ... 268
11.7.3 Multicriteria Decision-Making Method ... 269
11.7.4 Application of AHP Method ... 271
References ... 273

Index ... 277

Foreword

Instrumentation is considered an art of science and technology approaching the measurement and control of process variables within a real-world plant, a manufacturing area, a laboratory, or maybe a very diverse and sparse array of sensors for all kinds of environments, such as Earth, atmosphere, ocean, forests, space, and so on. Any instrument measuring a physical quantity, such as flow, temperature, level, distance, angle, pressure, or electrical parameter, could be as simple as a direct reading gauge, a little more complex for multivariable process analyzers, or even totally spread in a large geographical area with networking capabilities. The devices, or instruments, can be part of a control system in refineries, factories, vehicles, or power systems. Traditional mechanical devices, such as the ones used in boilers, electrical generators, transformers, and transmission lines, have been replaced by advanced electronic and digital devices. The new era of renewable energy and smart grid systems, with pervasive communication protocols, based on many technologies such as Bluetooth, Wi-Fi, and Internet of Things, has made necessary an overall discussion of technology, computer science, computer engineering, networking, and real-time and software-based designs that can encompass the new age of modern automation and control system, composed of computers, electronic instrumentation, communication, and control loops, with interacting smart grids and digital circuitry. This book is a bridge between industrial processes represented by the distributed generation and the analog/digital paraphernalia commonly used in modern power systems. The required sensing variables will need accuracy, fast response, and adequate industrial process control decision-making.

This book reviews electronic instrumentation for modern distributed generation systems. The authors discuss the fundamentals of electronic instruments, analog amplifiers for instrumentation, digital and analog sensors, basis of the electronic instruments for electromechanical measurements, signal simulators, power systems signal analysis, and data acquisition systems. They also describe how to apply artificial intelligence techniques in instrumentation for distributed generation and how smart grid can be enhanced with such contemporary instrumentation paradigm.

I recommend this book for readers who want to understand how modern photovoltaic panels, wind power, fuel cells, energy storage, and electrical machines can be monitored, managed, and controlled with digital electronic instrumentation. The chapters contain content related to design, analysis, and circuit implementation for electronic instrumentation. The field of distributed power generation is very important for making possible renewable energy systems to be integrated into the modern utility power grid. The discussion includes intelligent alarms, smart loads, system design, and integration. The book can be recommended for senior undergraduate or first-year graduate courses. Also, industry-oriented professionals and practicing engineers will benefit from reading this book.

Dr. Bimal K. Bose
University of Tennessee
Knoxville, Tennessee

Preface

Instrumentation is a very broad field. Modern electrical, mechanical and industrial engineering are totally dependent on variables measured by sensors, processing units and actuators based on *instrumentation* circuitry. To adapt and migrate from the traditional mechanical devices used in electrical power systems, such as those used in boilers, electrical generators, transformers, and transmission lines, and replace them with electronic and digital devices, an in-depth revision of the basics of instrumentation is necessary. With this book, the readers will understand the physical fundamentals guiding the electrical and mechanical devices that form the modern automation and control system, which is widely composed of computers, electronic instrumentation, communication, control loops, smart grids, and digital circuitry.

This book covers a link between the industrial process represented by the distributed generation and the analog/digital paraphernalia commonly used in modern power systems. This link deals with the collection and conditioning of electrical and mechanical signals so as to introduce them to computer systems to find an advanced and fast response to the question, and then give a solution back to the industrial process. Therefore, the electronic instrumentation in this book will be mostly related to solving problems in electrical engineering of small power generation plants, energy processing, and electronic drivers.

This book presents approaches to electronic instrumentation for modern distributed generation and covers the fundamentals of electronic instruments, analog amplifiers for instrumentation, digital and analog sensors, basics of the electronic instruments for electromechanical measurements, signal simulators, power systems signal analysis, data acquisition systems, use of instrumentation software, and artificial intelligence in instrumentation for distributed generation. Artificial intelligence has been recently regarded by Stephen Hawking, Bill Gates, and Elon Musk as so important that proper development and ethical design must be considered. Fuzzy logic and neural networks and how they can be used in practice for electronic instrumentation of distributed generation and power systems are discussed in this book. The authors have not come across any other books dealing with intelligent instruments.

Therefore, it is hoped that this book will bring extensive knowledge on power systems, electronic devices, modern photovoltaic panels, wind power, fuel cells, energy storage, and electrical machine needs regarding electronic instrumentation. It covers the basics of how to deal with signals from most power generation engineering and convert them into digital computer signals useful in the processing, automation, and control fields.

The book provides basic and advanced knowledge on design, analysis, and circuit implementation for electronic instrumentation; how to get the best out of the analog, digital, and computer circuitry design steps; and how the new trends are more and more related to computer processing, digital circuitry, and control. The scope of this book is to approach electronic instrumentation for interfacing power systems and computers with analog and digital circuits. Our intention is to discuss electronic instrumentation for low and medium electrical power applications, mostly related to photovoltaic, wind turbines, and small electrical power plants in general, because these matters are very important for distributed power generation applications using analog and digital electronic instrumentation. We cover intelligent alarm and smart loads and systems plus distributed generation and smart grids. Also, we intended to present the reader with user-friendly technical content that can be useful in advanced undergraduate courses as well as graduate ones. In addition, we hope the book will be a good reference for practicing engineers who want to understand the foundations of this important topic in electrical engineering.

Felix Alberto Farret
Santa Maria, Brazil

Marcelo Godoy Simões
Denver, Colorado

Danilo Iglesias Brandão
Belo Horizonte, Brazil

MATLAB® and Simulink® are registered trademarks of The MathWorks, Inc. For product information, please contact:

The MathWorks, Inc.
3 Apple Hill Drive
Natick, MA, 01760-2098 USA
Tel: 508-647-7000
Fax: 508-647-7001
E-mail: info@mathworks.com
Web: www.mathworks.com

Acknowledgments

We are very thankful to many colleagues, professionals, and students, through suggestions and, sometimes, extra duties. The manuscript took many years to mature and be written down. During this period, we interacted with so many people who understood the importance of this work and firmly supported us. Special thanks to Adriano Longo, Antônio Ricciotti, Carlos De Nardin, Ciro Egoavil, Diogo Franchi, Emanuel Vieira, Felipe Fernandes, Frank Gonzatti, Fredi Ferrigolo, Lucas Ramos, Luciano de Lima, Luis Manga, Maicon Miotto, Marcio Mansilha, and Vinicius Kuhn, to whom we offer our utmost recognition and respect. We also thank the staff at CRC Press for their full dedication to this project; they encouraged us and filled us with enthusiasm, in particular Nora Konopka and Kyra Lindholm for their special support in moments we really needed help. Thanks to our families and friends, who were deprived of our company and attention for many months; they never complained but rather gave us their help and understanding. Finally, we express our gratitude to our schools: the Federal University of Santa Maria, the Colorado School of Mines, and the Federal University of Minas Gerais. Their institutional support encouraged cooperation in a very creative environment.

Authors

Felix Alberto Farret received his BSc from the Federal University of Santa Maria, Brazil, in 1976; specialized in electronic instrumentation from Osaka Prefectural Industrial Research Institute, Japan, in 1975; and received his MSc from the University of Manchester, England, in 1981 and PhD in electrical engineering from the University of London, England, in 1984. He was a visiting professor at the Colorado School of Mines in the Division of Engineering, Golden, Colorado, in 2002–2003. He published several international papers and books on renewable sources of energy and power electronics and successfully completed different projects sponsored by different government and nongovernment bodies related to these subjects.

Marcelo Godoy Simões is the director of the Center for Advanced Control of Energy and Power Systems (ACEPS) at the Colorado School of Mines, Golden, Colorado. He was a U.S. Fulbright Fellow at Aalborg University, Institute of Energy Technology (Denmark). He is an IEEE Fellow, with the citation "for applications of artificial intelligence in control of power electronics systems." Dr. Simões was a pioneer in applying neural networks and fuzzy logic in power electronics, motor drives, and renewable energy systems. He is the coauthor of the book *Integration of Alternative Sources of Energy* (Wiley Blackwell, 2006, now in the second edition).

Danilo Iglesias Brandão received his PhD in electrical engineering from the University of Campinas, Brazil, in 2015. He was a visiting researcher at the Colorado School of Mines, Golden, Colorado, in 2009 and 2013, and he was a visiting PhD student at the University of Padua, Italy, in 2014. He is currently a professor at the Department of Electrical Engineering of the Federal University of Minas Gerais, Brazil. His main research interests include active power filter, power quality, distributed compensation strategies, and microgrids.

1 Computer Interface and Instrumentation Electronics

1.1 INTRODUCTION

This chapter is based on the assumption that the reader is familiar with basic electronics and understanding of analog circuit–based amplifiers. However, even if this is the first time for anyone with regards to this topic, the authors believe that the following sections show how electronics support and are key in the signal conditioning instrumentation design.

In order to start, we can note that electronic amplifiers are used mostly to increase or change the electrical properties of measured signals (voltage, current, impedance, or power). Such signals would hardly be useful to drive functions such as displays, signal processing blocks, or electric drives, and they must be processed by some kind of controller (analog or digital). Specific features such as differential input, feedback, and cancellation of common errors in signal acquisition will make such electronic amplifiers a more suitable device in interfacing electronic circuits or processes, monitored by microcontrollers, or microprocessors or computers. This chapter shows that the differential amplifier is an important system for distortion cancellation of electromagnetic common-mode noises induced in conductors and instrument signal cables, typical of industrial environments [1–3].

In order to better understand the nonideal behavior of an instrumentation amplifier, this chapter describes the most relevant aspects of basic electronics in applications that would affect from their operational qualities.

1.1.1 Operational Amplifiers

An operational amplifier is typically available and ready for use, but it is important to understand its constraints and limitations in the operation. The current technology of operational amplifiers makes simpler the design trade-offs such as gain variations by temperature and humidity, avoiding considerations of aging of components, radiation effects, and differences between the discrete components that comprise the system.

The standard operational amplifier typically has a parallel voltage feedback. As such, it can be used to perform almost instantaneously various mathematical functions, such as signal inversion, scaling, phase shift, and general operations of signal integration and differentiation. In addition, it can meet the most diverse ranges of applications, such as of high-frequency, high-gain, and low-noise amplification of signals.

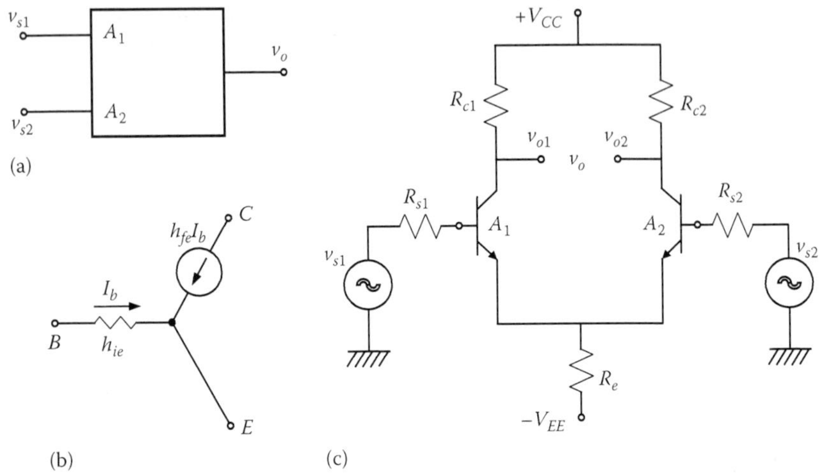

FIGURE 1.1 Differential amplifier: (a) differential signal block, (b) hybrid model of an NPN transistor, and (c) basic circuit of a differential discrete amplifier.

Most commercial operational amplifiers can be modeled as differential amplifiers, consisting of two inputs and one output, as depicted in Figure 1.1a. One input port is considered the signal-inverting input and the other the noninverting one. These inputs allow the designer to offset common signals in those conductors connecting the amplifier to the external environment, which are often affected by electromagnetic noises (due to radiation effects) and ripples of power sources (due to conduction effects) [4–6].

1.1.2 Gains of Differential Amplifiers

Figure 1.1a shows a signal block that relates the input and output signals for a differential amplifier. One can assume that the signal sources may have identical resistances. Such approximation is very realistic because the current manufacturing techniques of integrated circuits achieve matched parameters for transistors and resistors. Figure 1.1b and c shows the equivalent models required to understand such a basic differential amplifier. The fundamental relationships are given by Equations 1.1 through 1.3:

$$v_d = v_{s1} - v_{s2} \tag{1.1}$$

$$v_c = \frac{1}{2}\left(v_{s1} + v_{s2}\right) \tag{1.2}$$

$$v_o = v_{o1} - v_{o2} = A_1 v_{s1} + A_2 v_{s2} \tag{1.3}$$

In the equations above, the subindices "d," "c," and "o" refer, respectively, to the differential, common, and output signals and A_1 and A_2 are the respective voltage gains of each transistor.

A common signal is when it is equally received by inputs 1 and 2 of the differential amplifier. The differential signals are input at the amplifier and their difference (with respect to a ground point or to some other common point) must be amplified, at the same time rejecting the common signal. Therefore, the signals at each amplifier input in Figure 1.1c can be given as a function of the common and differential signals. Adding Equations 1.1 and 1.2 and rearranging them yield:

$$v_{s1} = v_c + \frac{1}{2} v_d \tag{1.4}$$

Taking the difference between Equations 1.1 and 1.2 results in:

$$v_{s2} = v_c - \frac{1}{2} v_d \tag{1.5}$$

Replacing Equations 1.4 and 1.5 in Equation 1.3 and rearranging, we arrive at:

$$v_o = \left(A_1 + A_2\right) v_c + \frac{1}{2}\left(A_1 - A_2\right) v_d \tag{1.6}$$

or, in a more compact form:

$$v_o = A_c v_c + A_d v_d \tag{1.7}$$

where

$A_c = A_1 + A_2$ is the common gain

$A_d = \frac{1}{2}\left(A_1 - A_2\right)$ is the differential gain

If Equation 1.7 is written by explicitly showing the differential values, the following results:

$$v_o = A_d \left(v_d + \frac{A_c}{A_d} v_c \right) \tag{1.8}$$

where the relationship A_d/A_c is defined as the common-mode rejection ratio (*CMRR* or ρ).

In the ideal case, the common-mode rejection ratio should ideally be infinite. For this condition, the entire received common-mode signal by the two inputs would be eliminated. In practical cases, a $CMRR = 1000$ is sufficient for most applications [7,8].

The following analysis assumes the hybrid common-emitter model of a transistor amplifier, as shown in Figure 1.1b, where definitions for such equivalent model are as follows: h_{fe} for the forward alternating current (AC) transfer ratio between collector and base and h_{ie} as the AC input resistance of the transistor base.

1.1.3 Calculation of A_d

It is possible to calculate A_d from its definition in Equation 1.7, making $v_c = 0$, i.e., the condition:

$$A_d = \left.\frac{v_o}{v_d}\right|_{v_c=0} \tag{1.9}$$

Furthermore, considering $v_c = 0$ in Equation 1.2, $v_{s1} = -v_{s2}$, and from Equation 1.1, it follows that $v_{s1} = v_d/2$. Therefore, as the differential current through the common-emitter resistor becomes null, there is no voltage drop across the emitter resistor, which makes it negligible in the equivalent differential input circuit, as shown in Figure 1.2a. By using the superposition theorem, both AC and DC (direct current) signals in linear circuits can be analyzed independently for the conditions of forward current and alternating current. In these circumstances, when AC signals are analyzed, the forward current sources are considered in short-circuit and vice versa. It can then help to reconstruct Figure 1.2a in its simpler form, as shown in Figure 1.2b [9–12].

The equivalent circuit simplifies the analysis of the differential input gain defined by Equation 1.9, giving the following expressions:

$$v_o = -h_{fe} I_b R_c \quad \text{and} \quad \frac{v_d}{2} = I_b\left(R_s + h_{ie}\right)$$

Replacing these expressions in Equation 1.9 results in:

$$A_d = \left.\frac{v_o}{v_d}\right|_{v_c=0} = -\frac{h_{fe} R_c}{2\left(R_s + h_{ie}\right)} \tag{1.10}$$

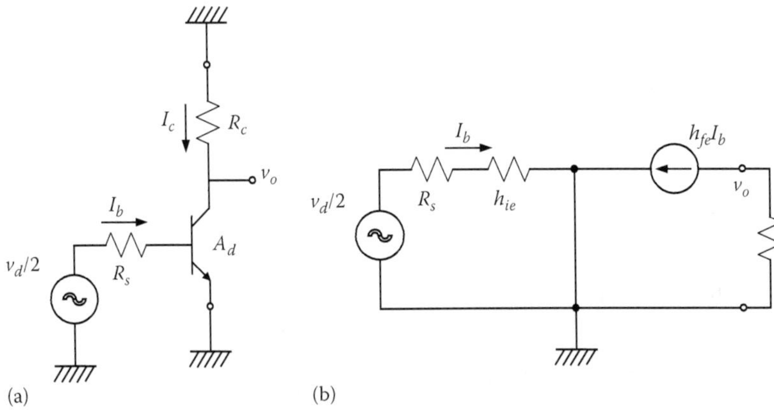

FIGURE 1.2 Equivalent circuits of the differential input: (a) equivalent differential input and (b) equivalent circuit for small signals.

1.1.4 Calculation of A_c

With a similar reasoning for the common input gain, A_c can be evaluated from its definition in Equation 1.7, but now making $v_d=0$, i.e., the conditions:

$$A_c = \left.\frac{v_o}{v_c}\right|_{v_d=0} \tag{1.11}$$

Considering $v_d=0$ in Equation 1.1, $v_{s1}=v_{s2}$, and from Equation 1.2, it follows that $v_{s1}=v_c$. So the common current through the emitter resistor is twice that of the emitter of each transistor, and the voltage drop across this resistor is $2I_cR_e$, as represented in Figure 1.3a. Based on these considerations, Figure 1.3a can be reconstructed in its simpler form, as shown in Figure 1.3b.

The equivalent circuit of Figure 1.3 simplifies the analysis of the common input gain given by Equation 1.11, using the following expressions:

$$v_o = -h_{fe}I_bR_c \quad \text{and} \quad v_c = I_b\left(R_s + h_{ie}\right) + \left(1 + h_{fe}\right)I_b\left(2R_e\right)$$

Replacing these expressions in Equation 1.11 results in:

$$A_c = \frac{-h_{fe}R_c}{\left(R_s + h_{ie}\right) + \left(1 + h_{fe}\right)\left(2R_e\right)} \tag{1.12}$$

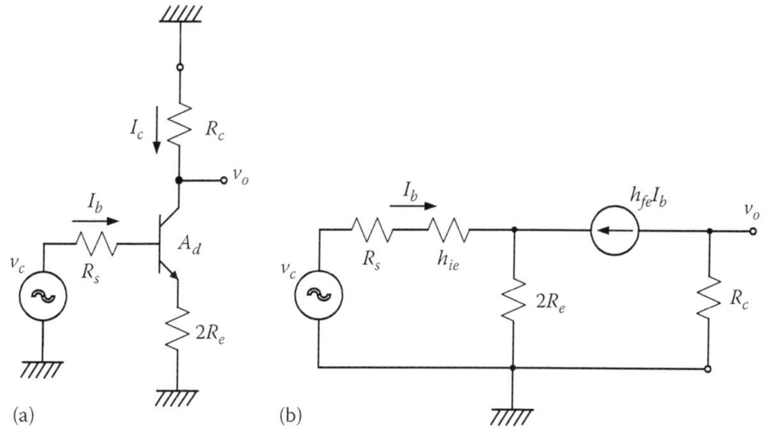

FIGURE 1.3 Equivalent circuits of the differential amplifier: (a) equivalent common input of the differential amplifier and (b) equivalent circuit for small signals.

1.1.5 Calculation of Common-Mode Rejection Ratio (*CMRR*)

From Equations 1.10 and 1.12, the common-mode rejection ratio can be obtained as:

$$CMRR = \frac{A_d}{A_c} = \frac{R_s + h_{ie} + 2R_e\left(1+h_{fe}\right)}{2\left(R_s + h_{ie}\right)} = \frac{1}{2} + \frac{R_e\left(1+h_{fe}\right)}{R_s + h_{ie}} \tag{1.13}$$

Equation 1.13 leads to an important conclusion about the desired value for the common-emitter resistance R_e. Given the (h_{fe}, h_{ie}) for the transistor and the (V_s, R_s) for the sensor, the *CMRR* must have a value as high as possible [13–16]; an ideal infinite R_e. However, in differential amplifier applications, it is common to replace R_e by a circuit with an ideal infinite impedance of the current source $(h_{fe}i_b)$ of a transistor collector, or by an infinite current mirror. Figure 1.4 shows the emitter resistance replaced by a transistor to increase its effective value.

1.1.6 Current Mirror Circuit

There are many possible versions of the current mirror circuit. Some of them are portrayed in Figure 1.5. It is very common to use bipolar transistors in order to take advantage of a constant base-emitter voltage. One can easily note that the current mirror has a behavior that depends only on the mirrored current parameters, such as:

$$V_{CC} - I_x R = V_{BE} \tag{1.14}$$

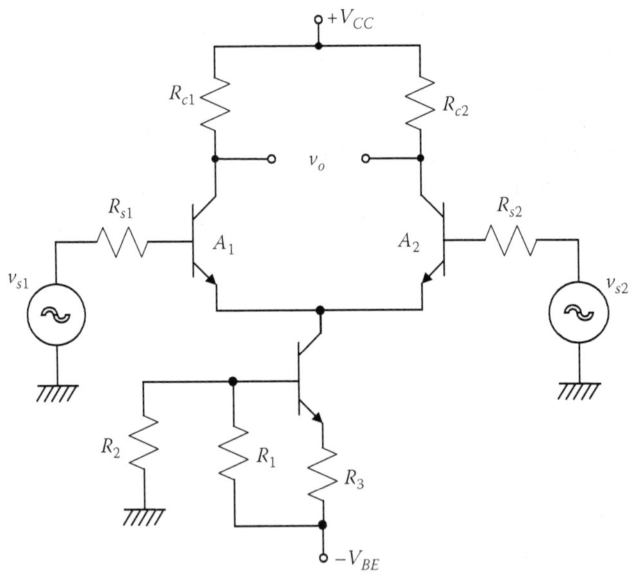

FIGURE 1.4 Emitter resistor replaced by a transistor.

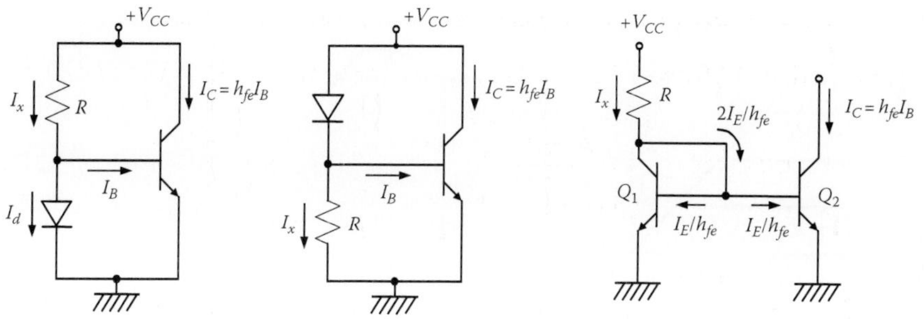

FIGURE 1.5 Examples of current mirror circuits.

where V_{BE} is the base-emitter voltage drop, or yet:

$$I_x = I_d + I_B = \frac{V_{CC} - V_{BE}}{R} \tag{1.15}$$

1.1.7 Parameters of Differential Amplifiers

A representation of a differential amplifier is detailed in Figure 1.6. The input resistance is represented by R_i, the output resistance R_o, open-loop gain A, and the gain-bandwidth product (GBP).

Unlike the ideal amplifier, a realistic operational amplifier (amp-op) circuit may not have either infinite gain nor infinite passband gain. For example, looking at the open-loop gain, in Figure 1.7, it can be noted that this gain depends on the amp-op frequency, in this case the old device 741. It can be observed that at very low frequencies, the open-loop gain is initially constant, but begins to roll at a rate of −6 dB/octave or −20 dB/decade (an eighth is the folding frequency and a decade is its multiplication by 10). This reduction in gain continues until the linear gain becomes unity, or 0 dB, and the frequency for this unity gain is called *unity gain frequency*.

The basic parameters of an ideal amplifier and two practical examples of actual amplifiers are shown in Table 1.1 for both inverting and noninverting signal operations. Usually, the first factor to be considered for an amp-op is its GBP.

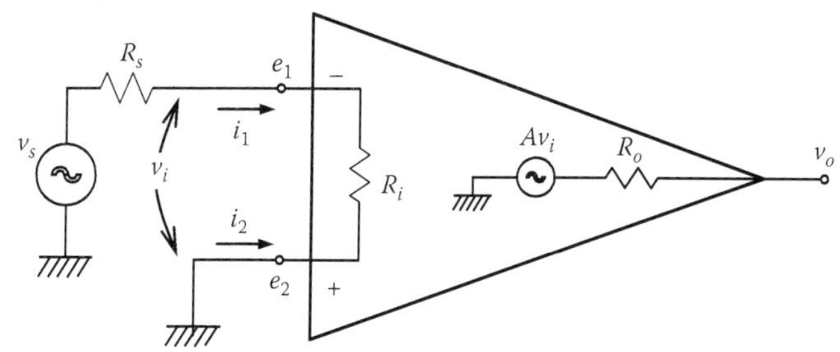

FIGURE 1.6 Simplified representation of a differential amplifier.

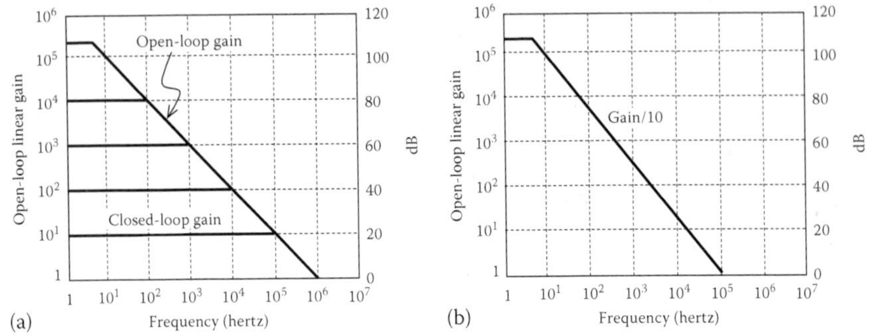

FIGURE 1.7 Gain-bandwidth product (GBP) of amp-op 741: (a) band pass ($GBP = 1\ MHz$) and (b) practical gain of 1/10.

TABLE 1.1
Ideal and Real Parameters of Operational Amplifiers

	Parameter				
Amplifier	**Input Impedance (Z_i)**	**Output Impedance (Z_o)**	**Gain (G)**	**GBP**	**Slew Rate**
Ideal	∞	0	∞	∞	∞
741C-BJT	2 MΩ	75 Ω	200 kV/V	1 MHz	High
3140-BI-MOS	1.5 TΩ	60 Ω	100 kV/V	4.5 MHz	High

In Figure 1.7a, the product of the open-loop gain A_{ol} and the frequency is constant at any point of the curve such that $GBP = A_{ol} \cdot BW$. Graphically, the bandwidth is the point at which the closed curve intersects the gain of the open gain curve for a family of closed-loop gains. In a practical situation, the amp-op operates correctly without the distortion amplifier when the designer considers a closed-loop gain of about 1/10 to 1/20 of the open-loop gain at a given frequency. As an example, using the gain response of Figure 1.7a, the closed-loop gains at 10 kHz should be around 100/10 = 10 or 100/20 = 5 since the open-loop gain is 100, which corresponds to $20 \log 100 = 40\,\text{dB}$.

An additional parameter of amp-ops is the transient response, which can be given by the rise time for the output signal switching from 10% to 90% of its rated value when a step function pulse is applied. This measurement must be conducted with the input signal with a specified closed-loop condition. In electronic circuit theory, the rise time t_r is related to the bandwidth (BW) of the amp-op through the following relationship:

$$t_r = \frac{0.35}{BW}(\text{slew rate}) \tag{1.16}$$

1.2 PROCESSING OF ANALOG SIGNALS

Operational amplifiers may be used to process either analog signals or digital signals. A comparison between such signal processing techniques can be made in view of data processing features: speed, precision, and flexibility, as shown in Table 1.2.

Many other criteria could be taken into consideration, such as cost, size, commercial availability, efficiency, noise generation, susceptibility to noise, integration means, data transmission, and changes in technology, among others. Once the criteria is selected, we must set up the required processing configuration, such as defining the circuits for proper amplification, summing of variables, inversion, integration, differentiation, multiplication by a constant, multiplication of two signals, variable phase control, and signal generation. The amplifiers can be inverters or noninverters, depending on whether the input signal is applied, i.e., by inverting or noninverting the input port with respect to the output signal.

1.2.1 Inverting Amplifier

An inverting amplifier will need a closed-loop gain defined by external components (two resistors) if the internal gain of the amp-op is very high. The determination of the closed loop can be made as follows: an amplifier circuit as shown in Figure 1.8 can have input impedance supposedly infinite. Therefore, the input current should be ideally zero, and then $i = i_f$. As can be seen from the figure, a virtual short circuit between the input terminals can be considered, and then the amplifier gain becomes:

$$i = \frac{v_i}{Z_1} = i_f = \frac{0 - v_o}{Z_2}$$

TABLE 1.2
Selection Criteria for Digital and Analog Processing

	Criterion		
Processing	**Speed**	**Precision**	**Flexibility**
Analog	Quasi-instantaneous	Very limited	Stiff
Digital	Fast	Expandable	Very flexible

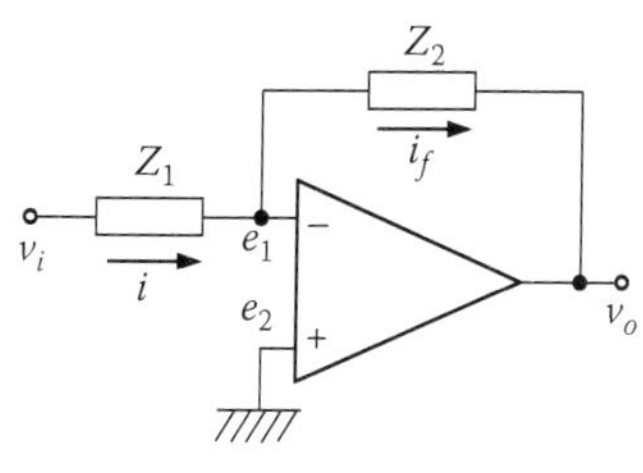

FIGURE 1.8 Feedback inverter amplifier.

i.e., the general feedback gain of the inverting amplifier is given by:

$$A_f = \frac{v_o}{v_i} = -\frac{Z_2}{Z_1} \tag{1.17}$$

The feedback amplifier of Figure 1.8 can also be considered as a constant multiplier or signal inverter, as in Equation 1.18:

$$v_o = \left(-\frac{Z_2}{Z_1}\right) v_i = k v_i \tag{1.18}$$

If $Z_1 = Z_2$, the amplifier is a signal inverter.

For all negative feedback loops around the amp-op circuits, it is possible to make this kind of analysis, where the current inside the amp-op is zero at the same time that the noninverting and inverting terminals keep a virtual zero voltage across them.

1.2.2 Summing Amplifier for Inverting Variables

As an extension of what is explained for the signal amplifier (Section 1.2.1), one can make a summing amplifier with "n" input currents (Figure 1.9), namely:

$$i_f = \sum_{k=1}^{n} i_k = \sum_{k=1}^{n} \left(\frac{v_k}{Z_k}\right) = \frac{0 - v_o}{Z_f} \tag{1.19}$$

or yet:

$$v_o = -Z_f \sum_{k=1}^{n} \left(\frac{v_k}{Z_k}\right) = -\sum_{k=1}^{n} \left(\frac{Z_f}{Z_k} v_k\right) \tag{1.20}$$

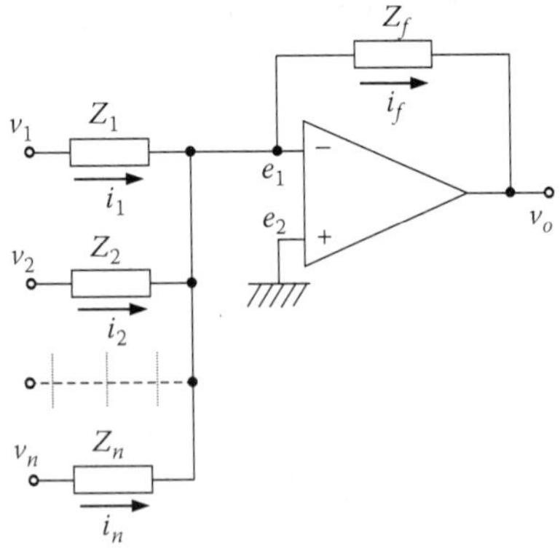

FIGURE 1.9 Summing amplifier.

This circuit allows the weighted sum of input signals, where each input k is weighted by a factor $-Z_f/Z$. If all impedances of the signal sources are equal, then:

$$v_o = -\frac{Z_f}{Z}\left(v_1 + v_2 + \cdots + v_n\right) \tag{1.21}$$

1.2.3 Integration Amplifier

The feedback amplifier operates as an integrator if the feedback impedance is a capacitor and the input impedance is a resistor, as shown in Figure 1.10. In this case, using Equation 1.17, the amplifier gain will be given by:

$$A_d = -\frac{Z_2}{Z_1} = -\frac{1/j\omega C}{R} \tag{1.22}$$

which can be placed in the form of a Laplace variable, as follows (i.e., $j\omega = s$):

$$A_d\left(s\right) = -\frac{1}{RC}\frac{1}{s} \tag{1.23}$$

Another mathematical way to interpret this is by considering the instantaneous values of current:

$$i = \frac{v_i}{R} = i_f = -C\frac{dv_o}{dt} \tag{1.24}$$

i.e.:

$$v_o = -\frac{1}{RC}\int v_i dt \tag{1.25}$$

where $1/RC$ is the integration constant.

The initial integration values can be added as the initial charge of the feedback integration capacitor. In practical circuits, it is sometimes necessary to have a way

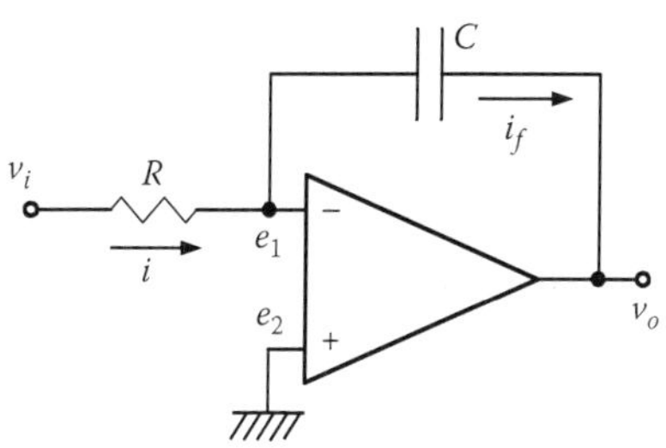

FIGURE 1.10 Integration amplifier.

to reset the capacitor, with a J-FET (junction field-effect transistor) transistor to discharge the capacitor voltage.

1.2.4 Derivative Amplifier

Analogous to the same concept of integration amplifier, one can devise an equation for the derivative amplifier (Figure 1.11) as follows.

Once again, using the general equation of the amp-op gain, we derive:

$$A_d = -\frac{Z_2}{Z_1} = -\frac{R}{1/j\omega C} \tag{1.26}$$

rewritten in the form of a Laplace variable, it becomes:

$$A_d\left(s\right) = -RCs \tag{1.27}$$

or even using the instantaneous values:

$$i_f = \frac{0 - v_o}{R} = i = -C\frac{dv_i}{dt} \tag{1.28}$$

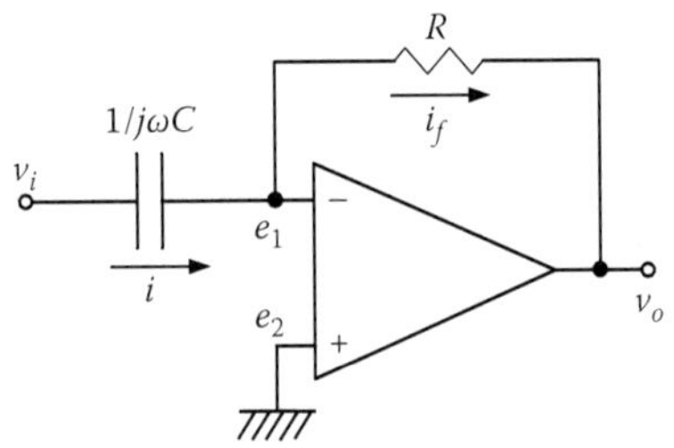

FIGURE 1.11 Derivative amplifier.

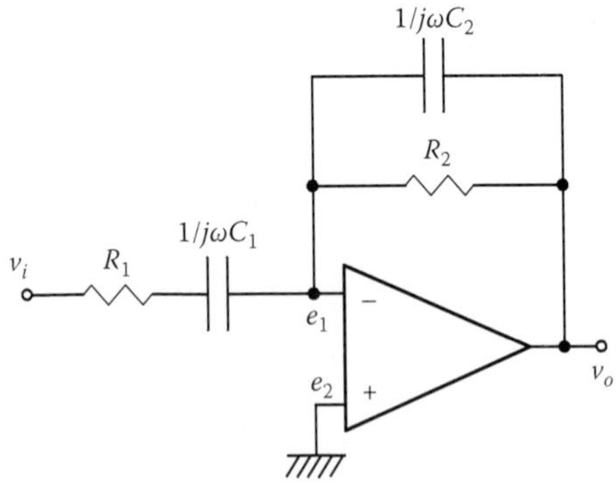

FIGURE 1.12 Modified derivative amplifier.

or:

$$v_o = -RC\frac{dv_i}{dt} \tag{1.29}$$

As can be seen from Equations 1.27 through 1.29, this amplifier can be unstable because its gain increases as the frequency increases. Therefore, any sudden transition in the input signal (frequency or transition), which is a time-domain signal with an ideally infinite frequency, will lead the output to saturation. In order to cope with this practical problem, it is necessary to place a series resistor with an input capacitor and a capacitor in parallel with the feedback resistor, as shown in Figure 1.12, in order to create a more stable system by reducing the high-frequency sensitivity, which could be triggered by noisy input signals. The poles are placed sufficiently high in frequency to prevent significant phase-shift error in the high-frequency range.

1.2.5 Noninverting Amplifier

The determination of the closed-loop gain of a noninverting amplifier for a circuit depicted in Figure 1.13 can again be considered by assuming the amp-op input impedance to be infinite. Then, the input current will be zero and $i=i_f$. Considering a virtual short circuit (i.e., zero voltage across inverting and noninverting inputs for negative feedback around the amp-op), we make $v_i=e_1$ and the amplifier gain becomes:

$$v_i = e_1 = \frac{v_o}{Z_1 + Z_2}Z_1 \tag{1.30}$$

i.e., the feedback gain of the noninverting amplifier is given by:

$$A_f = \frac{v_o}{v_i} = \frac{Z_1 + Z_2}{Z_1} = 1 + \frac{Z_2}{Z_1} \tag{1.31}$$

The smallest gain of the noninverting amplifier is 1 and positive. It will be negative only if one of the impedances is capacitive and the other is not. In most practical applications, the signal-inverting amplifier is preferable to the noninverting amplifier, because the inverting amplifier allows signal changes, the gain can vary from zero, and the amplifier allows negative feedback during state changes [17,18].

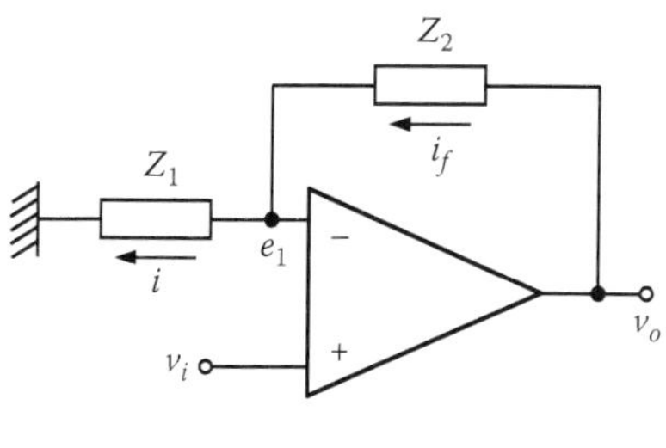

FIGURE 1.13 Noninverting feedback amplifier.

1.2.6 Buffer Amplifier

A direct application of the noninverting amplifier of Figure 1.13 is the buffer, i.e., a circuit that provides a high impedance at the input, but capable of driving a higher current, i.e., a low impedance at the output. The circuit is shown in Figure 1.14, consisting of an open circuit to the ground at the inverting input ($Z=\infty$) and a short circuit between the amplifier output terminal and the inverting input ($Z_f=0$) [19,20]. The output signal v_o of this amplifier is the exact duplication of the input signal v_i, but then the load current is fed by the amplifier voltage supply from the amp-op output. In this way, the amplifier can be used to reproduce the input signal without loading the previous stage (this is the reason this circuit is called buffer), because the load becomes an infinite impedance for the amplifier input and a zero output impedance for the buffer amplifier.

1.2.7 Level Detector or Comparator

Level detectors or comparators can be built by using the configuration examples of Figure 1.15. The Zener diode feedback serves to limit the level of the saturating

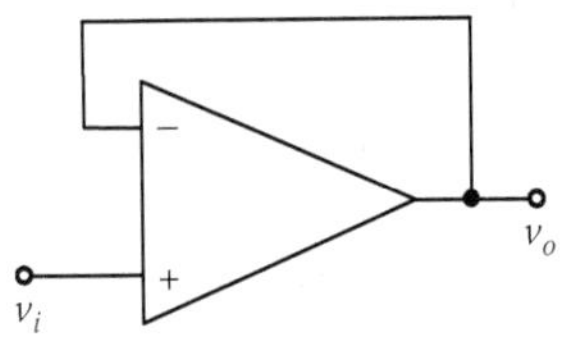

FIGURE 1.14 Buffer amplifier current.

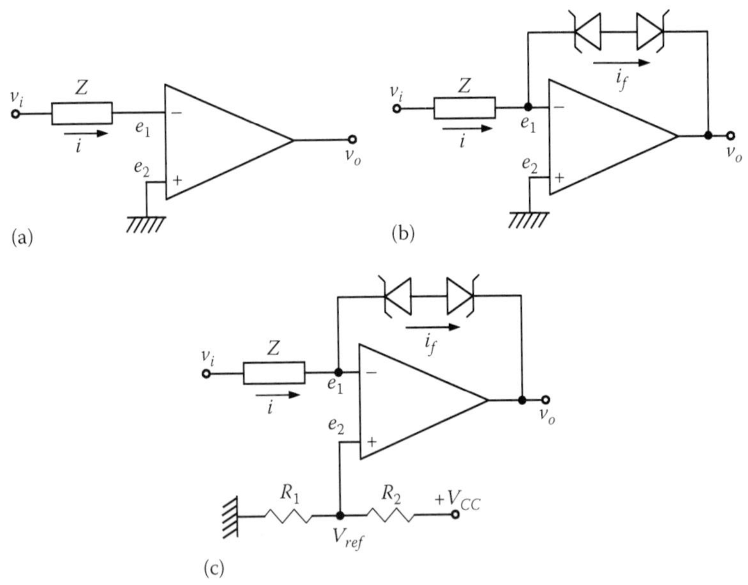

FIGURE 1.15 Level detectors: (a) zero crossing level detector, (b) zero crossing detector with voltage limiter, and (c) nonzero level detector.

voltage at the operational amplifier output. These detectors may have distinct or equal Zener levels.

1.2.8 Precision Diode, Precision Rectifier, and Absolute-Value Amplifier

Precision diodes can cancel out the knee voltage of diodes within the operational amplifier feedback. In the circuit shown in Figure 1.16, the positive half-cycle of the input signal is applied directly to the output port through the input virtual short circuit and feedback loop. The amp-op feeds the load as if the diode were a short circuit between input and output. In the negative half-cycle of the input signal, the feedback loop is seen as an open circuit and does not provide voltage to the load; therefore, the gain is zero. An adaptation of this circuit can be used as a peak detector (see Figure 1.17):

$$Z_f = \frac{Z_1Z_2 + Z_1Z_3 + Z_2Z_3}{Z_2} \tag{1.32}$$

The circuit shown in Figure 1.18 illustrates a precision half-wave rectifier with the possibility of voltage gain, while the circuit shown in Figure 1.19 illustrates the full-wave rectifier or absolute-value amplifier of the input signal based on this configuration.

A full-wave precision rectifier circuit with only two diodes and some extra components is shown in Figure 1.20. This version is not as accurate as the one in Figure 1.19 since it relies only on the accuracy of a resistive divider for the positive half-cycle at the amp-op output and depends on the load. Its accuracy is impaired by the need of extreme-precision components; otherwise, it distorts the output signal.

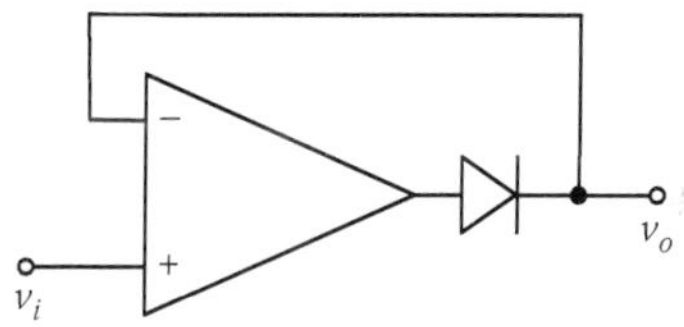

FIGURE 1.16 Precision diode (unit gain).

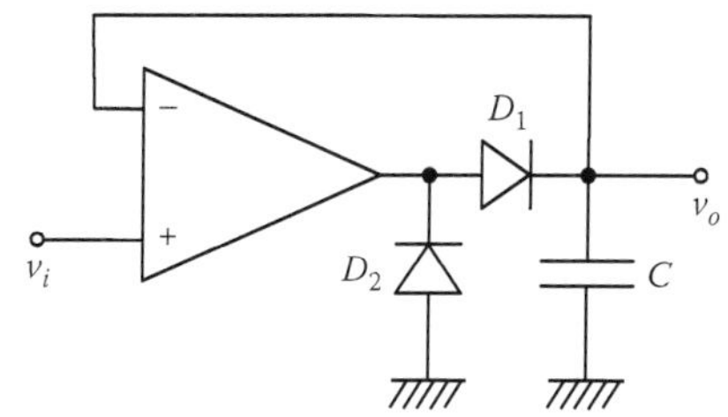

FIGURE 1.17 Peak detector.

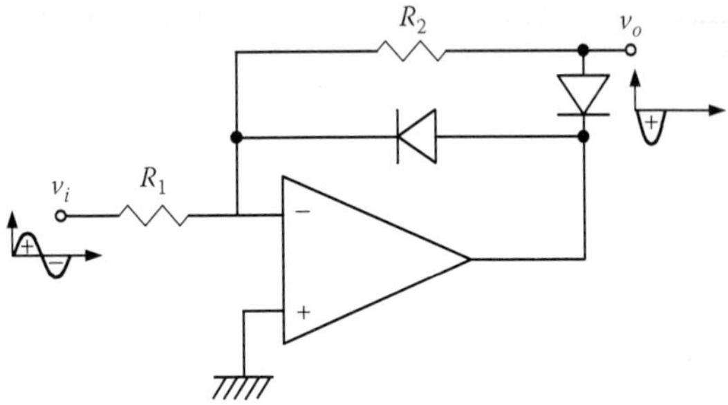

FIGURE 1.18 Precision half-wave rectifier with gain.

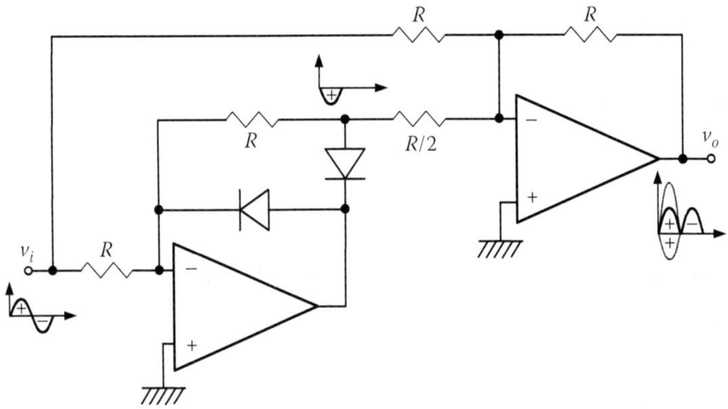

FIGURE 1.19 Precision full-wave rectifier with two amp-ops.

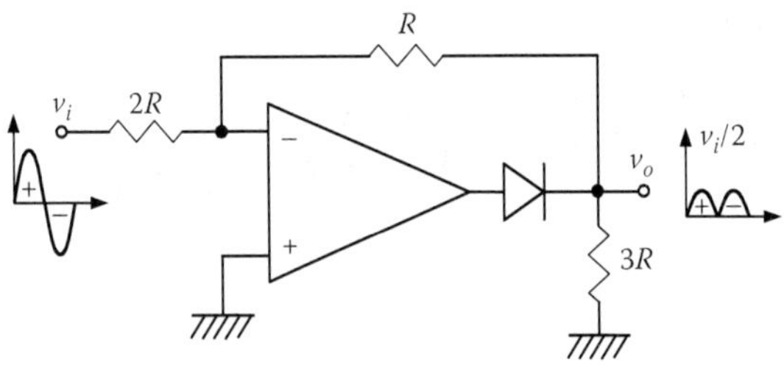

FIGURE 1.20 Full-wave rectifier with only one precision amp-op.

1.2.9 High-Gain Amplifier with Low-Value Resistors

To obtain high gains from an operational amplifier, it is often necessary to use high-value resistors, in the order of 1 MΩ or more. It turns out that, at higher frequencies, the stray capacitances of these resistances become significantly high, and thus, a way must be found to compensate them. One way of reducing the resistor values is considering the classical star-delta circuit transformation (Equation 1.32); the equivalent circuits are represented in Figure 1.21 [17–20].

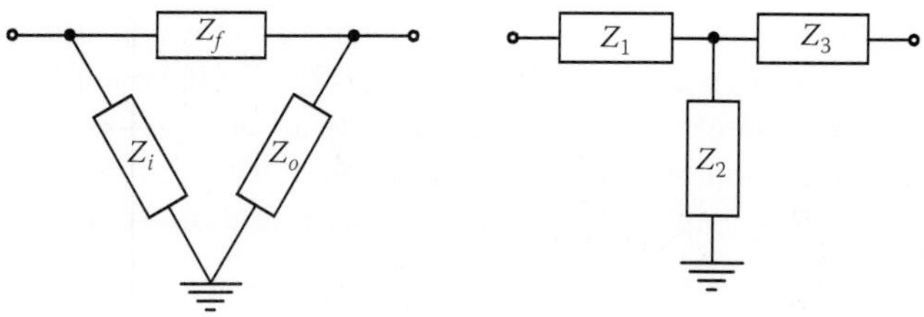

FIGURE 1.21 Star-delta equivalent circuits.

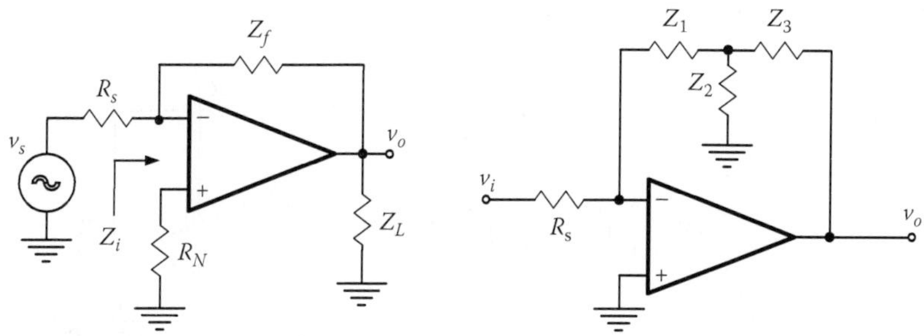

FIGURE 1.22 Equivalent star-delta circuit applied to operational amplifiers.

This transformation can be applied to the feedback amplifier, as shown in Figure 1.21.

1.2.9.1 Example of a High-Gain Amplifier with Low-Value Resistors

In the configuration of a feedback inverter amplifier with a gain equal to 100 and an input impedance of 10 kΩ, calculate the feedback resistor with low values.

Solution: To let an ordinary inverting amplifier have a gain of 100 and an input resistance of 10 kΩ, the feedback resistor value would have to be 1 MΩ; the default values using the equivalent circuit given in Figure 1.22 for the voltage gain, by means of Equation 1.32, are as follows:

$$10^6 = \frac{10^4 R_2 + 10^4 \cdot 10^4 + 10^4 R_2}{R_2} \tag{1.33}$$

Isolating R_2 and solving the above equation give: $R_2 = 102\ \Omega$.

1.3 THE ERROR EQUATION

The imperfections usually found in commercial differential amplifiers applied to instrumentation can add errors to signals if certain precautions are not taken at the design phase. These errors concern the drift and offset values of currents and voltages found, to a greater or lesser degree, in all commercial amplifiers. In the case of a nonzero voltage offset, i.e., $v_1 \neq v_2$, one would expect $v_o \neq 0$. It is assumed that the manufacturer has placed external pins to allow an external compensations of the offsets. In the case of an offset input current, a general equation has been established that allows the evaluation of errors due to differences between the excitation input currents of differential amplifiers, i_1 and i_2, in addition to those due to the voltage offset e_{os}.

As an example, suppose the inverting amplifier shown in Figure 1.8, where offset and drift voltages and currents are considered DC values and the impedances are replaced by resistors. By shorting the input signal source in Figure 1.8, one obtains the Thévenin equivalent circuit for the inverting amplifier terminals. The supply voltage and the Thévenin resistance, respectively, are as follows:

$$V_{Th} = \frac{R_1 V_o}{R_1 + R_2} = \beta_{DC} V_o \tag{1.34}$$

$$R_{Th} = R_1 // R_2 = R_I \tag{1.35}$$

Figure 1.23 represents, in general terms, the offset current and voltage conditions in differential amplifiers due to a nonzero internal source of signal at the inverting input with a resistor R_I, a signal-inverting amplifier with a resistance R_N at the noninverting input. Resistors R_{10} and R_{20} represent the offset and drift path currents connecting the input and output circuits as well as offset ground currents. Therefore, one can define, respectively, the offset voltage and current as follows:

$$e_{os} = e_2 - e_1 \tag{1.36}$$

$$i_{os} = i_2 - i_1 \tag{1.37}$$

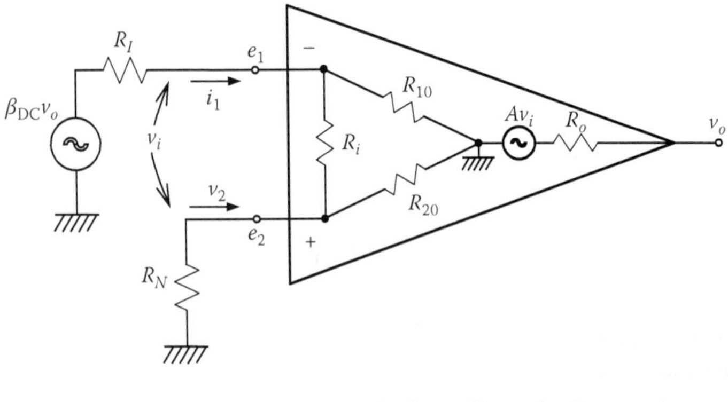

FIGURE 1.23 Simplified amp-op diagram including offsets of voltage and current.

From Figure 1.23, it can also be established that:

$$e_1 = \beta_{\text{DC}} v_o - i_1 R_I \tag{1.38}$$

$$e_2 = -i_2 R_N \tag{1.39}$$

Replacing Equations 1.38 and 1.39 in Equation 1.36, we derive:

$$e_{os} = -i_2 R_N - \beta_{\text{DC}} v_o + i_1 R_I \tag{1.40}$$

which may be represented in terms of v_o as:

$$v_o = \frac{1}{\beta_{\text{DC}}}\left(-e_{os} - i_2 R_N + i_1 R_I\right) \tag{1.41}$$

From Equation 1.37, one can state that $i_2 = i_{os} + i_1$, and Equation 1.39 can be rewritten as the general error equation:

$$v_o = \frac{1}{\beta_{\text{DC}}}\left[-e_{os} + i_1\left(R_I - R_N\right) - i_{os} R_N\right] \tag{1.42}$$

The largest value of the offset error will be obtained when all terms in Equation 1.40 have the same signal expressed in absolute values, which provides:

$$V_o = \frac{1}{\beta_{\text{DC}}}\left(\left|e_{os}\right| + \left|i_1\left(R_I - R_N\right)\right| + \left|i_{os}\right| R_N\right) \tag{1.43}$$

Two cases can be considered:

1. If $R_N = 0$, the maximum error is as follows:

$$V_o = \frac{1}{\beta_{\text{DC}}}\left(\left|e_{os}\right| + \left|i_1\right| R_I\right) \tag{1.44}$$

2. If $R_N = R_I$, the maximum error is as follows:

$$V_o = \frac{1}{\beta_{\text{DC}}}\left(\left|e_{os}\right| + \left|i_{os}\right| R_I\right) \tag{1.45}$$

Adopting typical values of commercial amplifiers, listed in Table 1.3, for offset errors in bipolar amplifiers at 25 °C, it can be concluded from Equation 1.44 that if $i_{os} \ll i_1$ or i_2 and $R_N = R_I$, the offset error will be minimized. The offset current usually doubles for every 10 °C due to temperature effects on the bipolar junctions of the transistors.

TABLE 1.3
Typical Deviations for Bipolar Discrete Amplifiers at 25 °C

Type of Error	Magnitude
e_{os}	±0.2 mV
i_1 or i_2	±20 nA
i_{os}	±2 nA
$\partial e_{os}/\partial T$	±0.5 μV/°C
$\partial i_1/\partial T$	±0.2 nA/°C
$\partial i_{os}/\partial T$	±0.02 nA/°C

1.4 LARGE-SIGNAL AMPLIFIERS

Large-signal push-pull amplifiers are mostly used in instrumentation for remote communications and automation of power plants and bioengineering since they are able to reproduce quite involving wave shapes with many distortions [14–16]. Most of the distortions introduced by nonlinearities in the dynamic transfer characteristics of operational amplifiers can be minimized by the push-pull amplification. A well-balanced push-pull circuit cancels out all even harmonics and the DC current at the output port leaving the third harmonic and its multiples as well as the distortion on the main sources.

The advantages of push-pull amplifiers are related to the lower level of harmonic distortions for a given high power transistor. These distortions include the DC source differences and the smaller distortion possibility from transformer cores. Power source ripples can also be canceled out. The electronic hum is not eliminated by the push-pull configuration.

The input voltage feedback of a **Class B push-pull amplifier** can minimize the distortions previously discussed due to the base-emitter voltage zero crossing (crossover effect). Figure 1.24 illustrates a case where the feedback crossover distortion in the pair of complementary unmatched transistors is minimal. A more detailed design of an amplifier like this can be obtained from Reference 12.

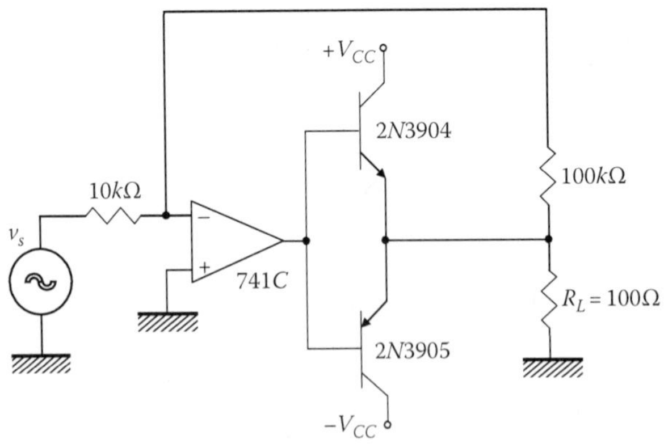

FIGURE 1.24 A Class B feedback push-pull amplifier.

EXERCISES

1.1 Specify the circuit components in the figure below for a feedback gain = 10 in direct current and a gain = 5 at $f = 1000\,\text{Hz}$. If the amplitude of the input signal is 2.6 V, what would be your recommendation for the values of $\pm V_{DC}$?

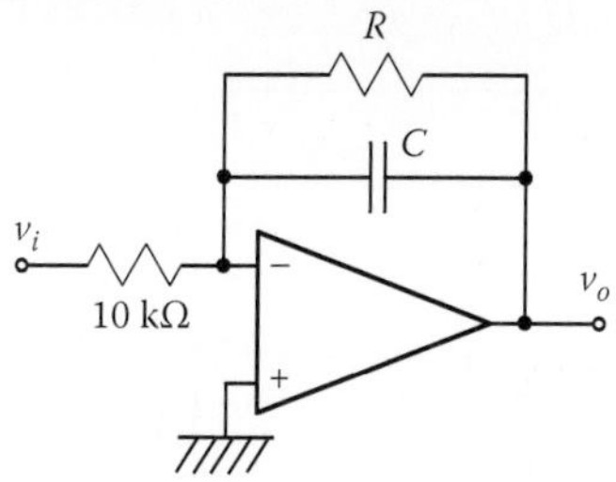

1.2 What is the minimum expected error in the electrical measurement using an instrument that has an input resistance of 500 kΩ when it is used to measure a resistive divider with a 250 kΩ load and high levels of voltage? Make a circuit sketch that could reduce this error using an operational amplifier without changing anything inside the instrument.

1.3 Using the table of typical values of an operational amplifier (Table 1.3), calculate the error in volts in the amplifier output voltage shown below, with and without the resistance in the noninverting input.

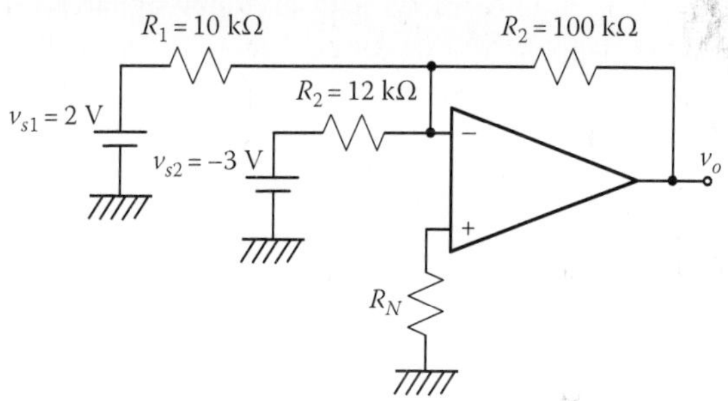

1.4 Propose an analog circuit to solve the equation $dy/dx = -10y + 12$ using operational amplifiers.

1.5 Calculate the percentage error in the measurement of the output voltage of a resistive divider, such as in the figure below, when it is not used as a buffered output.

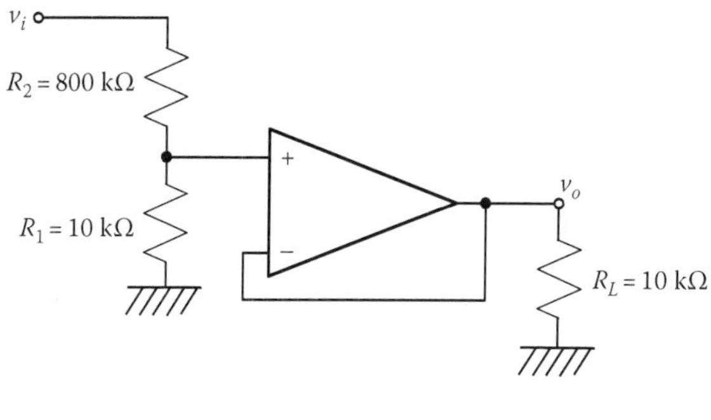

1.6 Compare the expected percentage error of an instrument based on an amplifier with the following offset values: $e_{os} = -0.2\,\text{mV}$, $i_1 = 2.2\,\text{nA}$, and $i_{os} = 2\,\text{nA}$. Calculate the amplifier error in the configuration given in the figure below, with and without the compensation resistance in the noninverting terminal.

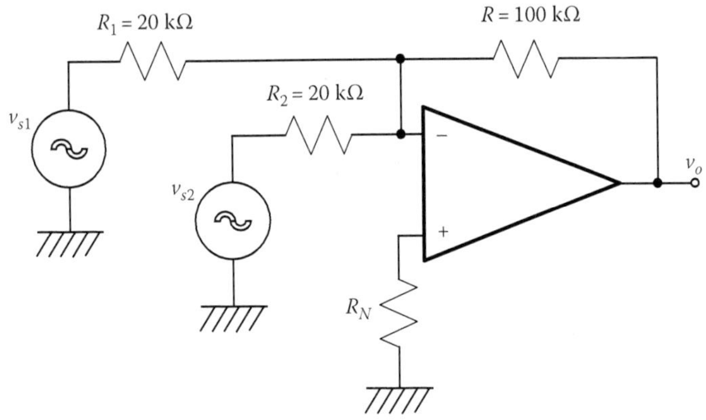

1.7 Calculate the parameters R_N, e_{os}, i_{os}, and β_{DC} of the amplifier in Figure 1.6 when the offset values are as follows: $i_1 = 14\,\text{nA}$, $i_2 = 16\,\text{nA}$, and $e_1 = -0.26\,\text{mV}$.

1.8 Calculate the amplitude of the output signal of the instrumentation amplifier presented in the figure below, with an input signal $v_i = 0.005 \sin \omega t$ at $f = 60$ kHz. What happens if a direct current signal is used with the same amplitude input signal? What is the instrument's input impedance for $f = 60$ kHz? What is the recommended value for $\pm V_{DC}$?

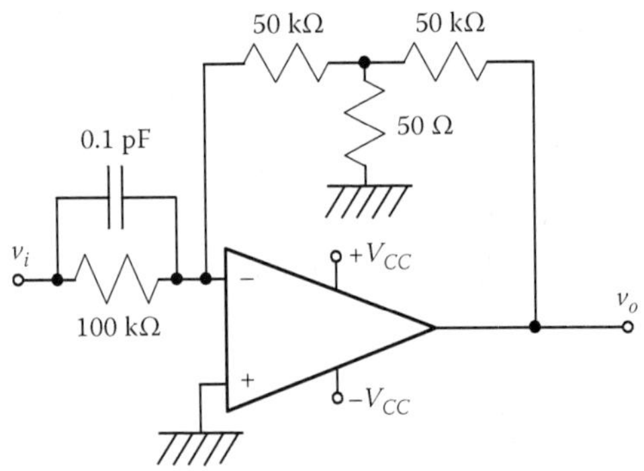

1.9 What is the resonant frequency of a Wien bridge using ideal operational amplifiers when the resistors used in the RC-circuit are 100 Ω each and the capacitors are 0.3 μF each?

1.10 Calculate the passband of the first-order filter below to be used as input of a digital multimeter, with an RC input series impedance of $10^4 - j10^6\ \Omega$ for a frequency of 10 kHz, and where the gain product versus the amplifier passband is as shown in Figure 1.6.

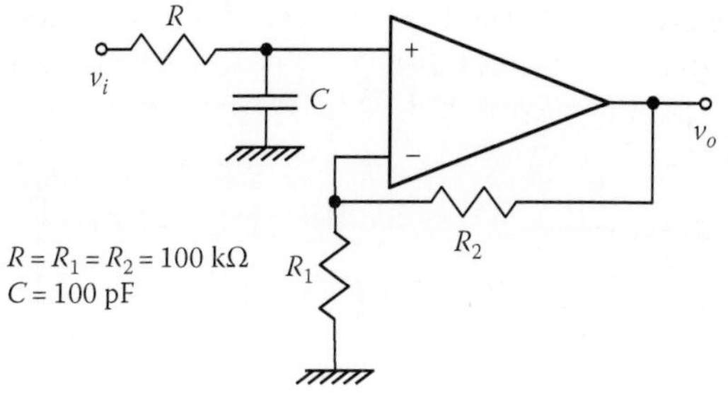

1.11 Give five of the main characteristics of instrumentation amplifiers, discussing how each one can affect the measurement quality.

REFERENCES

1. Silveira, P.R. and Santos, W.E., *Automation and Discrete Control (Automação e Controle Discreto)*, Editor Érica, São Paulo, Brazil, 229pp., 1998.
2. Diefenderfer, A.J., *Principles of Electronic Instrumentation*, W.B. Saunders Co., Philadelphia, PA, 667pp., 1972.
3. Slotine, J.J.E. and Li, W., *Applied Nonlinear Control*, Prentice Hall International, Englewood Cliffs, NJ, 461pp., 1991.
4. Graemme, J.G., Tobey, G.E., and Huelsman, L.P., *Operational Amplifiers: Design and Applications*, Burr-Brown Ed., Tokyo, Japan, 473pp., 1971.
5. Zelenovsky, R. and Mendonça, A., *PC: Um Guia Prático de Hardware and Software*, 2nd edn., Editors MZ, Rio de Janeiro, Brazil, 760pp., 1999.
6. Eveleigh, V.W., *Introduction to Control Systems Design*, TMH Edition-Tata McGraw-Hill Publishing Company, New York, 624pp., 1972.
7. Bowron, P. and Stephenson, F.W., *Active Filters for Communications and Instrumentation*, Editors McGraw-Hill Book Co., Berkshire, U.K., 285pp., 1979.
8. Auslander, D.M. and Agues, P., *Microprocessors for Measurement and Control*, Editor Osborne/McGraw-Hill, Berkeley, CA, 310pp., 1981.
9. Bolton, W., *Industrial Control and Measurement*, Editor Longman Scientific and Technical, Essex, England, 203pp., 1991.
10. Wait, J.V., Huelsman, L.P., and Korn, G.A., *Introduction to Operational Amplifier Theory and Applications*, International Student Edition, McGraw-Hill Kogakusha, Tokyo, Japan, 396pp., 1975.
11. Boylestad, R. and Nashelsky, L., *Electronic Devices and Circuit Theory*, 11th edn., PHB, Delhi, India, 2012, ISBN: 10-0132622262, ISBN: 13-978-0132622264.
12. Horowitz, P. and Hill, W., *The Art of Electronics*, 3rd edn., Cambridge University Press, London, U.K., 2015, ISBN: 10-0521809266, ISBN: 13-978-0521809269.
13. Regtien, P.P.L., *Electronic Instrumentation*, Editor VSSD (Vereniging voor Studie en Studentenbelangen te Delft), Delft, the Netherlands, 2005, ISBN: 978-90-71301-43-8.
14. Dyro, J.F., *Clinical Engineering Handbook*, Elsevier Academic Press, Boston, MA, 2004, ISBN: 0-12-226570-X.
15. Karunakaran, C., Bharkava, K., and Benjamin, R., *Biosensors and Electronics*, Elsevier, Amsterdam, the Netherlands, 2015, ISBN: 978-0-12-803100-1.
16. Valentinuzzi, M.E., *Cardiac Fibrillation and Defibrillation: Clinical and Engineering Aspects*, World Scientific, Hackensack, NJ, 2011, ISBN: 13-978-981-4293-63-1.

17. Mirri, D., Filicori, F., Iuculano, G., and Pasini, G., A nonlinear dynamic model for performance analysis of large-signal amplifiers in communication systems, *IEEE Transactions on Instrumentation and Measurement*, 53(2), 341–350, April, 2004, DOI: 10.1109/TIM.2003.822714.
18. Rybin, Y.K., *Electronic Devices for Analog Signal Processing* Springer Series in Advanced Microelectronics, 33, Berlin, Heidelberg, Germany, eBook Kindle, ASIN: B007EMAP98, 2012.
19. Tietze, U., Schenk, C., and Gamm, E., *Electronic Circuits: Handbook for Design and Application*, Springer-Verlag, London, U.K., 2008, ISBN: 978-3-540-78655-9.
20. Kitchin, C. and Counts, L., *A Designer's Guide to Instrumentation Amplifiers*, 3rd edn., Analog Devices Co., Norwood, MA, 2006.

2 Analog-Based Instrumentation Systems

2.1 INTRODUCTION

It is not currently usual to implement linear amplifiers based on discrete components, except for high-frequency or high-power-level customized designs. Such linear amplifiers are commercially available as integrated circuits and are ready for immediate operation and able to efficiently attend standard problems, such as polarization component, cancellation of common-mode signals, thermal stability, and parametric stability effects caused by component aging, board humidity, component replacements, and sources of imbalances, noise, and other necessary compensation for their proper functioning. The integration of linear amplifiers as operational amplifiers are presently manufactured to allow their use in amplifiers and electronic instrumentation. of the foundation for such circuits based on the *differential amplifier* to be discussed in this chapter [1–3].

By understanding the ideal behavior of the differential amplifier, it is possible to establish a set of characteristics that would be desirable, in particular, for instrumentation amplifier purposes, such as those listed in Table 2.1, including the common-mode rejection ratio (CMRR), as defined in Section 1.1.2. These are the features to be considered when selecting a differential amplifier for instrumentation.

2.1.1 VOLTAGE FEEDBACK FOR DIFFERENTIAL AMPLIFIERS

Although there are many topologies for differential amplifiers, this book discusses two differential feedback amplifiers that seem to be the most useful in a basic course of electronic instrumentation: standard amplifiers and instrumentation amplifiers. To do this, let us obtain the differential gain of a standard differential amplifier, such as the one in Figure 2.1, representing their simplest version. In the theory of differential amplifiers in Figure 2.1 is defined:

$$v_o = A(e_1 - e_2) \quad \text{or} \quad \frac{v_o}{A} = e_1 - e_2 \tag{2.1}$$

where A is the gain of the differential amplifier, which is very high (usually $A \gg 10^5$); thus, we can conclude that $e_1 \cong e_2$, which means a virtual short circuit between the negative and positive terminals. The individual values of e_1 and e_2 can be obtained from the input circuit shown in Figure 2.1:

$$e_1 = v_1 - i_1 R_1 = v_1 - \left(\frac{v_1 - v_o}{R_1 + R_2} \right) R_1 \tag{2.2}$$

TABLE 2.1
Basic Features of an Instrumentation Amplifier

Characteristic	Desirable	Ideal
Input impedance	High	$\to\infty$
Output impedance	Low	0
Gain	High and adjustable	$\to\infty$
Coupling	DC	Direct
Linearity	Low	0
Input signal	Differential	Differential
CMRR	High	$\to\infty$
Noise	Low	$\to 0$
Offset and drift	Low	$\to 0$
Slew rate ($\lvert dv/dt\rvert_{max}$)	High	$\to\infty$

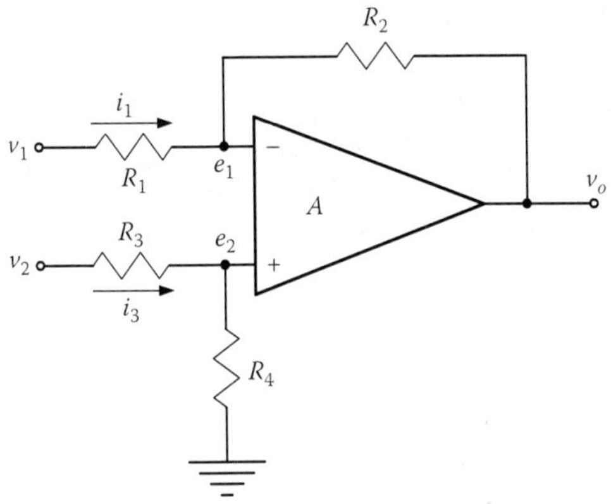

FIGURE 2.1 Standard differential amplifier.

$$e_2 = i_3 R_4 = \left(\frac{v_2}{R_3 + R_4}\right) R_4 \tag{2.3}$$

Subtracting Equation 2.1 from (2.2), equating e_1 and e_2 due to the virtual short circuit, and isolating v_o, we derive:

$$v_o = \frac{R_4}{R_1}\left(\frac{R_1 + R_2}{R_3 + R_4}\right) v_2 - \frac{R_2}{R_1} v_1 \tag{2.4}$$

The necessary condition to amplify the difference between v_2 and v_1 is that:

$$R_3 = R_1 \quad \text{and} \quad R_4 = R_2 \tag{2.5}$$

Under these assumptions, Equation 2.4 becomes:

$$v_o = \frac{R_2}{R_1}\left(v_2 - v_1\right) \tag{2.6}$$

The previous assumptions create three non ideal situations of a differential amplifier applied to instrumentation purposes:

1. It is difficult in practice to perfectly match resistor values, as requested by Equation 2.5.
2. Each amplifier input impedance, R_1 and R_3, has an inherently very low value (e.g., for biological instrumentation amplifiers).
3. It is not easy to vary the overall amplifier gain, because it depends on the simultaneous and accurate variation of the two matched potentiometers: $R_3 = R_1$ and $R_4 = R_2$ (Equation 2.5).

In practical terms, the differential gains are specified in decibels (dB). In applications using very small signals, such as those generated in strain gauges and thermocouples, a 40 dB gain is not very significant, as they usually need gains as high as 100 dB or 120 dB [1–3].

Wound wired resistors have greater stability with respect to temperature, aging, and humidity, but have some stray inductance, which causes CMRR degradation. A good solution is to use the Ayrton-Perry resistors, which are noninductive. If the capacitive effect is a problem, then a possible solution to compensate it is to place the components in a symmetric mirror position with respect to each other.

2.1.2 Standard Differential Amplifier for Instrumentation

There are many ways to achieve a good-quality differential amplifier, but in practice, it seems a norm for instrumentation purposes that the amplifier configuration shown in Figure 2.2 is the most appropriate. Figure 2.2 neglects the excitation currents of the two amplifier inputs, so initially that the currents flowing through resistors R_1, R_2, and R_3 are almost the same and equal to i_1. Moreover, v_{o1} and v_{o2} are, respectively, the signals at the two inputs of the standard differential amplifier, according to Figure 2.1 and Equation 2.6. Therefore, one can obtain the overall gain from individual gains of the two amplifier stages. The values v_{o1} and v_{o2} can be obtained from the input signal, v_1 and v_2, by means of the gain of the first stage, given as:

$$A_{v1} = \frac{v_{o1} - v_{o2}}{v_1 - v_2} \tag{2.7}$$

Thus, the common gain of this amplifier (for $v_1 = v_2$) can be established by making $v_{o1} \approx v_1$ and $v_{o2} \approx v_2$, and therefore $i_1 \to 0$:

$$A_c = \frac{v_{o1} - v_{o2}}{v_1 - v_2} \tag{2.8}$$

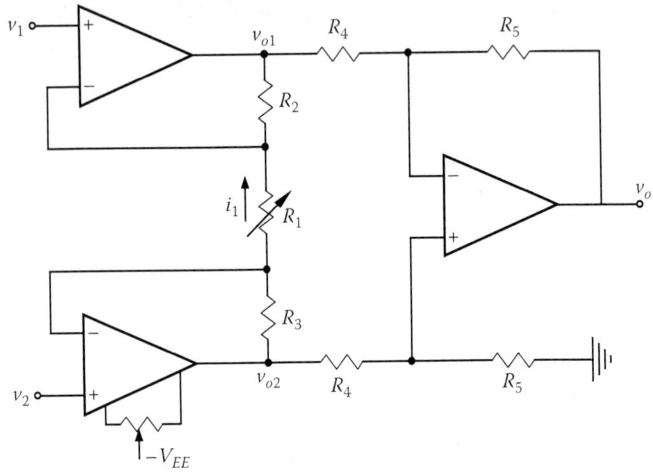

FIGURE 2.2 Standard instrumentation amplifier.

The differential gain (for $v_1 = -v_2$) can be established from:

$$i_1 = \frac{v_{o1} - v_{o2}}{R_1 + R_2 + R_3} = \frac{v_1 - v_2}{R_1}$$

which rearranged gives:

$$A_d = \frac{v_{o1} - v_{o2}}{v_1 - v_2} = \frac{R_1 + R_2 + R_3}{R_1} = 1 + \frac{R_2 + R_3}{R_1} \tag{2.9}$$

Since the common gain of the input stage is unity, by definition, it will have the same differential gain expression given in Equation 2.9, which is as follows:

$$A_1 = \frac{v_{o1} - v_{o2}}{v_1 - v_2} = 1 + \frac{R_2 + R_3}{R_1} \tag{2.10}$$

From Equation 2.6, for a single differential amplifier, the second-stage output voltage is:

$$v_o = \frac{R_5}{R_4}\left(v_{o2} - v_{o1}\right) \tag{2.11}$$

Rearranging Equation 2.11, the gain of the second stage may be given by:

$$A_2 = \frac{v_o}{v_{o2} - v_{o1}} = \frac{R_5}{R_4} \tag{2.12}$$

Combining the gains of the first and second amplifier stages in order to avoid the mirror effect, i.e., multiplying Equation 2.10 by (2.12), and simplifying, we derive:

$$A = A_1 A_2 = \frac{v_o}{v_2 - v_1} = \left(1 + \frac{2R_2}{R_1}\right)\frac{R_5}{R_4} \tag{2.13}$$

As can be inferred from Equation 2.12, the input stage can provide a high differential gain controlled only by R_1, keeping the first stage unity and common-mode gains without any resistor matching. Thus, the differential output signal represents a substantial reduction in the comparative signal in common mode and is used to drive conventional differential amplifiers (the second stage), which compensate for any residual common-mode signal. Commercial examples of instrumentation amplifiers are LH0036 and AD522 and the high-precision differential amplifier 3630. The component sometimes called "differential instrumentation amplifier 725" is just a very good conventional amplifier and should considered as with an instrumentation amplifier.

2.1.3 Programmable Gain Amplifier

In a general data acquisition system with several inputs, the multiplexer is typically followed by an amplifier, which should supply the digital-to-analog (D/A) converter with a signal at specific levels, e.g., between −5 and +5 V. To do so, it is important to have a digital control over the gain amplifier. An amplifier example with digital programmable gain is shown in Figure 2.3. The resistance R_1 of this amplifier has the

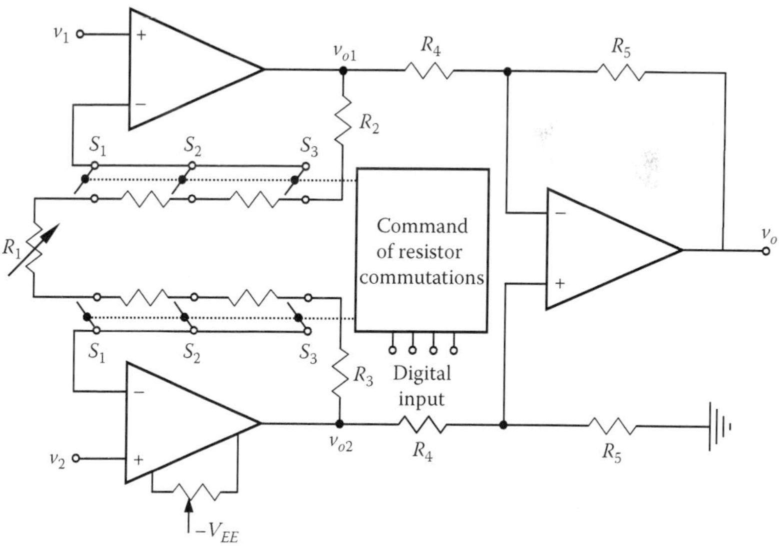

FIGURE 2.3 Differential amplifier with a digitally programmable gain.

same function of the ordinary gain control circuit shown in Figure 2.2, except that this one can be digitally varied by a counter output or directly from a computer data bus or yet from some other digital encoding.

2.2 CUSTOMIZED INSTRUMENTATION AMPLIFIER

In addition to the standard amplifiers described in the previous sections, an instrument may need many other topologies of amplifiers for specific applications. In particular, small-signal amplifiers are usually placed under a high level of electromagnetic noise environments. Therefore, the selected instrumentation amplifier must be compatible with its technical purpose, suitability, relative cost and overall characteristics.

2.2.1 Isolated Instrumentation Amplifier

Typically, the common-mode voltage v_c specified for instrumentation amplifiers is very low (e.g., 15 V). On the other hand, the voltage to be measured can be high and must be isolated from the data acquisition system for security reasons, or to attend to electromagnetic compatibility levels of electronic data processors, generally limited between 3 and 30 V. An example is the case when someone needs a common-mode voltage of a few kilovolts, as in pole isolation or coupling of the high-power converter bridges in distributed generation. Furthermore, isolated amplifiers are needed in places where there exists the possibility of electrocution, e.g., in the case of hospital instruments or between equipment panels and operators. Therefore, according to the intended purpose, it is possible to use optical, magnetic, or capacitive isolation amplifiers (with linearity compensation or isolation only in one of the isolated stages).

2.2.2 Magnetic Isolation

Magnetic isolation is performed through an isolation transformer, as shown in Figure 2.4. This sort of isolation needs a modulator with its own floating power

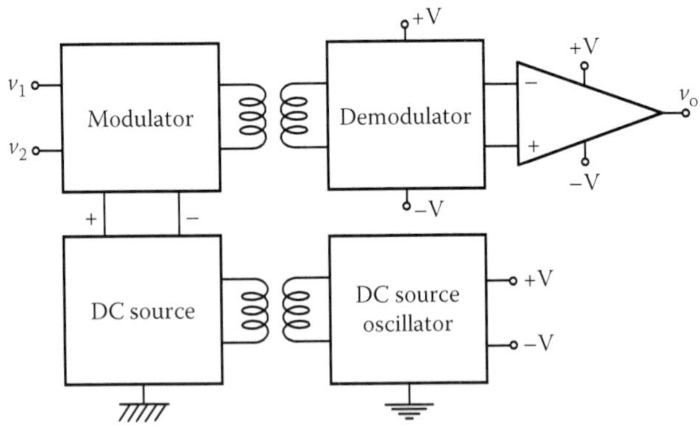

FIGURE 2.4 Amplifier with magnetic or galvanic isolation.

source excited by a few hundred kilohertz direct current (DC) oscillator. The signal can be modulated by pulse width modulation (PWM); pulse frequency modulation (PFM); (hardly) pulse amplitude modulation (PAM), usually applied to multilevel power converters; or pulse density modulation (PDM), which avoids amplitude distortions due to transformer nonlinearity. The main disadvantages of this sort of isolation are as follows: cost, sensitivity to parasitic electromagnetic radiation, and large response times. Magnetic isolation can be easily adapted for large-distance systems, such as in a power substation.

Commercial examples of amplifiers with magnetic isolation are ADuM4190 for frequencies up to 400 kHz, AD210 for frequencies up to 20 kHz, and AD215 for higher frequencies from the Analog Devices Co. used for biological processing signals and to prevent the ignition of explosive gases by sparks and for protection of patients from electric shocks. The amplifier AD210 has a nonlinearity of 0.012%, a gain from 1 to 100, a common-mode isolation voltage of 2500 V_{rms} and a continuous off of ±3500 V, a CMRR greater than 130 dB at 60 Hz, with the possibility of a resistance imbalance between sources of 5 kΩ, unity gain and PWM modulation, isolated input and output power supplies, ±15 V, ±5 mA, and peak continuous wide bandwidth: 20 kHz (full power). However, such a configuration is mostly used in isolated environments for data transmission, and sometimes to supply the DC power supply and its oscillator.

2.2.3 Optical Isolation

Optical isolation is achieved through two pairs of photodiode-phototransistor, called generically as photocouplers, so that the reference pair compensates for the optical coupler's nonlinearities, as represented in Figure 2.5.

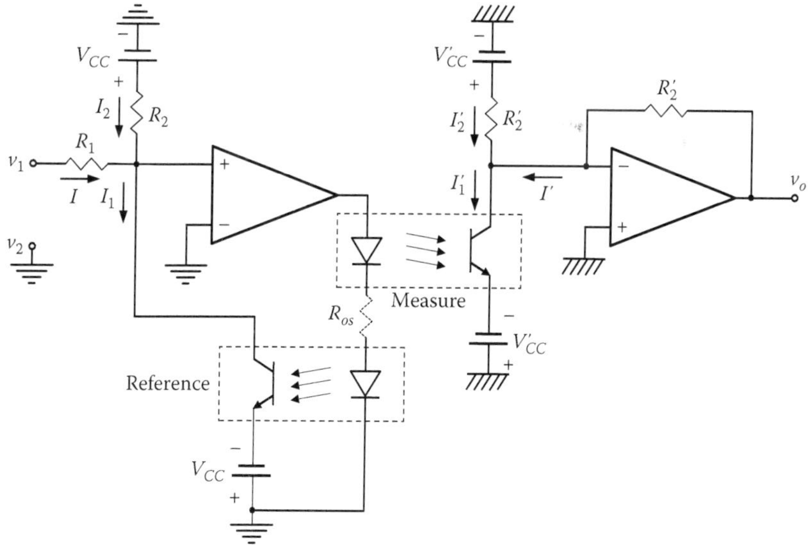

FIGURE 2.5 Amplifier with optical isolation.

The existing commercial optically isolated amplifiers are developed in integrated circuits, and they can achieve a very good matching between the photodiode and the phototransistor and among plus resistors. To represent the matched values of the output and input stages in Figure 2.5, the output-stage variables are differentiated by an apostrophe relative to the input-stage variables.

Note that in the circuit shown in Figure 2.5, the input stage acts as a noninverting amplifier. Still, in Figure 2.5, the feedback impedance of the optically isolated amplifier in the integrated loop is exerted by two diodes in series, with one cathode connected to the anode of the other one and then connected to the ground. The reference opto-transistor is optically coupled to the inverting output amplifier through the measure opto-transistor. The gain of this stage is unity because the biased diodes represent a short circuit to the operational amplifier output.

In order to show that there is nonlinearity compensation, the following relationships between the input and output stages are established:

$$I = I_1 - I_2 \tag{2.14}$$

where

$I = I'$

$I_1 = I_1'$ (matched photocoupler pair)

$I_2 = I_2' = \dfrac{V_{CC}}{R_2}$

If power supplies of the input stage and the output isolated amplifier are matched, one can consider with a good degree of approximation that the identities $I_1 \cong I_1'$ and $I_2 \cong I_2'$ and Equation 2.13 are true. Therefore, the voltage gain can be given from:

$$I' = \frac{v_o}{R_1'} = \frac{v_1 - v_2}{R_1} = I \quad \text{or, yet:} \quad A_v = \frac{v_o}{v_1 - v_2} = \frac{R_1'}{R_1} \tag{2.15}$$

In practice, the gain will never be much greater than unity to ensure that the photocoupler's quiescent values operate close to each other, and improved linearity is ensured, in addition to minimizing electrical field differences during the measurement.

The main advantages of optically isolated amplifiers are as follows: less influence on electromagnetic noise, lower weight and volume, and higher speed response. A commercial example of optically isolated amplifiers is the IL300 (Vishay). Such amplifiers exhibit a common-mode isolation voltage of 7500 V, a bandwidth of 200 kHz, a CMRR of 130 dB, a stability of ±50 ppm/°C, and a linearity of ±0.01%.

2.2.4 Capacitive Isolation

An amplifier with capacitive coupling converts direct or alternating voltage from the input into pulsed voltage, which is then amplified and reconverted to form an analog voltage similar to the input. This amplifier serves to minimize the input and output drift and offset values, and feedback signals, as shown in Figure 2.6. It reacts only to signal changes in v_i across the amplifier input and not to DC levels.

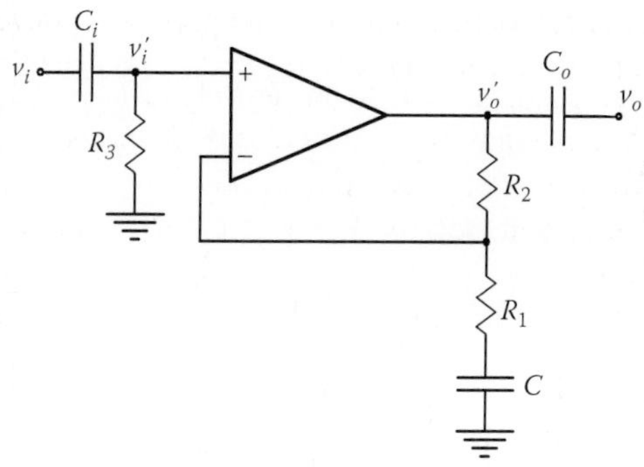

FIGURE 2.6 Amplifier with capacitive coupling.

The output voltage offset can be minimized by inserting a bypass capacitor in the feedback loop. Thus, the amplifier voltage gain becomes:

$$A_v = \frac{v_o'}{v_i'} = 1 + \frac{R_2}{R_1 + \dfrac{1}{j\omega C}} = 1 + \left(\frac{R_2}{R_1}\right)\frac{1}{1 + j\omega CR_1} \tag{2.16}$$

At the operating frequency, the capacitor appears as a short circuit, and the closed-loop voltage gain of the circuit is as follows:

$$A_{vm} = 1 + \frac{R_2}{R_1} \tag{2.17}$$

As mentioned above, drifts and offsets can be considered virtually as DC values. Thus, the capacitor appears as an open circuit for DC levels and the gain becomes unity. This fact reduces the output offset to a minimum, allowing maximum excursion of the output alternating current (AC) signals (maximum compliance). The bypass capacitor results in a cutoff frequency of:

$$f = \frac{1}{2\pi R_2 C} \tag{2.18}$$

A good commercial example of an amplifier with electrical coupling is ISO124 of Texas Instruments. This precision amplifier incorporates a modulation-demodulation operating cycle. The signal is transmitted digitally through a differential capacitive barrier of 2 pF. With a digital modulation, the barrier characteristics do not affect signal integrity and reliability, with good immunity to high-frequency transients through the barrier. The ISO124 does not need any external components, and offers a maximum non-linearity of 0.01%, bandwidth of 50 kHz, isolation mode, or voltage rejection of 60 dB at 140 Hz, 1500 V_{rms} isolation, voltage drift of 200 μV/°C, power supply range from ±4.5 to ±18 V, circuit quiescent currents of ±5.0 mA of input signal and ±5.5 mA of output signal for a wide range of applications.

2.3 VOLTAGE-TO-CURRENT CONVERSION (V-TO-I)

Most digital and analog signal processors are characterized by an output-based voltage. The signal sources may be of current type (e.g., inductors) or voltage type (e.g., capacitors). Therefore, for processing purposes, often, one type of signal must be converted into another and vice versa, whether the load amplifier is connected to the ground or is floating.

Basically, there are four different types of instrumentation based on voltage-to-current (V-to-I) converters:

1. Single input with ungrounded load
2. Single input with grounded load
3. Differential input with ungrounded load
4. Differential input with grounded load

2.3.1 Single-Input V-to-I Converter with Ungrounded Load

Figure 2.7 shows a single-input V-to-I converter with ungrounded load. As can be seen from the figure, the conversion between the load current and the input voltage is established as:

$$I_L \cong \frac{1}{R_f} v_i \tag{2.19}$$

2.3.2 Single-Input V-to-I Converter Load with Grounded Load

Figure 2.8 shows a single-input V-to-I converter with grounded load. As can be seen from the figure, the relationship between the load current and the input voltage is established from:

$$e_1 = v_i - \frac{v_i - v_o}{R_1 + R_2} R_1 \tag{2.20}$$

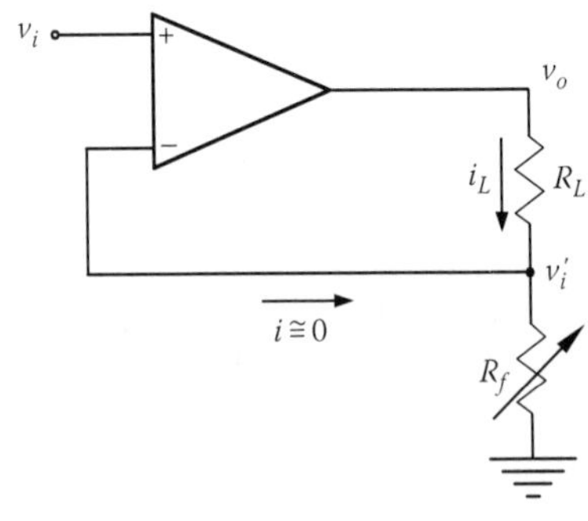

FIGURE 2.7 Single-input V-to-I converter with ungrounded load.

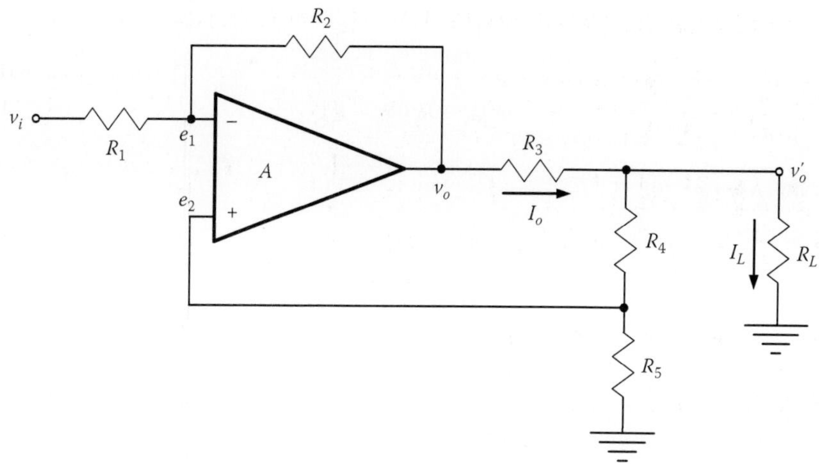

FIGURE 2.8 Single-input V-to-I converter with grounded load.

$$e_2 = \frac{v_o' R_5}{R_4 + R_5} \tag{2.21}$$

The virtual short circuit between the inputs of the differential amplifier allows $e_1 \cong e_2$, and therefore, combination Equations 2.20 and 2.21, results in:

$$v_i = v_o' \left[\frac{R_L R_5 (R_1 + R_2) - R_1 (R_3 + R_L)(R_4 + R_5) - R_L R_1 R_3}{R_2 R_L (R_4 + R_5)} \right] \tag{2.22}$$

The ratio between v_o' and v_o in Figure 2.8 may be established from:

$$i_o = \frac{v_o - v_o'}{R_3} = \frac{v_o'}{R_4 + R_5} + \frac{v_o'}{R_L}$$

or

$$v_o = v_o' \left(1 + \frac{R_3}{R_L} + \frac{R_3}{R_4 + R_5} \right) \tag{2.23}$$

As $v_o' = I_L R_L$, then combining Equations 2.22 and 2.23, and simplifying it, gives:

$$v_i = I_L \left[\frac{R_L \left[R_2 R_5 - R_1 (R_3 + R_4) \right] - R_1 R_3 (R_4 + R_5)}{R_2 (R_4 + R_5)} \right] \tag{2.24}$$

To make Equation 2.24 independent of R_L, discrete resistors should be chosen so that they can override the term containing R_L, namely:

$$R_2R_5 = R_1\left(R_3 + R_4\right) \tag{2.25}$$

After simplifying and replacing Equation 2.25 in Equation 2.24, we derive:

$$I_L \cong -\frac{R_2}{R_1R_3}v_i \tag{2.26}$$

The single-input V-to-I converter with grounded load is useful for a broad type of control signals with currents between 4 and 20 mA like those found in instrumentation for distributed generation.

2.3.3 V-to-I Converter with Differential Input and Ungrounded Load

Figure 2.9 shows a V-to-I converter with differential input and ungrounded load. The differential input stage is similar to those shown in Figures 2.1 and 2.2.

The relationship between the load current I_L and the second-stage input voltage v_i can be obtained from Figure 2.9 using the following identity:

$$I_L = I_3 - I_1 = \frac{v_o'}{R_3} - \frac{v_i - e_1}{R_2} \tag{2.27}$$

If $R_2 \gg R_3$, and taking into account Equation 2.27, this results in:

$$I_L \cong \frac{v_o'}{R_3} \cong -\frac{v_i}{R_3} \quad \text{i.e.:} \quad v_o' \cong -v_i \tag{2.28}$$

and the load current is almost independent of R_L.

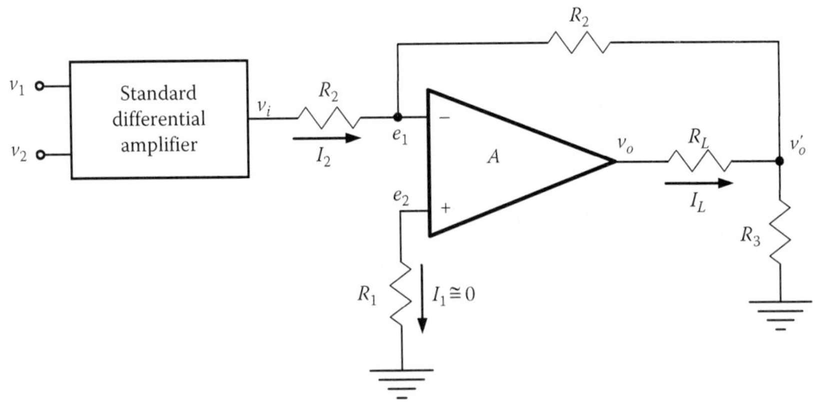

FIGURE 2.9 V-to-I converter with differential input and ungrounded load.

2.3.4 V-to-I Converter with Differential Input and Grounded Load

Figure 2.10 shows a V-to-I converter with differential input and grounded load. This differential amplifier is similar to the one shown in Figure 2.2, with the exception of the output feedback coming from the load. The relationship between the load current I_L and voltage $v_{o1} - v_{o2}$ across the input of the second stage can be obtained from Figure 2.10 by means of inverting e_1 voltages and noninverting e_2 voltages of the input of the second stage:

$$I_L = \frac{v_o - v_{ref}}{R} \tag{2.29}$$

$$e_1 = v_{o1} - \frac{v_{o1} - v_o}{2R_4} R_4 \tag{2.30}$$

$$e_2 = v_{o2} - \frac{v_{o2} - v_{ref}}{2R_4} R_4 \tag{2.31}$$

Subtracting Equation 2.30 from Equation 2.31 and simplifying, we derive:

$$v_o - v_{ref} = v_{o2} - v_{o1} \tag{2.32}$$

and thus, replacing Equation 2.32 in (2.29), we derive:

$$I_L = \frac{v_{o2} - v_{o1}}{R} \tag{2.33}$$

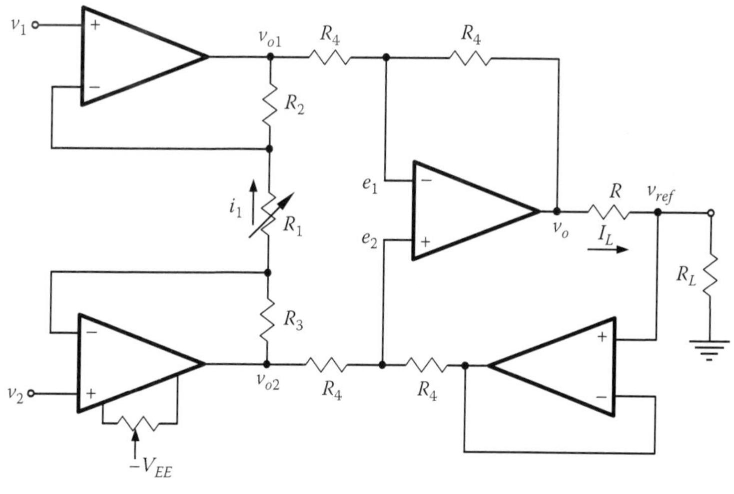

FIGURE 2.10 V-to-I converter with differential input and grounded load.

Therefore, load current is proportional to the differential voltage input and can be adjusted by varying the potentiometer R, regardless of the load.

2.4 ANALOG SIGNAL PROCESSING

Electronic and information technologies have been combined in modern instrumentation. They facilitate the treatment of signals, making them more directly useful for data monitoring, load drives, signal transmission, analysis, and data logging. This treatment can be very broad and some techniques are discussed in the following sections.

2.4.1 Signal Linearization

Signal linearization is a widely used technique for exponentially varying signals, facilitating their interpretation, though, of course, they introduce errors. Figure 2.11 illustrates the circuit and principle of linearization. The diode is applied in this case because its characteristic curve is approximately an exponential curve, as shown in Figure 2.11 [4,5].

The Shockley equation (Equation 2.34), represented in Figure 2.11 as a logarithmic relationship, can be used to linearize an operational amplifier since $v_o \cong -v_d$:

$$I_d = I_s\left(e^{v_d/\eta V_T} - 1\right) \tag{2.34}$$

where

i_d and v_d are, respectively, the diode direct current and voltage
I_s is the reverse saturation current of the diode (typically, from 10^{-15} A to 10^{-6} A)
η is an empirical constant (for the germanium diode $\eta=1$ and silicon diode $\eta=2$, ranging in practice from 1.1 to 1.8)
$q=1.6021\times10^{-19}$ coulombs (the electron charge)
$k=1.38\times10^{-23}$ J/K
T is the temperature in Kelvin
$V_T = kT/q$ (=25.8 mV at 20 °C)

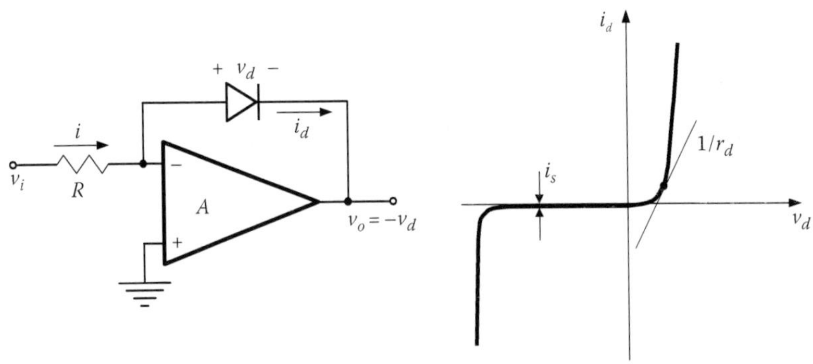

FIGURE 2.11 Implementing the Shockley equation with an operational amplifier.

For example, if $v_d \geq 0.1\,\text{V}$, and considering a germanium diode, Equation 2.34 results in $I_d = I_s(48.23 - 1)$. Thus, multiplying both sides of this equality by R, the approximation of the input voltage, with a 2.1% error, is given as:

$$v_i = RI_d \cong RI_s\left(e^{v_d/\eta V_T} - 1\right) \tag{2.35}$$

Accordingly, the voltage across the diode terminals has the following logarithmic relationship:

$$v_d = -v_o = \eta V_T\left(\ln v_i - \ln RI_s\right) \tag{2.36}$$

An example of this linearization application is a logarithmic potentiometer excited by a voltage source according to a cursor position, which results in:

$$v_i = ke^{\alpha\theta} \tag{2.37}$$

where

- θ is the angular position of the potentiometer cursor to be linearized with respect to v_o
- k is the device's input voltage
- α is a scaling factor

Replacing Equation 2.37 in (2.36) gives:

$$\begin{aligned} v_o &= \eta V_T\left(\ln RI_s - \ln ke^{\alpha\theta}\right) = \eta V_T \ln RI_s - \eta V_T\left(\alpha\theta + \ln k\right) \\ &= -\eta V_T \alpha\theta + \eta V_T\left(\ln RI_s - \ln k\right) \end{aligned} \tag{2.38}$$

Thus, Equation 2.38 has the form of a linear equation $y = ax + b$ if:

$$y = v_o, \quad x = \theta, \quad a = -\eta V_T \alpha, \quad \text{and} \quad b = \eta V_T\left(\ln RI_s - \ln k\right) \tag{2.39}$$

2.4.2 Phase-Shift Amplifier

A phase-shift amplifier can establish a phase shifting between input and output signals while maintaining the same amplitude, as shown in Figure 2.12. Equation 2.40 is the expression of the inverting and noninverting inputs:

$$\begin{aligned} e_1 &= v_i - i_1 R_1 = v_i - \frac{v_i - v_o}{R_1 + R_2} R_1 \\ e_2 &= \frac{v_i R}{R - \dfrac{1}{\omega C}} \end{aligned} \tag{2.40}$$

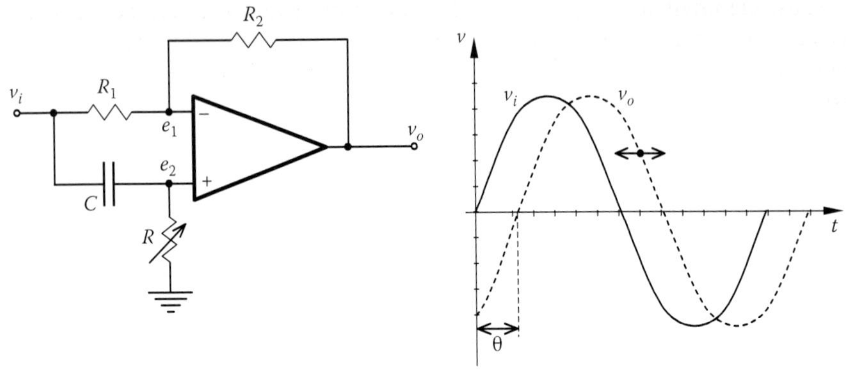

FIGURE 2.12 Phase-shift amplifier.

These equations are equivalent by the virtual short circuit, and then:

$$v_i - \frac{v_i - v_o}{R_1 + R_2} R_1 = \frac{v_i R}{R - \dfrac{1}{\omega C}}$$

or:

$$v_i \left(1 - \frac{R_1}{R_1 + R_2} - \frac{R}{R - \dfrac{1}{\omega C}} \right) = -v_o \frac{R_1}{R_1 + R_2} \tag{2.41}$$

Isolating the output/input signal ratio, we derive:

$$\frac{v_o}{v_i} = -\frac{1 - \dfrac{R_1}{R_1 + R_2} - \dfrac{R}{R - \dfrac{1}{\omega C}}}{\dfrac{R_1}{R_1 + R_2}} = -\frac{R_2}{R_1} + \frac{R}{R_1} \frac{R_1 + R_2}{R - j\dfrac{1}{\omega C}} \tag{2.42}$$

Assuming $R_1 = R_2$, then:

$$\frac{v_o}{v_i} = \frac{\omega RC + j}{\omega RC - j} \tag{2.43}$$

Transforming Equation 2.43 to its polar form, we derive:

$$\frac{v_o}{v_i} \underline{|\theta} = |1| \underline{\left| 2\tan^{-1}\left(\frac{1}{\omega CR} \right) \right.} \tag{2.44}$$

Thus, the phase-shift amplifier provides a unity gain and a desired phase shift between the input and output signals for a particular frequency. Therefore, if a variable shift phase is desirable, it can be achieved by adjusting the resistance with respect to capacitance, which is established as follows:

when $R_1=0$:

$$\frac{v_o}{v_i}=|1|\underline{|180^\circ} \tag{2.45}$$

when $R_1 \to \infty$:

$$\frac{v_o}{v_i}=|1|\underline{|0^\circ} \tag{2.46}$$

Summarizing, the phase shifting between the input and output signals may vary from 0° to 180° by just adjusting the noninverting input resistor (R_1) from zero (i.e., short circuit) to infinity (i.e., open circuit).

Similarly, in the circuit shown in Figure 2.12 and Equation 2.44, the same phase shifting can be obtained by adjusting the capacitor C as follows:

when $C=0$:

$$\frac{v_o}{v_i}=|1|\underline{|-180^\circ} \tag{2.47}$$

when $C \to \infty$:

$$\frac{v_o}{v_i}=|1|\underline{|0^\circ} \tag{2.48}$$

Both conditions shown above enable a phase-shift variation between the input and output signals while maintaining the same amplitude. In practice, the short circuit and the open circuit cause a small-amplitude deviation, since in these cases, the operational amplifier input terminals are actually shorted to the ground for the input signal and in this case it is not virtual short-circuit anymore, thereby changing the normal operation of an operational amplifier.

2.4.3 Schmitt Trigger Comparator

Figure 2.13 illustrates a Schmitt trigger comparator used in instrumentation to protect switching signal states when the closing voltage signal happens at a higher level with respect to the opened voltage level. The Thévenin equivalent voltage from the reference voltage point is defined as:

$$v_{Th}=\frac{R_2}{R_1+R_2}V_{DC} \tag{2.49}$$

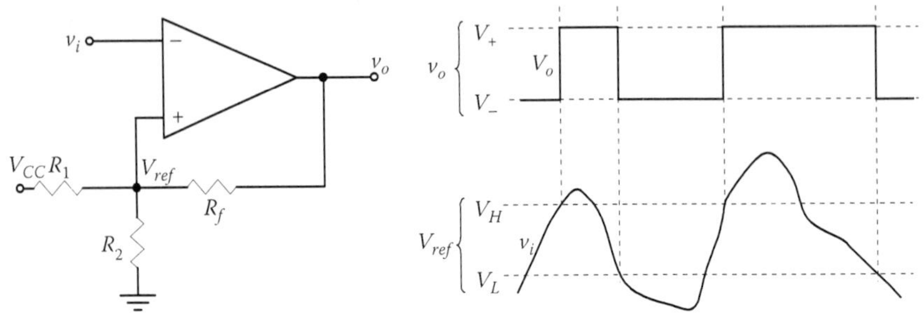

FIGURE 2.13 Comparator Schmitt trigger.

and the Thévenin resistance from this same point is as follows:

$$R_{Th} = \frac{R_1 R_2}{R_1 + R_2} \tag{2.50}$$

Therefore, the reference voltage can be defined as a function of the current through the Thévenin resistance, whenever the output voltage is at the positive V_+ or negative V_- level, which is about the same level of the biasing source V_{CC}. The feedback resistance R_f is in parallel either with R_1 or R_2. When v_i is higher than v_{ref}, then V_o operates at its positive level, like V_+, and we have:

$$R_{f+} = \frac{R_1 R_f}{R_1 + R_f} \tag{2.51}$$

$$V_H = \frac{V_{CC} R_2}{R_2 + R_{f+}} \tag{2.52}$$

Similarly, when v_i is lower than v_{ref}, then V_o operates at its negative level, like V_-, and we have:

$$R_{f-} = \frac{R_2 R_f}{R_2 + R_f} \tag{2.53}$$

$$V_L = \frac{V_{CC} R_1}{R_1 + R_{f-}} \tag{2.54}$$

2.4.4 Oscillators for Signal Generators

Electronic textbooks discuss in detail many of the following oscillator circuits: Hartley (L_1CL_2), Colpitts (C_1LC_2), Clapp ($C_1LC_2C_3$), crystal oscillators (with series or parallel feedback), oscillator Wien, and many others [6–8]. The Wien oscillator is the industry standard and has the basic topology shown in Figure 2.14.

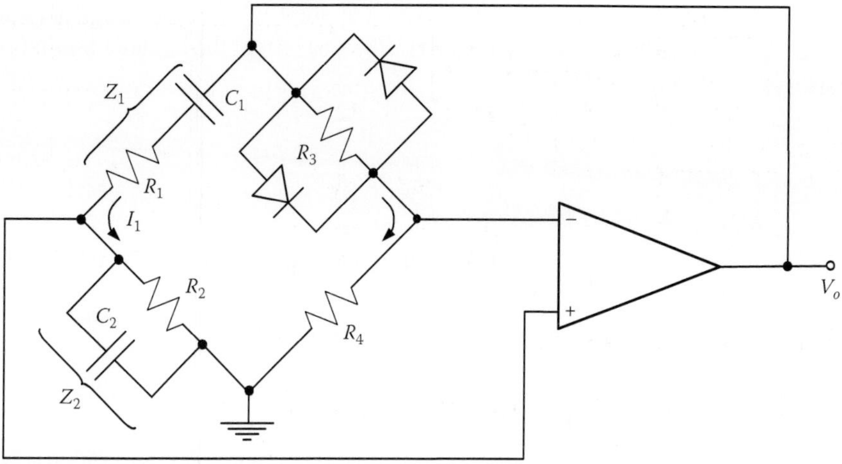

FIGURE 2.14 Wien oscillator.

2.4.4.1 Wien Oscillator

The equilibrium condition of the Wien oscillator, disregarding loading between the operational amplifier input and the bridge output, can be set from:

$$I_2R_3 = I_1\left(R_1 + \frac{1}{j\omega C_1}\right) \tag{2.55}$$

$$I_2R_4 = I_1\left[\frac{R_1\dfrac{1}{j\omega C_2}}{R_2 + \dfrac{1}{j\omega C_2}}\right] \tag{2.56}$$

Dividing Equation 2.55 by (2.56) and rearranging the terms, we derive:

$$\frac{R_3}{R_4} = \frac{\dfrac{1+j\omega R_1C_1}{j\omega C_1}}{\dfrac{R_2}{1+j\omega R_2C_2}} \tag{2.57}$$

Equation 2.57 can be simplified and replaced as follows:

$$\frac{R_3}{R_4} = \frac{\omega\left(R_1C_1 + R_2C_2\right) - j\left(1-\omega^2R_1C_1R_2C_2\right)}{\omega C_1R_2} \tag{2.58}$$

Observe that the left-hand side, R_3/R_4, is a real number, then, the imaginary part on the right-hand side must be zero, which allows us to find the circuit's oscillation frequency as follows:

$$f = \frac{\omega}{2\pi} = \frac{1}{2\pi\sqrt{R_1C_1R_2C_2}} \tag{2.59}$$

Furthermore, handling the real part on both sides of Equation 2.58, we find the relationship between the discrete component values, i.e.:

$$\frac{R_3}{R_4} = \frac{R_1}{R_2} + \frac{C_2}{C_1} \tag{2.60}$$

It is better to use resistors rather than capacitors because it is easier to vary a resistor than a capacitor; therefore, the former is preferred to select the frequency, although the use of capacitors with a double body is also common. So, the Wien bridge presents several problems. The first one imposes the need of a simultaneous value changes in the two control resistors to maintain the balance condition of Equation 2.60, if an adjustable frequency of oscillation is desired. Usually, double potentiometers are not well matched. The second problem is related to the use of the Wien bridge as a reference frequency by replacing the entries "+" and "−" of the operational amplifier with the output terminals of the excitation voltage that one wants to measure. This problem is related to the difficulty in balancing up the bridge in case the excitation voltage is not perfectly sinusoidal, since the harmonic content would cause a real change in the oscillating frequency of the bridge. The third problem is related to the stability of the components with regard to temperature, replacement, aging, and moisture.

Resistor R_1 is adjusted until the oscillations begin. In this condition, the inverting input of the operational amplifier will be around $v_o/3$. As the voltage across the diodes increases, they start to conduct and the diode impedance decreases, which therefore increases the negative feedback.

The adjustment of R_1 varies the amplitude of the output wave, which reaches a new stable amplitude. There is a slight interaction between the amplitude, stability of the amplitude, and distortion of the output waveform. The control amplitude is indirect since the potentiometer must be positioned such that distortion is minimized. This distortion limitation can also be achieved with thermal limiters, such as thermistors or incandescent bulbs. The greater is the distortion, the greater is the amplitude of the output waveform. Matched diodes minimize this distortion. The frequency stability depends on the quality stability of the components used in the circuit.

The amplitude of the generated signal can also be defined from the specification of amplifier compliance, i.e., the maximum peak-to-peak values of the input and output signals limited by the positive and negative levels of amplifier saturation.

In order to design components of the Wien oscillator, as depicted in Figure 2.14, one can select:

$$R_1 = R_2 = R \quad \text{and} \quad C_1 = C_2 = C \tag{2.61}$$

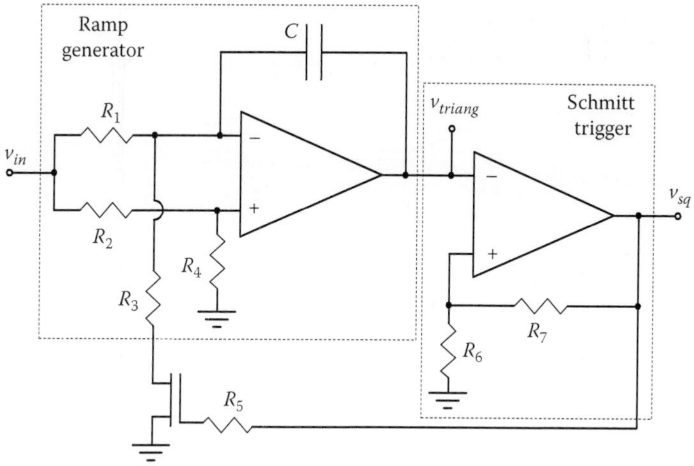

FIGURE 2.15 Voltage-controlled oscillator (VCO) (or voltage-to-frequency converter, VFC).

If these values are replaced in Equations 2.59 and 2.60, we derive:

$$f = \frac{1}{2\pi RC} \quad \text{and} \quad \frac{R_3}{R_4} = 2 \tag{2.62}$$

2.4.4.2 VCO Oscillator

The voltage-controlled oscillator (VCO) or voltage-to-frequency converter (VFC), shown in Figure 2.15, may be used as a signal modulator. Note that resistors R_1 and R_2 serve only to soften the effects of steady sources (low internal resistance) of the virtual short circuit across the operational amplifier inputs.

Resistor R_3 sets the duty cycle of the square wave. The limits of the triangular wave voltage (v_{triang}) are due to the connection of resistors R_6 and R_7, and in some configurations, the resistors are in series with another auxiliary resistor of the parallel power supply, both on the high level, $2V_{DC}/3$, and on the low level, $V_{DC}/3$, of the Schmitt trigger (switching boundaries) established at the square wave output (v_{sq}). The frequency of this circuit is calculated from the RC-ramp slope integrator, both in ascent and in descent, to come from voltage levels $V_{DC}/3$ to $2V_{DC}/3$. The input voltage level defines the output frequency of the VCO.

2.4.4.3 Relaxation Oscillator

An example of the relaxation oscillator is given in Figure 2.16, with a cycle continuously repeating with a period of approximately 2.2*RC* as discussed below, independent of the power supply level. If the relaxation oscillator is connected to an integrator circuit, it can obtain a triangular wave. In this case, the capacitor charge varies as a function of time *t*, from zero to *T*/2, under a load voltage $V_z + BV_z$ displaced vertically by $-BV_z$, i.e.:

$$v_c(t) = (V_z + BV_z)\left(1 - e^{-t/RC}\right) - BV_z \tag{2.63}$$

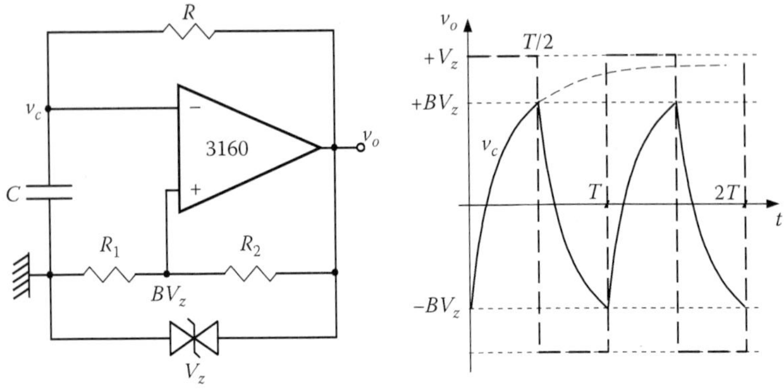

FIGURE 2.16 Relaxation oscillator.

where

$$V_{c+} = \frac{V_z R_1}{R_1 + R_2} = BV_z \tag{2.64}$$

$$B = \frac{V_c}{V_o} = \frac{R_1}{R_1 + R_2} \quad \text{for} \quad i = \frac{V_o}{R_1 + R_2} = \frac{V_c}{R_1} \tag{2.65}$$

Using Equations 2.64 and 2.65 for a 50% duty cycle (charging time = discharge time) of the capacitor voltage, one has:

$$v_c\Big|_{t=T/2} = BV_z = \left(V_z + BV_z\right)\left(1 - e^{-T/2RC}\right) - BV_z \tag{2.66}$$

or isolating the exponential term:

$$e^{-T/2RC} = \frac{1-B}{1+B} \tag{2.67}$$

Applying the logarithm on both sides of this identity and solving it as a function of T, we derive:

$$T = 2RC \ln \frac{1+B}{1-B} = 2RC \ln\left(2\frac{R_1}{R_2} + 1\right) \tag{2.68}$$

A more complete version of the relaxation oscillator shown in Figure 2.16 is presented in Figure 2.17, where the stiff voltage sources, represented by the voltage across capacitor C and the voltage divider $R_1 - R_2$, become more flexible with resistors R_3, R_4, and R_5.

$$f = \frac{1}{2RC \ln\left(1 + 2\frac{R_1}{R_2}\right)}$$

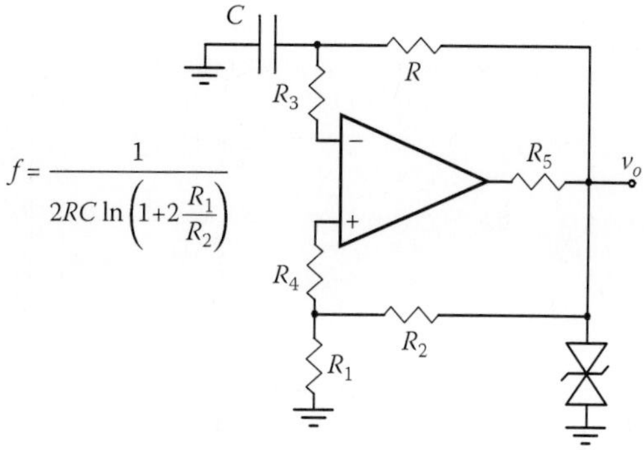

FIGURE 2.17 Example of relaxation oscillator.

A relaxation oscillator with buffer works well with the magnetic isolation amplifier shown in Figure 2.4 to transfer energy from the primary source to the isolated side amplifier.

2.4.4.4 Crystal Oscillators

The main feature of crystal oscillators is the stability of the oscillation frequency. Most of these oscillators are based on a quartz crystal, whose electrical equivalent is shown in Figure 2.18. The crystal suffers changes in its structure when subjected to an applied voltage across the plates containing it. The crystal quality factor is very high, varying typically from 20,000 to 1,000,000 [12]. The equivalent electrical model of the crystal is an RLC series circuit in parallel with the resulting capacitance of the crystal encapsulation, as illustrated in Figure 2.18. The crystal parameters R (internal friction of the crystal structure or viscous damping), C (ductility or reciprocal of the spring constant), and L (mass) are highly stable with respect to the mechanical capacitance across its terminals. A crystal-based 90 kHz oscillator has typically the following parameters: $L=137\ H$, $C=0.0235$ pF, $C_M=3.5$ pF, with $C_M \gg C$, and

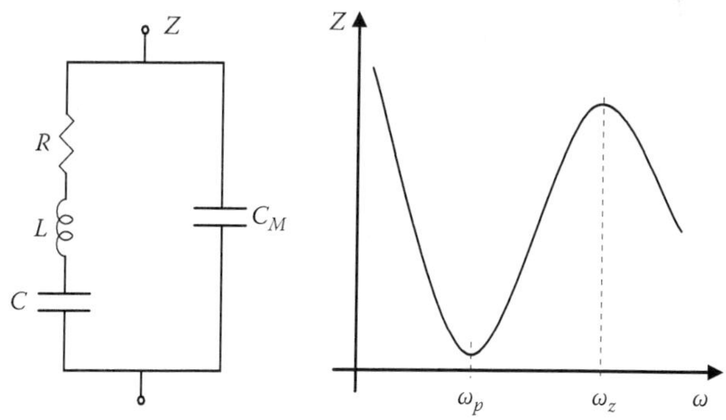

FIGURE 2.18 Equivalent circuit of a crystal.

$R = 15\ \text{k}\Omega$, corresponding to a $Q = 5500$. Therefore, neglecting the internal resistance of the crystal, impedance across the crystal terminals is:

$$Z = j\left(\omega L - \frac{1}{\omega C}\right) // \left(-j\frac{1}{\omega C_M}\right) = \left(-j\frac{1}{\omega C_M}\right)\frac{\omega L - \dfrac{1}{\omega C}}{\omega L - \dfrac{1}{\omega C} - \dfrac{1}{\omega C_M}} \tag{2.69}$$

Multiplying the numerator and denominator of the above equation by ωL, we derive:

$$Z = \left(-j\frac{1}{\omega C_M}\right)\frac{\omega^2 - \dfrac{1}{\omega C}}{\omega^2 - \dfrac{1}{L}\left(\dfrac{1}{C} + \dfrac{1}{C_M}\right)} \tag{2.70}$$

Equation 2.70 shows that there is a series resonance frequency (frequency of zero impedance) $\omega_z^2 = 1/LC$. Actually, C_M establishes an antiresonance frequency f_p or parallel resonance $\omega_p^2 = \dfrac{1}{L}\left(\dfrac{1}{C} + \dfrac{1}{C_M}\right)$ (frequency of the pole), as shown in Figure 2.18. As $C_M \gg C$, these frequencies are about the same and the circuit will oscillate at a frequency between zero and that of the pole, but close to the frequency of parallel resonance.

The series capacitance C has a much lower value than the parallel capacitance C_M, which makes the frequency of the crystal oscillator independent of it and of this series capacitance originated by the crystal envelope. Therefore, C_M is negligible for the purpose of establishing a crystal resonance frequency of $X_L - X_C - X_{CM} = 0$. That is, if we neglect R, the crystal impedance across the crystal terminals will be:

$$\omega L - \frac{1}{\omega C} = 0 \tag{2.71}$$

i.e.:

$$f = \frac{1}{2\pi\sqrt{LC}} \tag{2.72}$$

2.4.4.5 Types of Crystal Oscillators

Figure 2.19 shows the types of crystal oscillators, every one with a particular specificity. Figure 2.20 shows a series resonant crystal oscillator using an operational amplifier. The output wave shape is approximately square and the wave uses a high gain with a pair of diodes in its output [9–11].

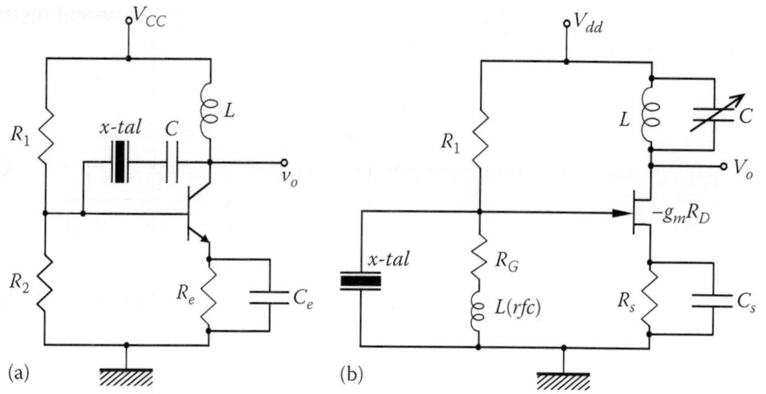

FIGURE 2.19 Types of crystal oscillators: (a) resonant series and (b) resonant parallel (Miller oscillator).

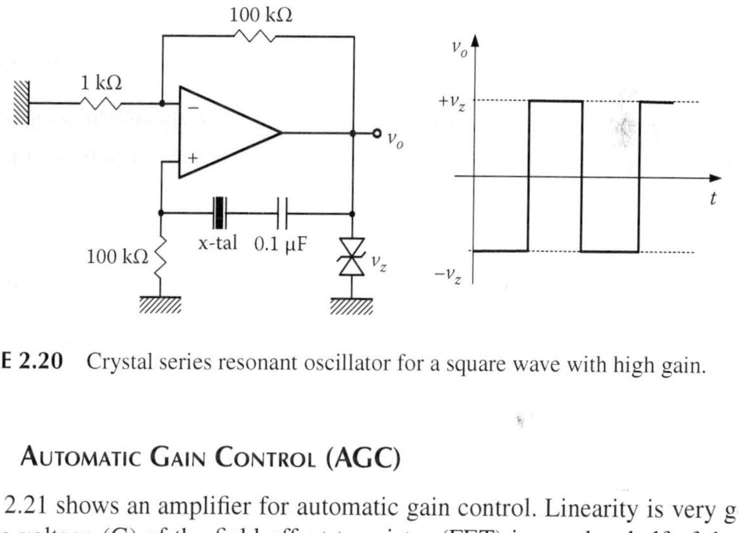

FIGURE 2.20 Crystal series resonant oscillator for a square wave with high gain.

2.4.5 Automatic Gain Control (AGC)

Figure 2.21 shows an amplifier for automatic gain control. Linearity is very good if the gate voltage (G) of the field-effect transistor (FET) is equal to half of the drain source (DS) voltage. The amplifier gain is adjustable from 1 to 1000 by the voltage control V_c. The two resistors connected to the FET gate improve linearity while remaining always $V_{GS}=V_{DS}/2$, and there is no distortion up to 8.5 V of the output signals.

2.5 SIGNAL CONDITIONING

The use of attenuators for signal conditioning helps to match system voltage levels to the instrumentation levels. The attenuator may be a resistive divider usually adapted to a buffer amplifier output. The attenuator has the purpose of matching the instrument impedances and measuring quantities, such as voltage and current levels [13,14].

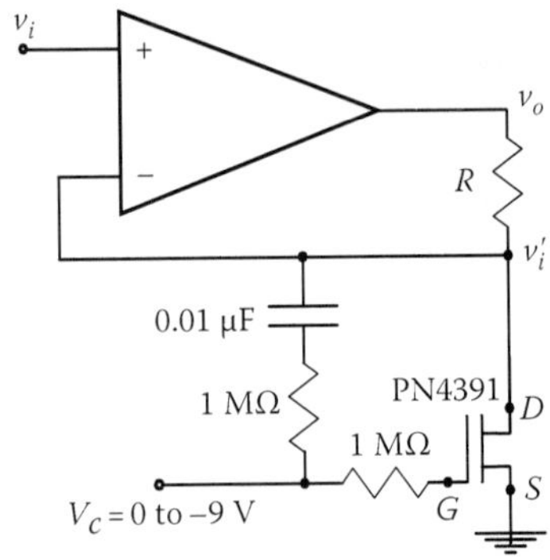

FIGURE 2.21 Automatic gain control (AGC).

Besides impedance matching, there are undesired levels of direct current (offset and drift) that must be mitigated in data acquisition. One way to eliminate offsets and drifts in the transmission of low-level DC signals is signal modulation (see Section 2.2.4). Furthermore, an AC signal helps to eliminate external interferences in the transducer signal. A usual application for compensating offsets and drifts is to make use of a chopper converter (DC–DC converter) along with a signal modulation technique, such as PWM, PFM, PDM, or PAM. By definition, for AC signals, PWM, PFM, PDM, or PAM, all can be applied. Figure 2.22 shows an example of how the signal feeding the amplifier will look like.

The treatment of these PWM, PFM, PDM, and PAM signals use widespread techniques in signal conditioning field, especially in triggering processes by the computer, known as D/A and analog-to-digital (A/D) conversions. Other forms of conversion are digital-to-digital (e.g., binary code to the Gray code and from digital NTSC to PAL systems) and analog-to-analog (e.g., transformers and voltage dividers). With the advent of digital high-definition (HD) TV, the difference between NTSC and PAL is practically limited to the frame rate of the display, which is 23,975p/30p/60i in NTSC and 25p/50i in PAL standards. In modern sets, as the digital decoder is

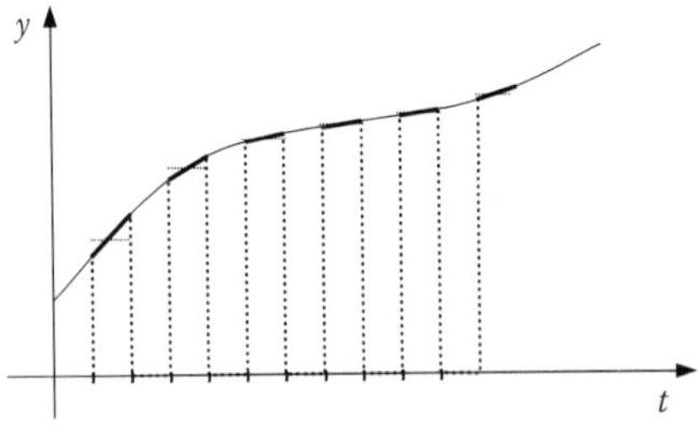

FIGURE 2.22 Pulsed signal.

running on DC digital TV sets, the voltage frequency is not an issue; thus, the voltage frequency is irrelevant, since there is an AC/DC converter.

2.5.1 Digital-to-Analog Conversion

There exist a variety of techniques for converting digital signals into analog signals found in basic texts and electronic signal processing. However, the most well-known techniques are the adder inputs type and pondered R-2R ladder type, both discussed in this section.

Figure 2.23 shows an adder type with binary inputs whose weighted resistor values are in a binary sequence 2^n. If the reference level V_{ref} is used as the logic "1" and the ground is the virtual logic "0," then the output analog voltage will be given by:

$$V_a \cong V_{ref}\left(\frac{b_0}{2^0}+\frac{b_1}{2^1}+\cdots+\frac{b_{n-1}}{2^{n-1}}\right) \tag{2.73}$$

where

b_i is the logic state of bit "i" ("0" or "1")
n is the number of converter input bits

Equation 2.73 is not exact because there is always switching distortion and the need for filtering the amplifier output voltage. Moreover, the number of digital bits n will determine the converter resolution (the number of possible discrete values), which will always be less than or equal to $v_a/2^n$, i.e., the number of possible discrete v_a values. The nonzero minimum variation in the level of v_a is $V_{ref}/2^{n-1}$ (least significant bit, LSB), and $2V_{ref}(1-2^{-n})$ is the maximum value, which occurs when all bits are set at level "1" relative to level "0" (see Equation 2.73). This expression results in the sum of terms of the geometric series equation (Equation 2.73), calculated by:

$$S_n = \frac{a_1\left(1-q^n\right)}{1-q} \tag{2.74}$$

where the first term is $a_1=1$ and the ratio is $q=1/2$.

Converter limitations of weighted inputs are the following: the diversity of the needed resistors values, the large increase in the number of resistors needed to attend to the large number of bits, the high accuracy of the input resistors for larger numbers of bits to avoid undesired effects on conversion precision, and load surges on the reference source as the bits are simultaneously turned on or off.

Figure 2.24 shows a D/A converter using a R-$2R$ ladder network. If the output of this circuit is loaded with $2R$, the Thévenin impedance seen by each key (bit) of the ladder and the ground is $3R$, either going directly to the ground or through the source and then going from there to the ground. For this reason, these inputs are typically supplied by constant-current sources and the result is the sum of the currents feeding the output amplifier.

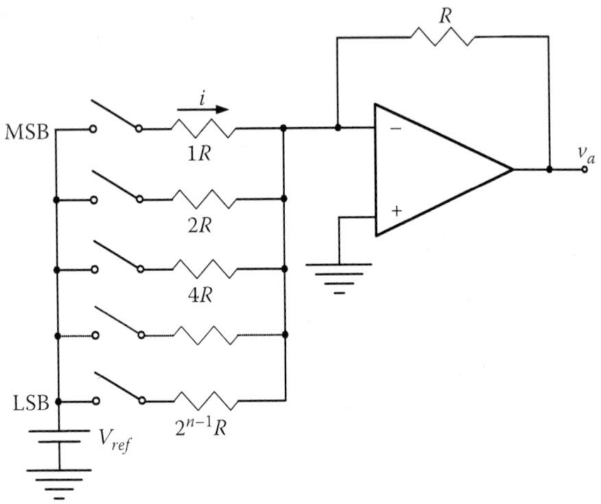

FIGURE 2.23 D/A converter of the pondered adder type.

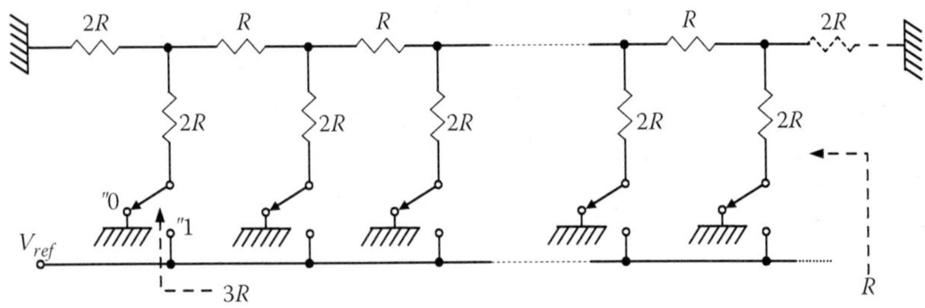

FIGURE 2.24 R-$2R$ ladder network for a D/A converter.

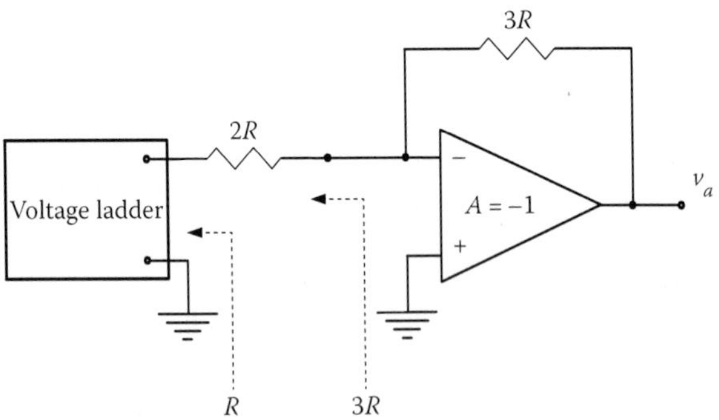

FIGURE 2.25 Gain and impedance of D/A converter of ladder type.

In terms of load, one can use the equivalent Thévenin circuit, assuming $V_{ref}=0$, as shown in Figure 2.25. The input resistance of this converter is connected directly to the ground or through the source resistance. In both cases, it is seen by the output amplifier just as R.

2.5.2 Analog-to-Digital Conversion

Another conditioner also very useful to control, drive, and monitoring industrial processes is the A/D converter. This converter is the core of data acquisition systems, making possible for the computer to interpret and process the data coming from physical processes.

The voltage-to-frequency conversion (V/f) may be considered an analog-to-digital conversion and is used in optical transmission or magnetic signals, thereby eliminating the nonlinearities of the components or opto-magnetic cores. This sort of converter can be transformed into a single-bit converter, and it is sometimes called 1-bit converter or delta-sigma. The delta-sigma circuit has two main sections: *Delta* receives the incoming digital signal and monitors the outgoing pulse train, then it creates an *error signal*, which is based on the difference between the binary signal coming in and the pulse train going out. *Sigma* adds up the results of the error signal created by delta and supplies this sum to a low-pass filter. It is called so because it uses only on and off states. The 1-bit converter can be used in voltage-to-frequency converter, as illustrated in Figure 2.26. It has a kind of frequency modulation produced by the input analog signal V_a. This frequency is read by the computer in a time window T_w. Slow conversion, low cost, and simplicity of implementation are its main features.

If a higher number of bits are required, it is preferable to use the standard A/D converter shown in Figure 2.27. This converter operates as follows. Whenever placing an analog value to be converted, the counter is started, which only stops when the level of the voltage ramp reaches the value of V_a. This value should be kept constant during the conversion (e.g., by using a circuit sample-and-hold operation). The counter also needs a signal to set or reset it. The conversion time varies in proportion to the signal level to be converted. Notice that the input clear (clr) performs as an enable action, and must be set so for a proper operation.

The fastest converter is the flash (parallel) or the comparator converter (flash encoding), as shown in Figure 2.28. This converter is commercially limited from

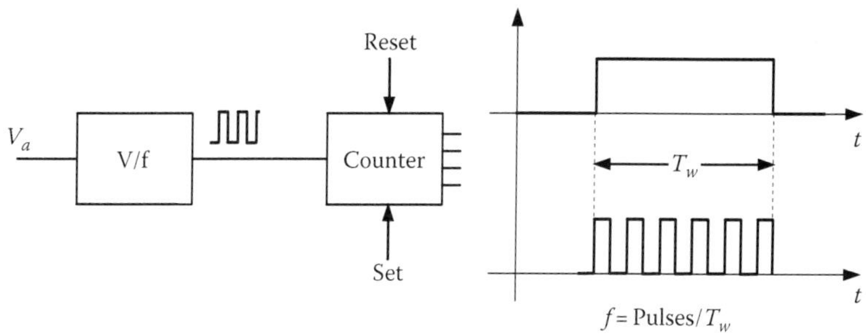

FIGURE 2.26 1-bit A/D converter using voltage-to-frequency conversion.

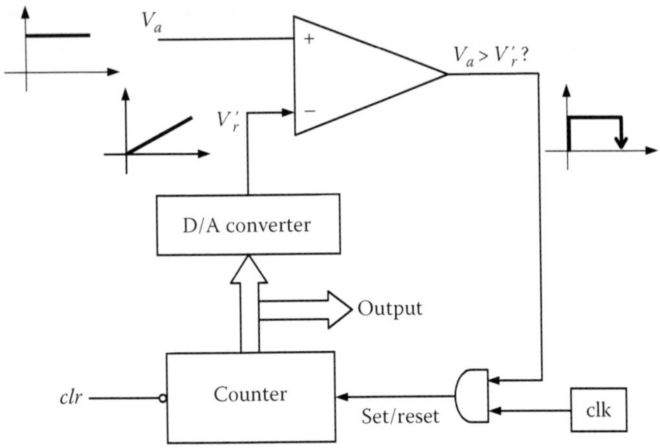

FIGURE 2.27 Standard D/A converter.

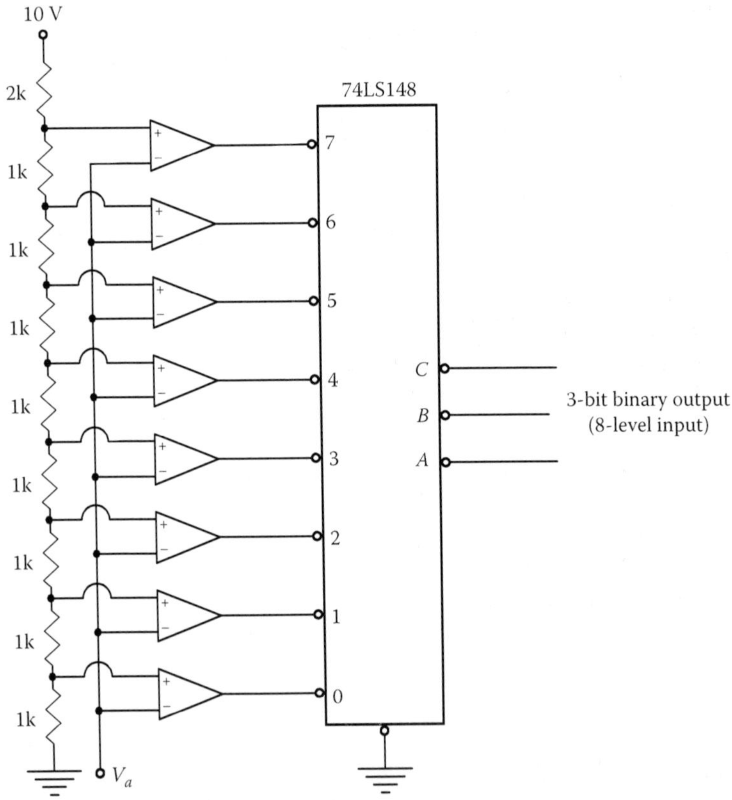

FIGURE 2.28 A/D flash converter of the parallel encoder type.

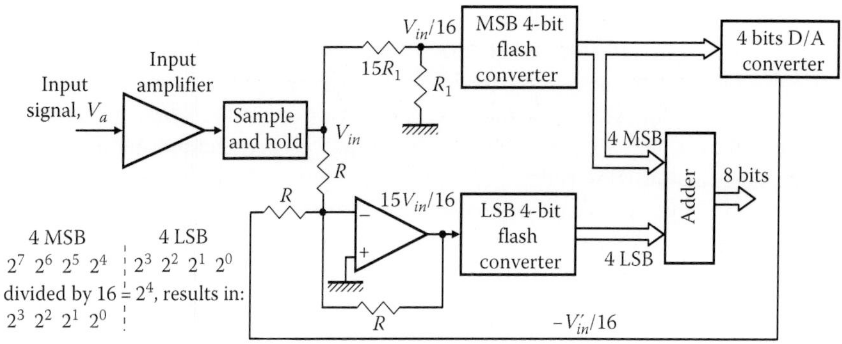

FIGURE 2.29 Schematic diagram of a double-flash converter.

16 to 256 levels (4–8 bits), because above these levels, it becomes expensive, large, and too complex. The resolution of this converter is given by the ratio between the number of output bits and the analog input levels, i.e.:

$$\text{Number of levels} = 2^{\text{bits}} \tag{2.75}$$

The 74LS148 IC (priority encoder) and NE521 are widely commercialized, and the latter has a conversion time of less than 20 ns.

To reduce the hardware of flash converters while maintaining the same conversion rate, one can use two flash converters, as shown in Figure 2.29. The input signal, after the sample-and-hold operation, follows two paths. The first one splits the analog input signal by $2^{n/2}$ through a resistive divider, where n is the number of bits at the desired output such that $1/2^{n/2}$. The second path supplies the remaining difference $(2^{n/2} - 1)/2^{n/2}$.

In the example illustrated in Figure 2.29, n is equal to 8. The resulting value $V_{in}/2^{8/2} = V_{in}/16$ of this path converts the most significant 4 bits by the flash converter, i.e., the number of most significant bits (MSB) is divided by 2^4 and it becomes equal to the number of less significant bits. Its output is reconverted back into an analog signal, which is subtracted from the input signal V_{in} in a relationship R_2/R_1, resulting in a fraction of 15 $V_{in}/16$. The resulting value of this subtraction is then converted by a second flash converter, which complements the least significance 4-bit series (LSBs).

In terms of hardware, the combined use of the two flash converters shown in Figure 2.29 reduces the number of comparators from 256 to 32 (i.e., eight times smaller) compared with the example discussed in Figure 2.28. In contrast, the conversion time is twice, which is the conversion time of the two flash converters, operating at one conversion after the other, added to the D/A converter time. Surely, this is the fastest conversion type and the most commonly used for processing digital signals, being a direct way of addressing all the bits at the same time.

The A/D converter by successive approximations is the most popular with regard to cost, good accuracy, and relative speed of conversion (1–50 μs). It uses a conversion method similar to that used by the digital voltmeter. It is based on the comparison of, one after the other, the MSB n-bits of the output code feeding a D/A converter. The result is compared with the analog input to be measured via the comparator. The undesirable features of this converter are the very low immunity to input spikes, which are disastrous for digital processing, and present unexpected nonlinearities and missing codes.

A variant of the successive approximation converter, uses up/down counters to generate the codes for the successive attempts. The method consists of an initial digital attempt, which should be above or below the analog value to be converted. The result of this comparison sets a flip-flop type D (when $V_r < V_a$) or restarts counting (when $V_r \geq V_a$). This converter is slow and responds well to signal discontinuities at the input, but works better when the variations are smooth; thus, it is known as the tracking converter (up-and-down counting).

Another well-known A/D converter uses single integration to estimate digitally the input value, as shown in Figure 2.30. An internal ramp voltage generator begins by comparing the conversion of a voltage with the output voltage of a counter. Tolerance of capacitance is this converter's main limitation.

Another method that eliminates A/D conversion is discussed in details in Chapter 4 and is illustrated in Figure 2.31 where a trace capacitor compares the

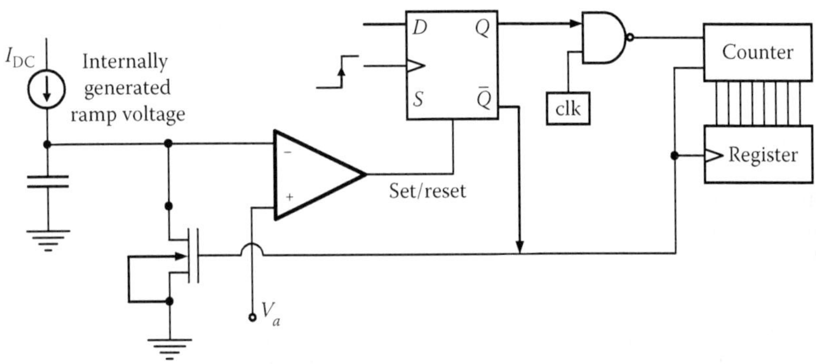

FIGURE 2.30 Single-ramp integration.

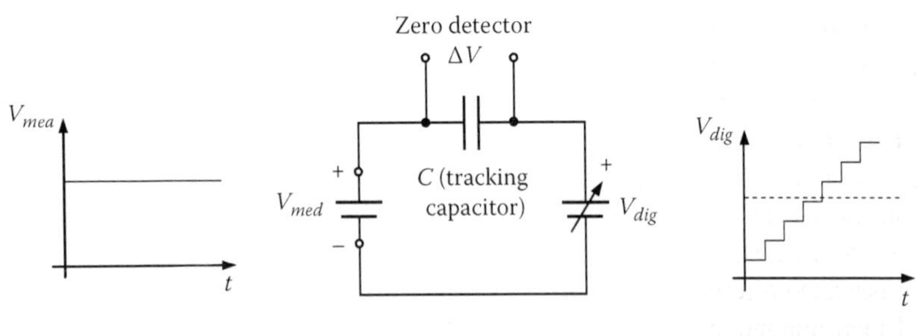

FIGURE 2.31 Analog-to-digital converter with load balance.

reference voltage V_{ref} generated by a digital counter with the value being measured. The current through or voltage across the capacitor is zero when the internal digital voltage and the analog measured voltage are equal, and the reading is registered on the counter.

There are several other A/D conversion techniques (around 12), some of them not very much conventional. However, these do not fall under the scope of this book, just to mention only some of them:

- Shaft encoder.
- Dual-slope integration.
- Delta-sigma converters: an internal power train of pulses is connected to a summing point with an analog voltage v_a. These pulses are counted up until the output level of an integrator (connected to the summation point) changes the state of the comparator.
- Switched capacitor with digital variation of the instrument internal voltage for comparison with the voltage being measured.

Most data acquisition boards include 8 or 16 channels from one analog input, and some more expensive options offer up to 128 channels for a single entry. Signal conditioners with internal or external multiplexing may become feasible to expand the capacity of data acquisition boards to treat large numbers of channels.

2.6 WORKING WITH CIRCUIT BOARDS FOR SIGNAL CONDITIONING

The practical implementation of circuit boards in instruments requires some practical measures in the components' layout, as well as in any other electronic circuit driving small signals:

1. Installation of coupling capacitors as close as possible to the integrated circuits subject to fast signal changes (conducted noise) as a means of locally supplying instant energy to these circuits if the main source is considerably apart from them. For very fast changing signals, it is used two parallel decoupling capacitors, one electrolytic of high capacity and another one of ceramic, tantalum, or polyester.
2. Use of cables and wires as short as possible, preferably integrated circuits; otherwise, such cables and wires will act as true noise pickup antennae (radiated noise).
3. Proper grounding signifies the existence of an only ground and a single connection to the ground; otherwise, parallel connections of different masses are at different grounding levels on the integrated circuit according to the resistance offered by the grounding path.
4. Electromagnetic shields are needed in electromagnetically noisy environments (radiated noise).

5. Mirroring effect of the component layout with respect to the ground point when there are significant electrical connections to the ground potential.
6. Signal noise introduces significant numerical errors in microprocessors, making misrepresentations of very small analog signals more frequent or whenever high gains become necessary.

EXERCISES

2.1 Design a good-quality differential isolated amplifier for a voltage signal $v = V_m \sin \omega t$ for a frequency range between 0 and 10 kHz, unity gain, linearity of 0.1%, selecting one among the following types of isolation: (a) magnetic, optical, and single-source electrical isolation for 50 mA, 12 V DC, continuous use, and/or alternating current; (b) voltage sources can be as follows—(i) conventional rectifier transformer, (ii) Zener splitter, (iii) high-frequency rectifier, (iv) battery, (v) capacitive divider, and (vi) resistive divider.

2.2 Give five main characteristics of instrumentation amplifiers, discussing how each of them can affect the measurement quality.

2.3 Establish the resistance values and the minimum operational amplifier supply power instrumentation for a voltage-to-current converter such that its transconductance gain can be –0.01 Siemens (S). From there, calculate what is the value of the output current for an input voltage signal $v_i = 0.5 \sin \omega t$ when $f = 1$ kHz and $R_L = 500\ \Omega$, 1 kΩ, and 2 kΩ. The operational amplifier can supply a maximum current of 20 mA.

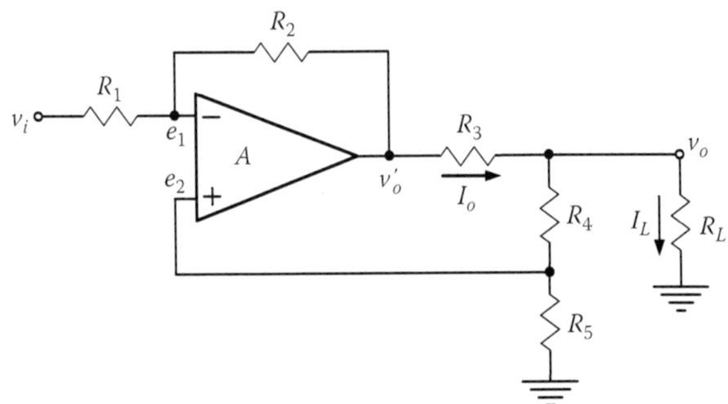

2.4 Assuming only resistors with 5% tolerance and an ideal operational amplifier, which would be the largest error you can expect from a 4-bit D/A conversion using the summation type converter. The minimum acceptable resistor to be used is: 10 kΩ and $V_{ref} = 1.2$ V. Determine the maximum voltage level that can be read and the resolution of such converter.

2.5 Calculate the maximum frequency that can be read, the resolution and the dynamic range of a 12-bit A/D converter for a signal sampled within a 2-s cycle at a 100 kHz sampling rate.

2.6 Explain the working principle of the D/A converter type ladder voltage for "*n*" bits, establishing which are the smallest and most significant values that they could register.

2.7 Describe in detail the operating principle of the fastest A/D converter commonly known in the technical literature.

2.8 Discuss the advantages of the A/D converter *R*-2*R* type with respect to the pondered adder type and when would the use of this latter type be justified.

2.9 Design a single-input V-to-I converter as in the following figure with an operational amplifier whose output current is in between 4 and 20 mA such that $0 < V_i < 5$ V and $V_{DC} = \pm 12$ V. Consider compliance limitations and $R_1 = R_2 = 100$ kΩ.

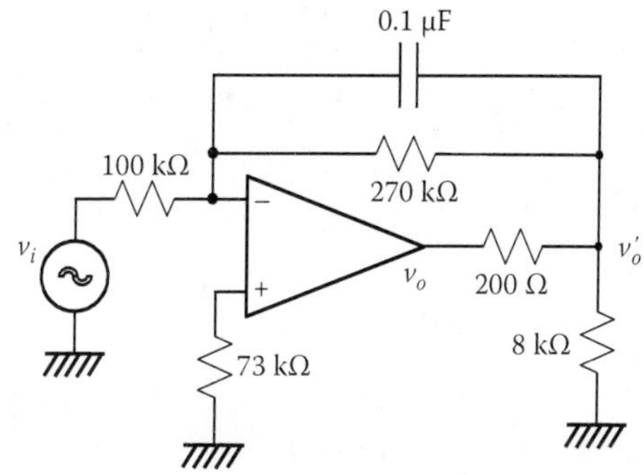

2.10 Calculate a reasonable scan frequency range and gain of a voltage-to-current converter to be used in the instrument amplifier suggested below.

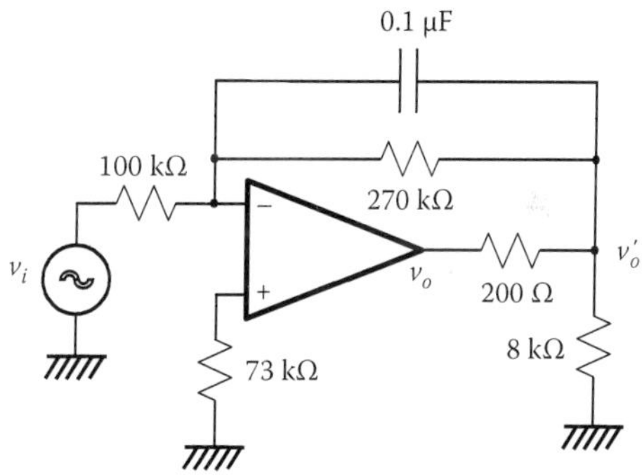

REFERENCES

1. Graemme, J.G., Tobey, G.E., and Huelsman, L.P., *Operational Amplifiers: Design and Applications*, Burr-Brown Publishers, Editors, Tokyo, Japan, 473pp., 1971.
2. Silveira, P.R. and Santos, W.E., *Discrete Automation and Control*, Érica Editors, São Paulo, Brazil, 229pp., 1998.
3. Diefenderfer, A.J., *Principles of Electronic Instrumentation*, W.B. Saunders Co., Publisher, Philadelphia, PA, 667pp., 1972.

4. Slotine, J.J.E. and Li, W., *Applied Nonlinear Control*, Prentice Hall International Publisher, Englewood Cliffs, NJ, 461pp., 1991.
5. Zelenovsky, R. and Mendonça, A., *PC: A Practical Guide of Hardware and Software*, 3rd edn., MZ Publishers, Rio de Janeiro, Brazil, 760pp., 1999.
6. Regtien, P.P.L., *Electronic Instrumentation*, VSSD Publisher (Vereniging voor Studie en Studentenbelangen te Delft), Delft, The Netherlands, 2015, ISBN: 978-90-71301-43-8.
7. Bowron, P. and Stephenson, F.W., *Active Filters for Communications and Instrumentation*, McGraw-Hill Book Company, Berkshire, U.K., 285pp., 1979.
8. Auslander, D.M. and Agues, P., *Microprocessors for Measurement and Control*, Osborne/McGraw-Hill, Berkeley, CA, 310pp., 1981.
9. Bolton, W., *Industrial Control and Measurement*, Longman Scientific and Technical, Essex, England, 203pp., 1991.
10. Eveleigh, V.W., *Introduction to Control Systems Design*, TMH Edition-Tata McGraw-Hill Publishing Company, New York, 624pp., 1972.
11. Wait, J.V., Huelsman, L.P., and Korn, G.A., *Introduction to Operational Amplifier Theory and Applications*, International Student Edition, McGraw-Hill Kogakusha, Tokyo, Japan, 396pp., 1975.
12. Boylestad, R. and Nashelsky, L., *Electronic Devices and Circuit Theory*, Pearson New International Edition, Essex, England, 2014, ISBN: 1-292-02563-8.
13. Horowitz, P. and Hill, W., *The Art of Electronics*, Cambridge University Press, London, U.K., 716pp., 2002.
14. Kitchin, C. and Counts, L., *A Designer's Guide to Instrumentation Amplifiers*, 3rd edn., Analog Devices Co., Norwood, MA, 2006.

3 Sensors and Transducers

3.1 INTRODUCTION

Any data coming from a physical phenomenon can be monitored by measurements via sensor elements so as to have information about the behavior of the quantities involved in the process. The sensor device contacting or approaching the physical phenomenon changes its physical characteristics according to a transformation law itself. Such events are most often unintelligible to human perception if not properly converted into measuring elements. To this end, transducers can translate to humans, or to an equipment, the data coming from the variable being measured. For example, a dilatation thermometer increases in volume a colored liquid within a capillary tube to indicate the temperature on a scale calibrated in degrees.

It can be seen then that a data acquisition system is part of a chain of transformations aimed to interpret the measurement in an intelligible form to be understood by humans or by other instruments or a measurement equipment. This instrumentation chain can be divided into four stages: the physical phenomenon, the interface, the measurement processor, and the display to interpret or process a solution, as illustrated in Figure 3.1.

The physical phenomenon is the real process, while the interface is the regulating signal part between generating data from physical processes and the required processor, such as voltage levels, data transmission standards, signal variation speed, and signal coupling or isolation, among others. The transducers are inserted in the interface block. The simplest example of an interface in electronic instrumentation can be the cables used to measure voltage and current by voltmeter and ammeter instruments. Interface compatibility between voltage or current levels can be exemplified by a current or voltage transformer, or a resistive divider, and so on. It is in this chain that transducers are inserted in. Interfaces are aimed at harmonizing the generating data processes and the required instrument [1,2]. Examples include voltage levels, data transmission standards, signal variation speed, and signal coupling or isolation, among others.

Transducers (or sensors) are used according to their affinity with the physical quantity being measured and with the instrument for signal or data acquisition. Many data acquisition systems may involve more than one transducer, e.g., the wattmeter.

Depending on their operating principles, mechanical or electrical sensors can be active or passive. A sensor is classified as passive whenever the interpretation of the measurement requires an external power source to represent it, e.g., a resistive divider. On the other hand, an active sensor is the one that measures physical variables with a topology that requires internal power. Table 3.1 gives examples of the two types, including other sorts of sensors.

On the other hand, transducers are seen as the simplest version of element combination to translate data from sensors about properties of the phenomenon under study

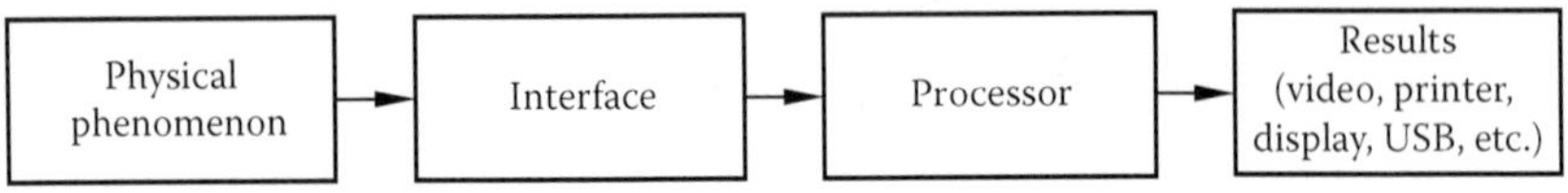

FIGURE 3.1 Basic elements of an instrumentation chain.

TABLE 3.1
Examples of Sensor and Transducers

Type/Description	Example	Physicochemical Principle	Input Signal	Output Signal
Electrical passive	Resistance thermometer	Temperature depending on the resistance	Temperature variation	Resistance change
Electrical active	Thermocouple	Thermoelectric effect	Temperature variation	Electromotive force (EMF)
Mechanic	Spring	Elastic expansion	Force	Length change
Electrochemical	Blood composition	Electrochemical effect	Alcohol content	EMF
Electrobiological	Electromyography	Electrobiological effect	Eye electrical signals	EMF
Radioactive	Geiger-Müller tube	Ionizing radiation	Radioactivity	Current

and therefore make it perceptible, graduated, and intelligible to human beings or to an equipment. Moreover, transducers can act directly or indirectly. However, sensors are defined as direct when their changes in properties due to the studied phenomenon can be directly sensed by human beings, without intermediate transformations to make them discernible, graded, and intelligible (e.g., temperature measurements using mercury dilatation over a graduated range). Transducers are indirect if changes in the properties of the studied phenomenon have to be processed in some way to be noticeable, graded, and understood by users or an equipment, e.g., the galvanometer mechanism that converters the measured electric current into a magnetic torque and hence moves a probe on a graduated scale.

According to the purposes of this book, sensor elements are understood as those elements having better sensitivity and representative properties related to the changes in the process being measured (mostly physical, chemical, electrical, mathematical, and biological) and manifesting themselves by modifying any of those properties in a more or less reversible way. Since sensors can be related to many fields of the human knowledge, any study about sensors has to be limited to some specific areas. In particular, this book is limited to electrical, electronic, and mechanical engineering, though it could serve as basis for many other fields of knowledge.

3.2 PASSIVE ELECTRIC SENSORS

Passive electric sensors can be resistive, capacitive, or inductive, according to their predominant composition. They have no source of energy in their composition.

3.2.1 Resistive Sensors

As a general rule, resistance changes are shown in the unknowns on the right-hand side of the following equation:

$$R = \rho \frac{\ell}{A} \tag{3.1}$$

where

ρ is the electrical resistivity of the sensor element (Ω-m)
ℓ is the physical component length (m)
A is the cross physical area for passage of the electric current (m^2)

The most common physical effects modifying the amounts of ρ, ℓ, and A in Equation 3.1 are temperature, moisture, compression/traction, material degradation, aging, biasing by electric or magnetic field, light, and radiation.

The effect of temperature on the electrical resistance of materials can be expressed as follows:

$$R_t = R_0\left(1 + \alpha \Delta t\right) \tag{3.2}$$

where

R_t is the electrical resistance at a particular temperature (°C)
R_0 is the electrical resistance at a reference temperature (°C), typically at 0 °C
α is the thermal coefficient of temperature
Δt is the temperature variation with respect to the same temperature reference of R_0

3.2.2 Resistive Sensors and Effects of Temperature on Measurements

Platinum is the most recommendable thermoresistive sensor because of its accuracy and stability. A thermoresistive sensor based on platinum consists of a resistance of 100 Ω at 0 °C, varying less than 0.004 Ω/°C. It is widely used in standard resistance thermometers, whose main characteristics are shown in Table 3.2. As platinum is very expensive, its use is directed for applications where stability and reading accuracy are essential, as in benchmarks, for example. In most practical cases, platinum is replaced by nickel and copper.

TABLE 3.2
Thermal Characteristics of Some Resistive Sensing Elements

Material	Linearity	Repeatability	Temperature (°C)	Deterioration	Cost	α (Ω/°C)
Platinum	Almost linear	Good	−200 to 850	Inert	Expensive	3.9×10^{-3}
Nickel	Reasonable	—	−80 to 300	Reasonable	Average	6.7×10^{-3}
Copper	Reasonable	—	−200 to 250	Sensible	Cheap	3.8×10^{-3}

For some metals arise nonlinearity temperature effects of higher orders, which can be expressed as follows:

$$R_t = R_0\left(1 + \alpha\Delta t + \beta\Delta t^2 + \gamma\Delta t^3\right) \quad \text{for } \alpha > \beta > \gamma \tag{3.3}$$

In the case of platinum:

$$\begin{aligned} \alpha &= 3.9 \times 10^{-3}\,°\text{C}^{-1} \\ \beta &= -5.9 \times 10^{-7}\,°\text{C}^{-2} \\ \gamma &\cong 0 \end{aligned}$$

Most fluids have a relative low electrical resistance. Hence, if the sensor is surrounded by a fluid, its resistance is affected as well as the measured values. Usually the sensor resistance is fully embedded in ceramic or epoxy in order to provide electrical insulation against the fluid environment, and also covered by a protective tube to let it contact the medium in which the temperature is being measured. For this reason, the measured response time can be very long (about several seconds).

Semiconductors, known as thermistors, may be sensible temperature sensors due to their high sensitivity to this phenomenon. Moreover, the temperature coefficient of a thermistor can be positive or negative (PTC or NTC). Typically, the variation of the sensor element resistance, as a function of temperature, is nonlinear and approximately −4%/°C, as represented in Figure 3.2. The NTC placed under fixed potential differences lets a small current flowing through it, which will, therefore, depend on the element temperature. Thus, a semiconductor can be used as a thermometer.

Another form of resistive sensor is known as a strain gauge, which is suitable for measuring the effort that a body is subjected to. Depending on British or American English, both gauge and gage are used. In this book, we adopted the word "gauge." This sort of sensor can be a wire, a blade, or a semiconductor strip. These sensors are fixed on the body surface to accompany its dimensional changes (Figure 3.3). The sensor consists of a very thin grid strip or wires in series whose electrical resistance varies linearly with the strength applied to the device. With this, the sensor resistance

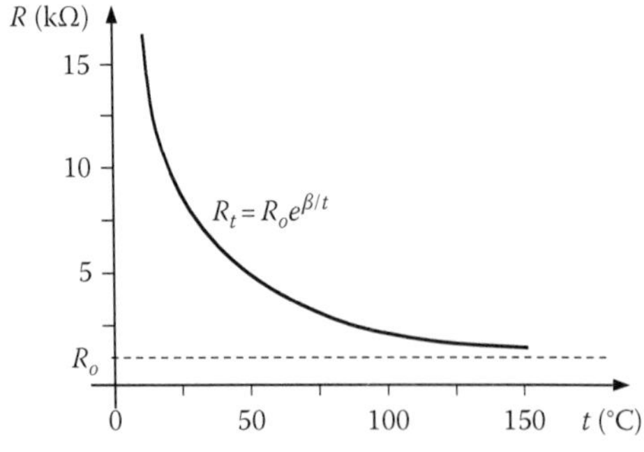

FIGURE 3.2 Variation of an NTC thermistor resistance with temperature.

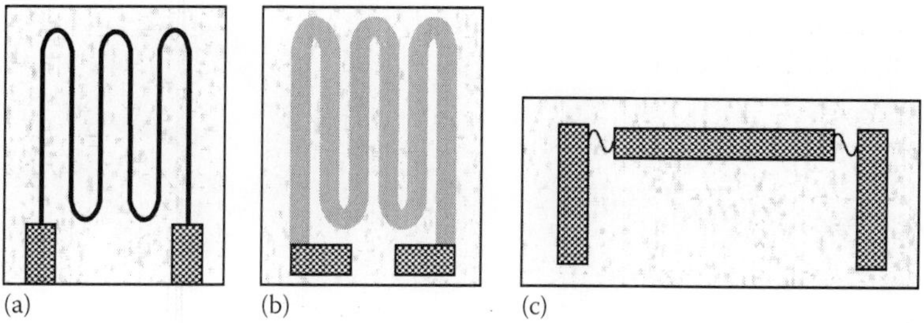

FIGURE 3.3 Strain gauges: (a) metallic, (b) metal blade, and (c) semiconductive tape.

also varies proportionally with the material stress (tension or compression) to which it is submitted.

Generally speaking, the resistance variation rate ΔR with stress ε (in Newtons) of a strain gauge is given by:

$$\frac{\Delta R}{R} = G\varepsilon \tag{3.4}$$

where G is the measurement constant (per Newton).

For silicon semiconductors, a value of G is in the range between 100 and 175 for type P, and −100 to −140 for type N. The negative sign means that the resistance decreases with an increasing effort. The main disadvantage of semiconductor strain gauges is their high-temperature coefficient.

For the alloy known as Advance, made of copper-nickel-manganese, the measurement constant is: $G = 200$/Newton. This alloy has low temperature and linear expansion coefficients.

Strain gauges can be used to sense linear, surface, or volume measurements of deformation and must be rigidly attached to the surface to be measured under some provisions, as illustrated in Figure 3.4.

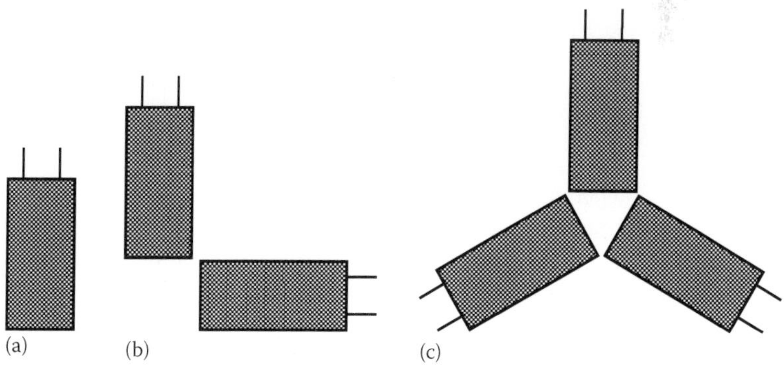

FIGURE 3.4 Assembly diagrams of strain gauges onto surfaces: (a) linear, (b) surface (x-y), and (c) volumetric (x-y-z).

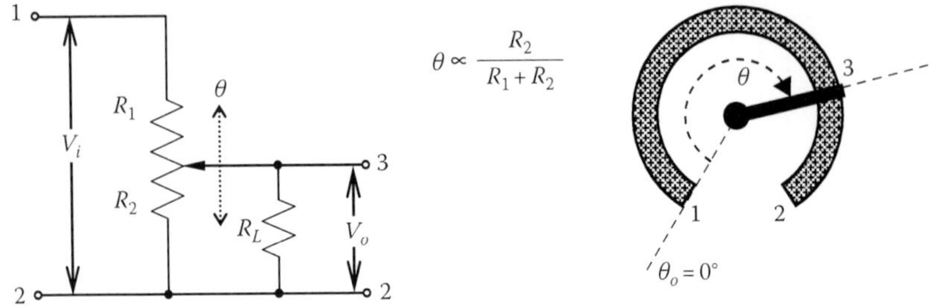

FIGURE 3.5 A potentiometer as a sensor.

Potentiometer sensors can be applied to sensing elements, e.g., to determine the angular position or linear displacement of a shaft bar. Figure 3.5 illustrates the relationship between the potentiometer resistance and the linear or angular position of the cursor. The output voltage v_o is related to the input voltage by the following equation:

$$v_o = \frac{x}{\frac{R_{pot}}{R_L} x(1-x)+1} v_i \quad \text{where}: x = \frac{R_2}{R_1 + R_2} = \frac{R_2}{R_{pot}} = \frac{v_o}{v_i} \tag{3.5}$$

The resistance of potentiometers varies between 20 Ω and 200 kΩ for wire and between 500 Ω and 80 kΩ for conductive plastic. The first type of sensor introduces a nonlinearity error and, in some cases, some discontinuity in resistance scanning, whereas the conductive plastic, although with no resolution problems, has higher α than metals. The accuracy of the potentiometer sensor based on conductive plastic varies from 0.1% to 1%, while for nonlinearity for wires wrapped with diameters varies for 0.5 mm < d < 1.5 mm.

Light-dependent resistors (LDRs) are conductive cells whose resistance varies with the lighting incidence. LDRs are made of cadmium sulfide and respond to the spectrum colors, just like the human eye. They are usually used for inexpensive light sensor devices, such as sunset relays for turning on streetlights.

3.2.3 Compensation of External Influences on Resistive Sensors

To compensate variations due to external influences or other means of physical quantities, generally a matched pair of sensors is used. One of the elements of the matched pair is used for active measurement, and the other one (dummy or ballast) is a reference of the measurement environment, with both placed very close to each other so as to be subjected to the same physical variable of the medium, say, temperature or humidity. The active sensor is subjected to the measured variable, while the dummy is not. A Wheatstone bridge, such as the one represented in Figure 3.6, can minimize or cancel out temperature or humidity effects on the resistance.

As is well known from circuit theory, to have a zero deflection on the current indicator in the central branch of a Wheatstone bridge, it is required that $i_1R_1 = i_2R_2$

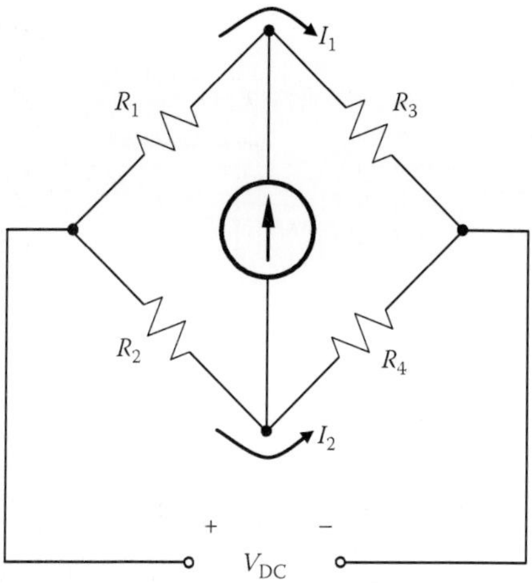

FIGURE 3.6 Wheatstone bridge.

and, similarly, $i_1R_3 = i_2R_4$, i.e.:

$$\frac{R_1}{R_2} = \frac{R_3}{R_4} \tag{3.6}$$

Suppose the temperature effects on the resistances are expressed by:

$$R_1' = R_1(1 + \alpha\Delta t) \tag{3.7}$$

$$R_2' = R_2(1 + \alpha\Delta t) \tag{3.8}$$

With these changes and the condition in Equation 3.6, for any $\alpha\Delta t$, we derive:

$$\frac{R_1'}{R_2'} = \frac{R_1}{R_2} = \frac{R_3}{R_4} \tag{3.9}$$

Another way of temperature compensation is the use of two active strain gauge sensors. For example, one is placed on the side under compression of an object being curved and the other one on the opposite side under tension. When these sensors are on different arms of the Wheatstone bridge, temperature effects are canceled out and the tension effects are doubled, since compression results in a decrease in strength while tension produces a corresponding increase in the sensor resistance. A typical example is the stress measurement used to evaluate the deformation of a diaphragm under pressure action to produce a sensor resistance variation related to this pressure (Figure 3.7). These sensors can also be used in load cells where their resistance is proportional to impressed strength and weight (Figure 3.8).

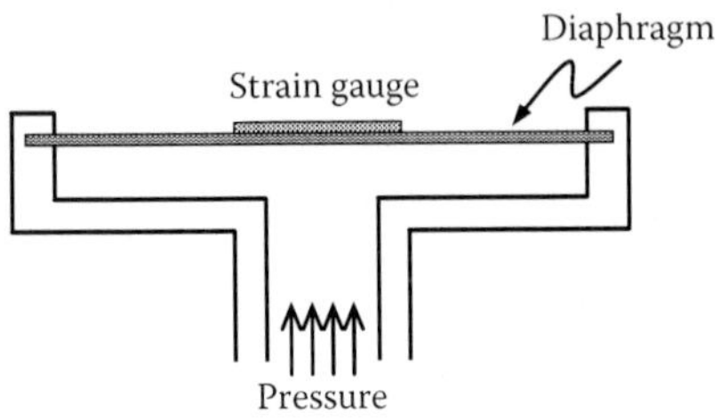

FIGURE 3.7 Pressure measurement in diaphragm.

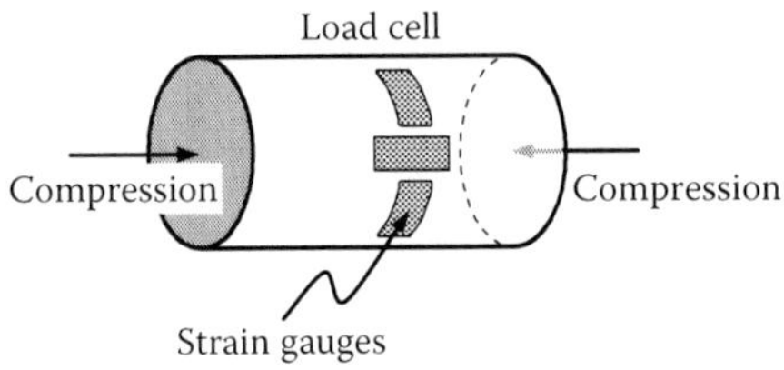

FIGURE 3.8 Load cell.

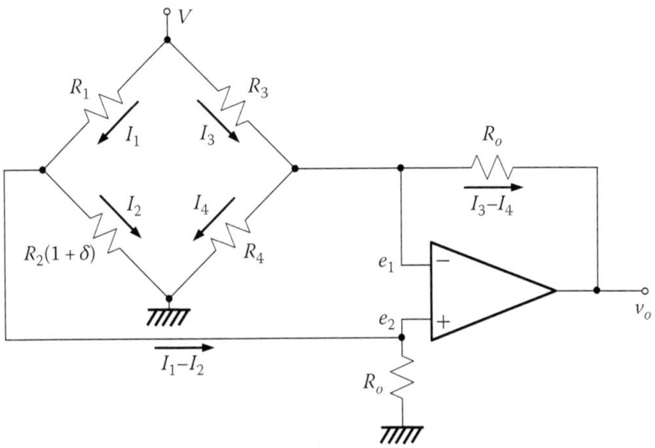

FIGURE 3.9 Bridge amplifier for strain gauges.

A good differential amplifier set up for strain gauges is shown in Figure 3.9. In this figure, the balance conditions of the Wheatstone bridge are established by Equation 3.6.

Note that $R_2(1+\delta)$ is in parallel with the noninvertible input resistor R_0, where δ is the incremental resistance variation caused by the tension or compression to which the strain gauge is subjected. If these resistors are selected to satisfy the conditions:

$$R = R_1 = R_2 = R_3 = R_4, \quad R_o \gg R(1+\delta), \quad \text{and} \quad \delta \ll 1 \tag{3.10}$$

then:

$$e_1 = R_o\left(i_3 - i_4\right) + v_o \tag{3.11}$$

$$e_2 = R_o\left(i_1 - i_2\right) \tag{3.12}$$

However, the bridge balance conditions are as follows:

$$i_3 R = i_1 R \quad \text{and, therefore,} \quad i_3 = i_1 \tag{3.13}$$

$$i_4 R = i_2 R\left(1+\delta\right) \quad \text{and, therefore,} \quad i_4 = i_2\left(1+\delta\right) \tag{3.14}$$

Replacing the conclusions of Equations 3.13 and 3.14 in Equation 3.11, we derive:

$$e_1 = R_o\left[i_1 - i_2\left(1+\delta\right)\right] + v_o \tag{3.15}$$

which equated with Equation 3.12 results in:

$$v_o = R_o \delta i_2 \tag{3.16}$$

As it was assumed that $R_o \gg R(1+\delta)$ and $i_1 \approx i_2$, then:

$$i_1 \approx i_2 \approx \frac{V}{R + R\left(1+\delta\right)} = \frac{V}{R\left(2+\delta\right)} \tag{3.17}$$

Substituting Equation 3.17 into Equation 3.16 yields:

$$v_o = R_o \delta i_2 = R_o \delta \frac{V}{R\left(2+\delta\right)} \tag{3.18}$$

As $\delta \ll 2$, then rearranging (3.18) leads to:

$$v_o \cong \left(\frac{V}{2}\frac{R_o}{R}\right)\delta \tag{3.19}$$

3.2.4 Capacitive Sensors

As a reference for a physical understanding of capacitive sensors, planar capacitors can be used, which apply the law:

$$C = \varepsilon \frac{A}{d} \tag{3.20}$$

where

$\varepsilon = \varepsilon_r \varepsilon_0$ is the electrical permittivity (F/m)
ε_r is the material relative permittivity referred to the air
$\varepsilon_0 \approx \varepsilon_{ar} = 8.854 \times 10^{-12}$ for the clean and dry air (F/m)
A is the physical actuation area of the electrical field over the plate (m^2)
d is the average distance between the capacitor plates (m)

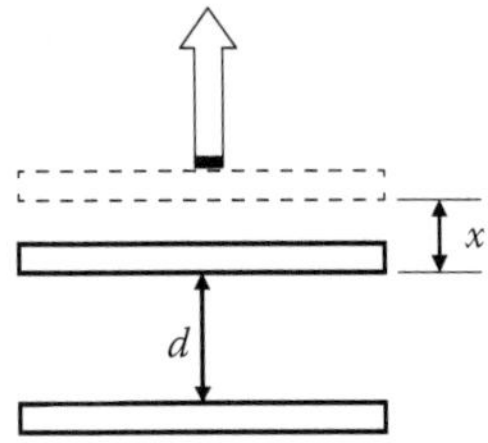

FIGURE 3.10 Displacement capacitive sensor.

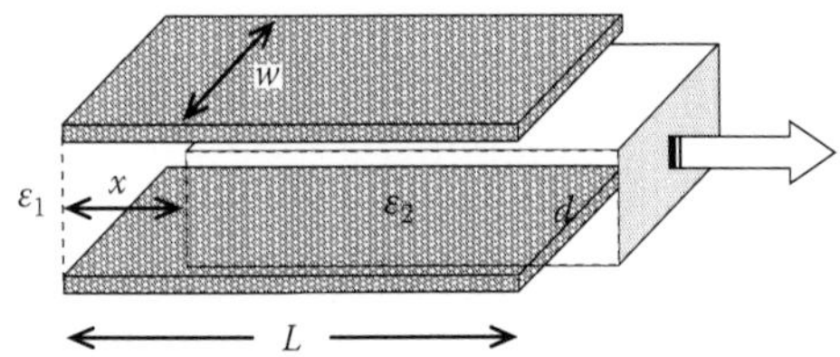

FIGURE 3.11 Capacitive insertion sensor.

The most common application of the capacitive sensor is to measure the displacement of one plate in respect to the other, with a dielectric in a capacitive based structure, as illustrated in Figure 3.10. In this case, Equation 3.20 is presented as:

$$C = \varepsilon \frac{A}{d + x} \tag{3.21}$$

where x is the displacement of one plate with respect to the other.

The measurement can be considered also for insertion of the material separating the two plates, as shown in Figure 3.11. In this case:

$$C = \varepsilon_1 \frac{wx}{d} + \varepsilon_2 \frac{w(L-x)}{d} = \frac{w}{d}\left[\varepsilon_2 L - (\varepsilon_2 - \varepsilon_1)x\right] \tag{3.22}$$

where

- x is the shifting move of one plate with respect to the other
- w is the width of the plates
- L is the length of the plates
- ε_1 and ε_2 are the dielectric constants of the two materials separating the two plates

Another way to change the capacitance is used in rotary shaft positioning, or in tuning circuits using reactive tanks. Figure 3.12 illustrates rotational displacement. The sector areas with radius r and R, respectively, can be obtained by:

$$A_R = \int_0^{\theta} \frac{R^2}{2} d\theta \quad \text{and} \quad A_r = \int_0^{\theta} \frac{r^2}{2} d\theta \tag{3.23}$$

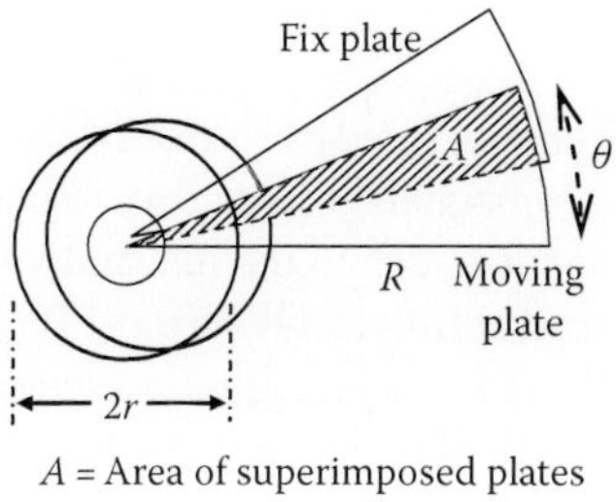

FIGURE 3.12 Capacitive rotation sensor.

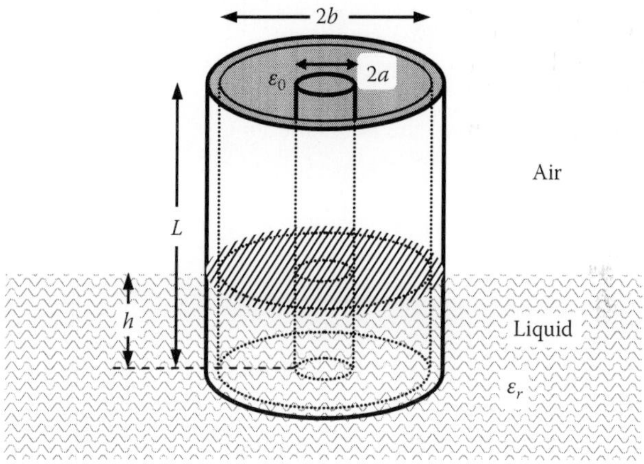

FIGURE 3.13 Cylindrical capacitive sensor.

An example of a liquid level being measured using a cylindrical capacitive sensor is shown in Figure 3.13, in which a solid conductive cylinder is placed concentrically in a cylindrical conductive tube. Both are then introduced into a vessel containing the liquid whose level is to be measured. The device operates as two capacitors with two concentrically cylinders. The individual capacitance/meter, considering radius "a" and "b" is given by:

$$C = \frac{2\pi\varepsilon_r\varepsilon_0}{\ln(b/a)}h \tag{3.24}$$

Equation 3.25 combines the two capacitances (parts immersed in and above the liquid) to describe the phenomenon:

$$C = \frac{2\pi\varepsilon_0}{\ln(b/a)}\left[L-(\varepsilon_r-1)h\right] \tag{3.25}$$

3.2.5 Inductive Sensors

The most common applications of inductive sensors are related to variations along the magnetic path d, sensed by magnetic reluctance, variations in the permeability of a magnetic inductance μ, and variations in the induced voltage.

The relative magnetic permeability ranges from 1.0 (vacuum or pure air permeability) to thousands (ferrite and other magnetic materials). Magnetic reluctance is the magnetic analog of electrical resistance, magnetomotive force (FMM) is the analog of voltage (electromotive force, EMF), and magnetic flux ϕ is the analog of electric current. Therefore, the average reluctance of a magnetic core can be given by:

$$\Re = \frac{FMM}{\phi} = \frac{Ni}{BA} = \frac{H\ell}{\mu HA} = \frac{\ell}{\mu A} \tag{3.26}$$

where

ℓ is the average path length of the magnetic flux
N is the number of turns
B is the magnetic density
H is the magnetic intensity
A is the coil area
μ is the material magnetic permeability

An example of a variable reluctance based inductive sensor is shown in Figure 3.14, in which the position of one plate alters the magnetic reluctance path of the magnetic cores: E- and I-shaped and wrapped with N turns. In this case, the change in magnetic reluctance is practically only due to the air gap.

As a reference for a physical understanding of inductive sensors, assume that an infinite cylinder is wrapped with wire windings uniformly covering its surface. The infinite cylindrical inductor will follow the general law:

$$L' = \mu\pi n^2 R^2 \tag{3.27}$$

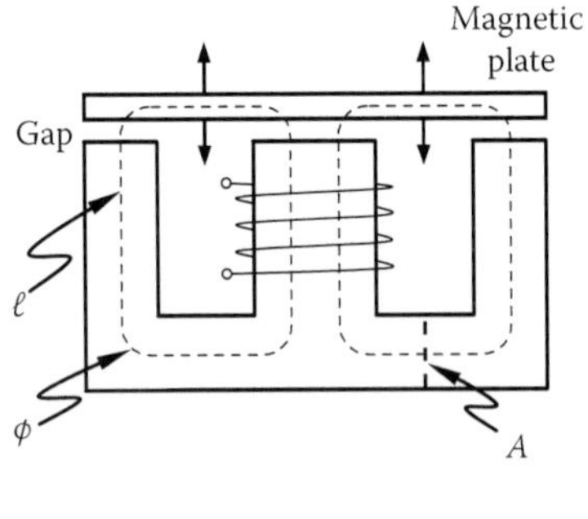

FIGURE 3.14 Sensor of reluctance.

where

L' is the inductance per meter of the coil
$\mu = \mu_0 \mu_r$ is the magnetic permeability of the material
$\mu_0 = 4\pi \cdot 10^{-7}$ is the magnetic permeability of vacuum (H/m)
μ_r is the relative permeability of the medium inside the cylinder
n is the number of turns per meter of the coil
R is the cylinder radius in meters

Another application of magnetic sensors is the variable differential inductor (VDI). There is a differential change in the magnetic inductance of coils A and B as a high-permeability core moves inside the coil set, as depicted in Figure 3.15. A bridge can be used to monitor this change. A simplified Equation 3.27 can be used to express a finite cylinder with a 10% margin of error:

$$L = \frac{\mu N^2 A}{\ell} \tag{3.28}$$

For higher-accuracy measurements, there is the linear variable differential transformer (LVDT), depicted in Figure 3.16a, using an active source in one primary and two secondary windings. The secondary coils are connected in antiseries. In balanced conditions, the potential difference between the two coils should be zero.

For analysis purposes, let us assume that a constant sinusoidal source is used to excite the primary coil. The voltages induced in the two secondary coils A and B can be expressed by:

$$\begin{aligned} v_A &= k_A \sin(\omega t - \phi) \\ v_B &= k_B \sin(\omega t - \phi) \end{aligned} \tag{3.29}$$

where k_A and k_B are coefficients dependent on the coupling degree between N_1 and $N_{2A} - N_{2B}$.

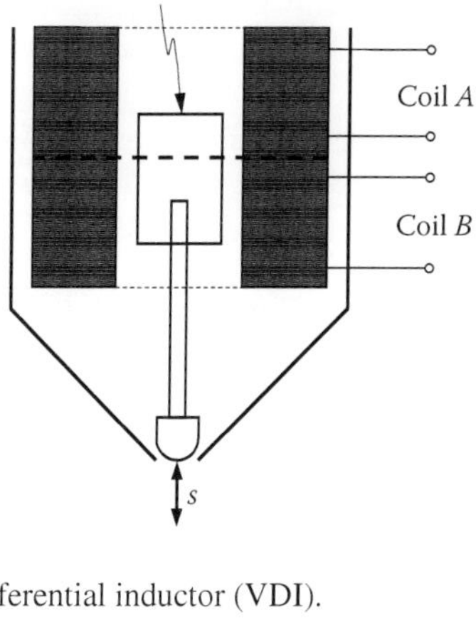

FIGURE 3.15 Variable differential inductor (VDI).

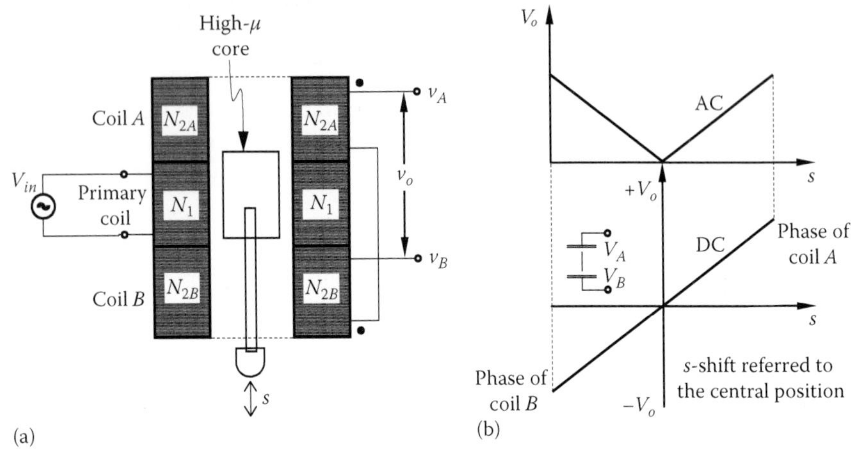

FIGURE 3.16 Linear variable differential transformer (LVDT): (a) electromechanical structure, (b) output voltage proportional to displacement.

The output voltage can be expressed then as:

$$v_o = v_A - v_B = (k_A - k_B) \cdot \sin(\omega t - \phi) \tag{3.30}$$

There is a phase shift between the induced voltages of 180°, such as that when the cursor is positioned in the middle of the path between coils, results in $k_A = k_B$, and the net induced output voltage is zero.

The output voltages can be independently rectified to direct current (DC) values to allow distinction of each coil phase producing the DC characteristic, as shown in Figure 3.16b. The two DC sources placed in opposition phase result in a difference v_o between them. Figure 3.17 illustrates this situation. Both features (AC [alternating current] and DC) are idealized because there will always appear some nonlinearity next to the core extremes (from 0.1% to 1%). LVDTs are used to measure displacements in the range of 0.25–250 mm.

As expressed by Equations 3.1, 3.20, and 3.26, passive electric sensors follow an acting law analog to each other, and their effects are proportional to the component area, inversely proportional to the length of the sensor, and dependent on the properties of the material of which they are made. Table 3.3 shows a summary list of passive electric sensors. In this context, a straight dependence of passive sensor characteristics on its dimensions can be observed [3–6].

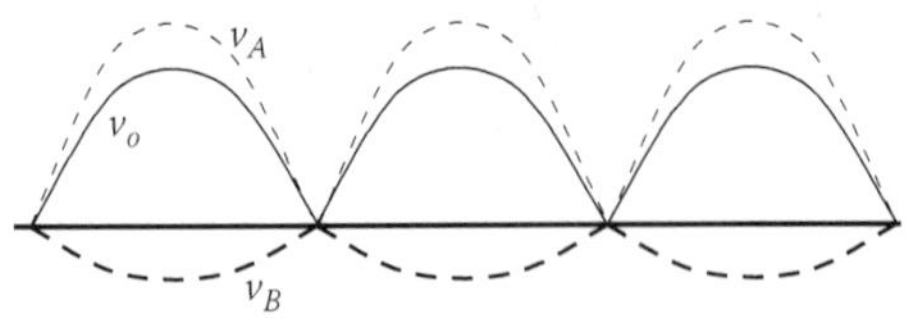

FIGURE 3.17 Representation of the induced and rectified voltages of coils *A* and *B*.

TABLE 3.3
Passive Electric Sensor Constructive Parameters

Type	Relationship
Resistive	$R = \rho \frac{d}{A}$
Capacitive	$C = \varepsilon \frac{A}{d}$
Inductive	$\Re = \frac{1}{\mu} \frac{d}{A}$

3.3 ACTIVE ELECTRIC SENSORS

Active electric sensors must necessarily contain a power source in their representation. The most common are thermocouple, photovoltaic (PV), piezoelectric, magneto-dynamic device (MHD), electrochemical, and Hall effect device (HED). Magnetic hydrodynamics refers to electrically conducting fluids. Examples of such fluids include plasmas, liquid metals, and salt water or electrolytes. The fundamental concept behind MHD is that magnetic fields can induce currents in conductive moving fluids, which in turn creates forces on the fluid, which changes the magnetic field itself [7–10].

3.3.1 Thermocouples

Sensors of the thermocouple type produce an EMF caused by a temperature difference at the junction of two distinct metals. There will be an electric current flowing through the junction every time there is heat flowing from the environment to this junction. Conversely, when passing a current through the junction, there will be a heat flow from it to the environment. This fact is based on three physical effects: Peltier, Seebeck, and Thompson.

The Peltier effect (due to Jean-Charles Peltier) was explained in 1834 as for how a joint (pair) of different materials can diffuse electrons on each other under the passage of electric current. Depending on the materials involved, there will be a temperature rise of the junction at one end and a corresponding reduction at the other end. If there is a stack of these materials, this effect can also be used for cooling or power generation. The temperature difference due to the Peltier element is limited by dissipation since the Peltier current is also limited. This temperature difference is determined by the Peltier coefficient, relating the heat flow to the current, the electrical resistance, and the thermal resistance between the hot and cold material faces.

The second physical effect is the Thompson effect, which explains how a differently heated rod (temperature gradient) at both ends develops an EMF by a density difference in the concentration of electrons caused by the temperature difference between these points. For example, metals such as cobalt, nickel, and iron, with the "cold" terminal connected to a higher electric potential, and the "hot" terminal connected to a lower electric potential, in which the electric current flows from the cold

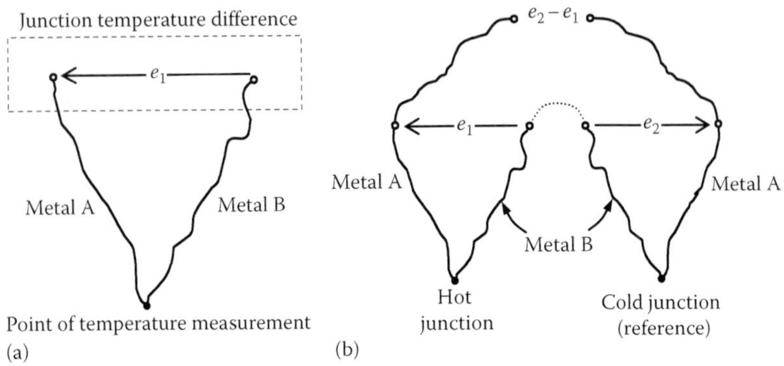

FIGURE 3.18 Thermocouple: (a) thermocouple junction and (b) thermocouple stack.

to the hot end. The electric current flows from the low thermal potential point to the higher thermal potential point. In this condition, there is no absorption of heat. Such an effect is called "negative effect of Thomson." Zinc and copper are typical metals with a positive Thomson effect.

The Seebeck effect, sometimes known as Peltier-Seebeck or thermocouple, is the association of the Peltier and Thompson effects (Sears, p. 195), and it is manifested as an EMF, as illustrated by the resulting thermocouple shown in Figure 3.18a.

The typical voltage variation of a junction of two metals with temperature is 50 μV/°C. As this value is too small to be measured, it is usual to choose a reference temperature of a different value from the value to be measured. Normally, this is 0 °C (ice in equilibrium with water in saturated air at 1 atm) or 100 °C (vapor in equilibrium with pure water at 1 atm).

The thermoelectric effect is therefore a result of three effects: the Peltier effect, which is the current flowing through a junction of two different materials, causing an environmental heat flow to the junction; the Seebeck effect, which is the rearrangement of carriers at the junction of two different materials at the same temperature; and the Thompson effect, which is the current flowing through an element when it is subjected to a temperature gradient. The thermoelectric effect to be measured must use a tiny reading current that does not interfere with the heating or cooling of the material. The overall relationship between the read values can be approximated by:

$$E = s\Delta T \tag{3.31}$$

where

E is the produced EMF (volts)
ΔT is the temperature difference between junctions (Kelvin or degree centigrade)
s is the thermocouple sensitivity due to temperature (volts per temperature degree)

For the purposes of measurement, the reference thermocouple can also be replaced by a source with an equivalent EMF, which would be generated in the same thermocouple model as if it was at the reference temperature.

TABLE 3.4
Thermoelectrical Seebeck Coefficient

Material	Potential (μV/°C)
Antimony	47.0
Bismuth	−72.0
Carbon	5.0
Copper	6.5
Constantan	−35.0
Nicromo	25.0
Platinum	0.0
Silver	6.5
Rhodium	6.0
Tungsten	7.5

The reference temperature is usually set at 0 °C ($E_{T,0}$), but for different intermediary temperatures ($E_{T,I}$), it is necessary to use the following law for interpolating and correcting the temperatures in tables ($E_{I,0}$):

$$E_{T,0} = E_{T,I} + E_{I,0} \tag{3.32}$$

There is a nonlinear relationship between EMF E and temperature T of the thermocouple junction, expressed as:

$$E = a_1T + a_2T^2 + a_3T^3 + \cdots \tag{3.33}$$

Differentiating Equation 3.33 on temperature, one can obtain the Seebeck coefficient of thermocouples, whose values are listed in Table 3.4 for some materials [2]. Other metals may be inserted in the circuit since they will have no effect on the EMF if their joints are at the same temperature; i.e., the first and last metals are the ones making the difference, as shown in Figure 3.18b. Note that for many pairs of metals, the terms a_2, a_3, etc. are sufficiently small to be ignored.

Table 3.5 shows the characteristics of the main types of thermocouples, having a 1% accuracy. Examples of alloys that can be used are constantan (an alloy of 55% copper and 45% nickel), chrome (an alloy of 90% nickel and 10% chromium), and alumel (an alloy of 96% nickel, 2% manganese, and 2% aluminum).

Thermocouples are usually placed in a sheath and wrapped with a mineral (thermal grease) which is a good conductor of heat and a bad conductor of electricity. The sheath has a faster response time than the thermal grease.

If the output EMF E needs to be higher, it is usual to have a series connection of devices, forming a thermocouple stack. However, the temperature increases near to the maximum, which reduces the useful life of the thermocouple.

TABLE 3.5
Main Types of Thermocouples

Type	Pair Material (Material$_1$-Material$_2$)	Operating Range (°C)	Sensit. (μV/°C)	Output 100 °C (mV)	Output 1000 °C (mV)	Resist. (Ω/m)
B	Plat./rhodium a 94%/6%–70%/30%	0 to 1800	0.00	0.033	4.83	3.15
E	Chrome-constantan	0 to 1000	60.48	6.32	76.36	12
J	Iron-constantan	−180 to 760	51.45	5.27	—	6
K	Chromel-alumel	−180 to 1370	40.28	4.10	41.27	10
R	Plat.-plat./rhodium a 87%/13%	0 to 1750	5.80	0.65	10.50	3.17
S	Plat.-platinum/rhodium a 90%/10%	0 to 1750	5.88	0.64	9.58	3.17
T	Copper-constantan	−180 to 370	40.28	4.28	—	5

Source: Adapted from Horowitz, P. and Hill, W., *The Art of Electronics*, Cambridge University Press, London, England, 2000, 716pp.

3.3.2 Piezoelectric Sensors

Piezoelectric sensors generate an EMF caused by pressure on crystals in the transverse shaft, where the pressure was applied. The basic principle is based on the concept that a crystal network under pressure moves ions from their normal positions. The most commonly used crystals are quartz, tourmaline, Rochelle salt, and some piezoelectric ceramics (e.g., zirconate and titanate).

Piezoelectric sensors are used to measure movements as acceleration, vibration, pressure, and impacts on the surface and floating forces. Among their applications are ultrasound, weighing scales, measuring the depth of rivers and streams (the echo method), potential piezoelectric transformers, and piezoelectric motors.

The inverse piezoelectric effect is also possible, i.e., by applying an EMF on the crystal faces, the crystal shrinks or expands; this effect is commonly used to produce ultrasound.

3.3.3 Photovoltaic Sensors

Photovoltaic sensors also generate EMF, but from a PN semiconductor junction. To do so, transistors and diodes are used, whose junctions are exposed to the light effects to produce photovoltaic currents following the full Shockley equation:

$$I_d = I_s\left(e^{V_d/\eta V_T} - 1\right) - I_\lambda \tag{3.34}$$

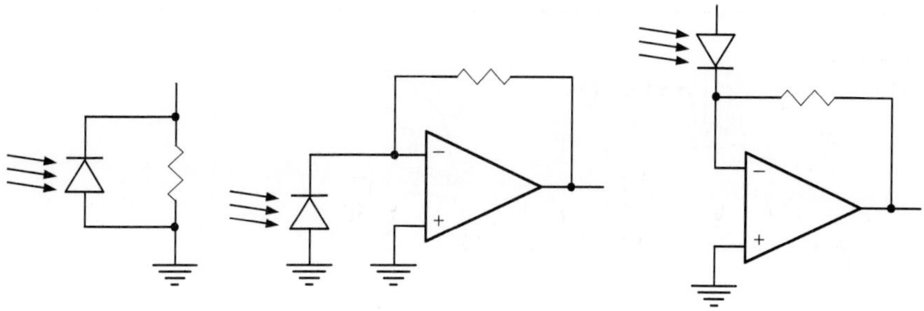

FIGURE 3.19 Examples of photosensitive sensors with diodes.

where

- I_d is the diode direct current
- V_d is the forward diode voltage
- V_T is the equivalent temperature voltage
- η is an empirical constant known as the emission factor, which varies between 1.1 and 1.8 for the diode of silicon and is 1.0 for germanium
- I_s is the reverse leakage or saturation current, which can be in the range of 10^{-6} to 10^{-15} A
- I_λ is the diode reverse current according to the light's wavelength, the temperature, and proportional to the incident light on the junction

Figure 3.19 shows some simple circuits using photodiodes.

3.3.4 Hall Effect Device

Precise measurements of magnetic fields with errors in the order of 1% are very important in physical sciences in connection with instrumentation. Examples are magnetic resonance (MRI), magnetrons, devices for magnetic focusing of electron beams, ultrasonic motors, the isolated measurement of DC current, in geology and prospecting, keyboards without contacts and panel keys.

The Hall effect is the production of a transverse voltage on a conductor (usually a semiconductor) carrying a current inside a magnetic field. Commercial magnetometers using the Hall effect cover the range from about 1 gauss to 10 kgauss.

The explanation of the Hall effect device (HED) is obtained through a simple application of a direct current measurement I_{DC}, as shown in Figure 3.20. A magnetic induction field B is established perpendicular to the HED, while a constant reference current I_{ref} is maintained, passing through the tablet in the direction from the inside to outside of the page (the left-hand rule). In classical physics, it is known that this field exerts a deflecting force F, pointing to the right, upon the electrons in the wafer (given by $F = BI_{ref}\ell$) in Figure 3.20. As the lateral force on the electrons of the wafer is due to lateral forces on the charge carriers (given by $F = qvB$), it follows that the electrons tend to move to the right, while draining over the wafer, to produce a transverse potential difference, such as that the left side will have a higher potential Δv than the right side.

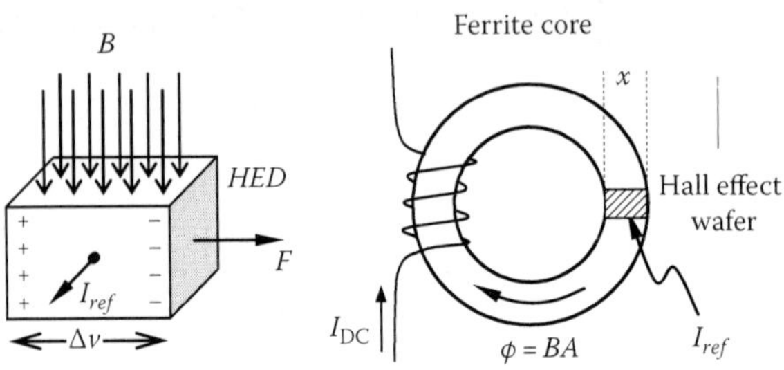

FIGURE 3.20 DC measurement using Hall effect.

The main advantages of HEDs are as follows: workable under AC or DC variations, low cost, simple, small volume, light weight, robust, and reliable. If the wafer has a width w, then:

$$\Delta v = \frac{I_{ref} B}{n \cdot e \cdot w} \tag{3.35}$$

where

n is the number of free electrons per cm^3

e is the electron charge (1.6×10^{-19} C)

An example of a transducer application to build up a HED is the differential amplifier shown in Figure 3.21a. This circuit is commonly used for general measurements

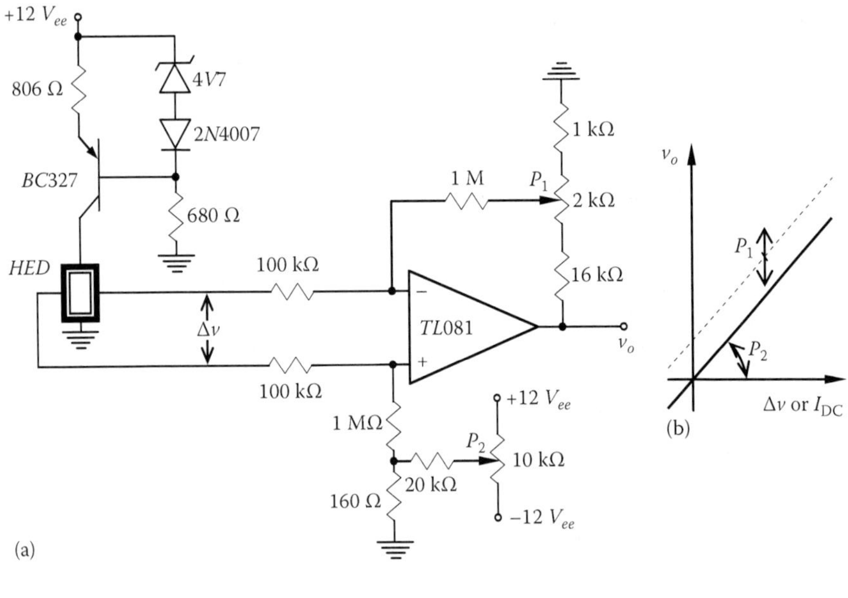

FIGURE 3.21 Amplifier for a Hall effect device: (a) differential amplifier with adjustments, and (b) input/output characteristic.

in concrete block deformations, shafts, metal structures, and overall weight sensors. In the case of current measurements, the HED is subjected to a uniform magnetic field generated by the DC I_{DC} to be measured through a coil wound around a toroid ferrite with a gap, where the Hall effect wafer is introduced. For measuring very small currents, it might be necessary to have several turns (n) of the looped wire, the overall gain will be then multiplied by n.

The calibration procedure starts by setting P_1 (fraction of v_o) to zero, after that adjusting P_2 (changing the ground level). A typical value of 0.1 V of the output voltage v_o) could correspond to a measured current of 1 A.

In Figure 3.21b, the current source is obtained by a fixed biased transistor. The voltage drop Δv across the HED terminals feeds a conventional differential amplifier with an offset adjustment potentiometer P_1, represented by the straight shift (dashed line) amplifier calibration. The gain adjustment is represented by the line slope of this calibration (full line), P_2, usually given in volt/ampere. The potentiometer P_1 provides a fraction of the output voltage as feedback to the noninverting input, while the pot P_2 changes the ground level.

3.3.5 Photomultipliers and Image Intensifiers

The signal levels used nowadays can be very small and with speed so high that only sophisticated sensors are suitable for measuring these variables. Among these sophisticated Photomultipliers are sophisticated sensors that can be used for measuring such small ultra-fast events. Photomultipliers measure the impact of a photon moving against an electron of an alkali photosensitive metal cathode. That happens in such a way that the photomultiplier amplifies this reduced current by accelerating the electrons on successive surfaces, known as dynodes, as shown in Figure 3.22. So, new electrons are displaced from these surfaces.

The photomultiplier shown in Figure 3.22 is a very-low-noise multiplier. It uses typically 100 V between successive dynodes to have a gain of about 10 in each stage,

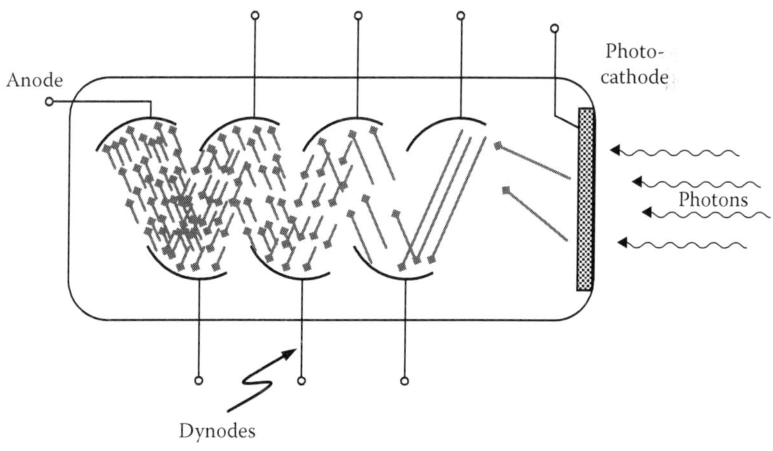

FIGURE 3.22 Process of photomultiplication.

resulting in a total of 10^6. The current collected at the anode is sufficiently high with respect to the generated noise, which is close to the ground potential. The photocathode voltage is at 1000 V. There is a need of an equalization loop with a potential between plates so that the linear distortion becomes minimal. A circuit comprised of 100 kΩ adjustable resistors will make possible to obtain the same voltage levels between the dynode pairs, such circuit with adjustable resistors is called as "equalization loop network".

Constructively, photomultipliers have a quantum efficiency of 25% and a huge gain provided by the dynodes. For low-light levels, the photomultiplier requires further amplifiers, load pulse integrators, discriminators, and counters. However, for higher-light levels, the individual photoelectron counting becomes too high (count of 10^6/s for an incident energy of 2 μW) such that the device can measure the anode current in a macroscopic way. The photomultiplier sensitivity is measured in A/μW, although the current limits are in mA or less. Therefore, the device may indicate current even for total darkness, because of the excitation of the photocathode electrons (typically 30 counts/s/cm^2 of the cathode area). To reduce these current levels, one can use a current compensation or cooling to about 25 °C below zero, or even a digital signal processing, reaching less than 1 count/s/cm^2. All this means that if exposed to light, the photomultiplier has to be cooled for a 24-hour period.

Typical applications for the photomultiplier, all for low-light levels, are related to photometry in spectrum photometry, espionage, war, astronomy, biology, and scintillators for particle detection, X-rays, and gamma rays. Relatively newer technologies, such as CCD (charged coupled device), light intensifier, SIT (silicon-intensified target), ISIT (intensified silicon-intensified target), and image dissectors, allow a light sensitivity much superior to the human eye, as is the case of the vidicon used in camcorders and digital photo cameras. These devices are two-dimensional targets sensitive to light that accumulate the image in such a way that it can be read electronically by scanning an electron beam or image shift, as the registers of analog displacement. This process is analogous to the one happening in photodiodes.

There are several generations of image intensifiers. The first generation consists of a sensitive surface as a photocathode tube with optical focusing on electrons and a phosphor screen placed behind it so that the photoelectrons from the cathode are accelerated by high voltages, reaching phosphorus with a sufficiently high energy to produce a glare light. The second generation uses plates with microchannels that help to achieve much higher levels of light in a single stage from very low levels, resulting in a signal reproduction of better quality through fewer transformation stages. The microchannel plate intensifier placed in the space between the cathode and phosphorus has microscopic holes whose interiors are coated with a sort of layer-multiplying dynodes.

Image dissectors are intensifiers comprising a photocathode area followed by a chain of photomultiplier dynodes. A small aperture is made, and some deflection electrodes are placed in order to make a photocathode active area that allows photomultiplication through the dynode system. It is like a photomultiplier with an electronically movable photocathode area. The image dissectors have efficiency and gains just like the conventional photomultiplier. However, they differ from the vidicon, CCD, and SIT, which are image integrator devices that do not accumulate the image within the entire field between readings.

3.3.6 Geiger-Müller Tube and Proportional Counter

Ionizing radiation from the energy released in nuclear disintegrations is high and can be detected by the Geiger-Müller (GM) tube. The problem in this sort of measurement is not the signal intensity, but on the contrary, the difficulty in finding materials that do not allow the passage of this intense energy and the voltage levels involved. For example, mica can absorb this energy, and low radiation can be used to manufacture the GM tube window and can easily support the vacuum before the cylindrical tube is filled with gas, as shown in Figure 3.23. The GM tube casing has a cathode made of metal with an insulated conductor, where the anode is in the center of the cylinder. The tube does not operate in a vacuum, but contains a gas under low pressure and ionized into electrons and positive ions by applying high voltage V_{DC}. This negative high voltage with respect to the central conductor attracts positive ions toward the cathode and pushes the electrons to the anode. A single particle entering the ionization chamber may cause complete ionization of the gas therein, generating a countable current pulse.

Note in the tube shown in Figure 3.23 that the difference between the outer areas of the driver and the tube causes the movement of electrons toward the anode much faster than of the positive ions toward the walls of the tube. This difference produces current pulses with a sharp decaying time. These current pulses associated with particles at different energy levels are equal due to the high internal amplification of the GM tubes. The count rate of these particles, however, is relatively low (somewhere around 1000 counts/minute) because the tube must extinguish the ionization each time a particle traverses through it. The voltage V_{DC}, in turn, applied between the anode and the cathode varies within a wide range, as shown in Figure 3.24. The levels of the voltage of most interest are described below.

The region of the ionization chamber is the first one to be reached by the particles. There is a collection of ions before any recombination happening. The current is very small because there is still no secondary ionization. The next region is the voltage range, where occurs the acceleration of ions toward the electrodes and the first levels of secondary ionization. Along with this occurs a linear current increase, proportional to the number of ions being counted, also causing an increase in the secondary gas

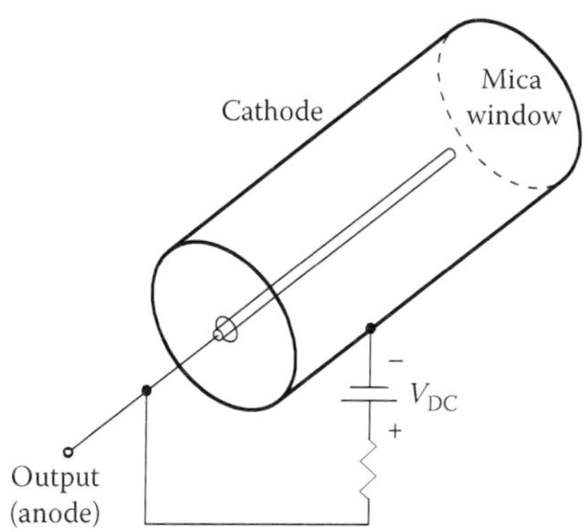

FIGURE 3.23 Simplified diagram of the Geiger-Müller tube.

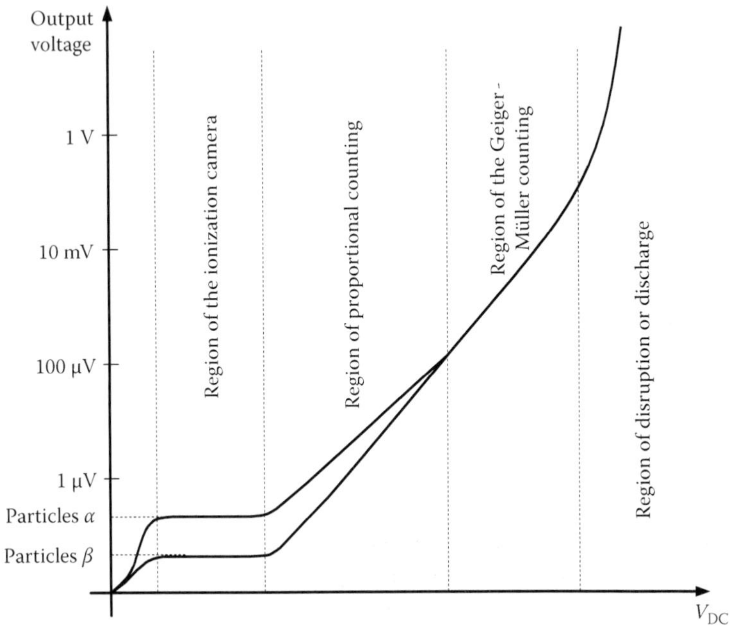

FIGURE 3.24 Transfer curves of a Geiger-Müller tube.

ionization. This increase in ionization, and thus in the internal amplification, makes the current appreciably greater as the output levels (anode) occur in the ionization chamber region. With a further increase in voltage, V_{DC}, there will be an avalanche effect.

The main characteristics of the GM tube are given in Table 3.6 in comparison with photovoltaic cells, phototransistors, thermocouples, and photoconductive cells.

At a lower voltage V_{DC} when an α or β particle forms an ion, it will soon accelerate, reaching a second molecule of the gas, ionizing it in the same way, and thus expanding rapidly throughout the gas contained in the tube. Therefore, the output current pulse has the same magnitude as that produced by any of the particles with the ability to produce the initial ion; the sensor must be at the proper voltage range. Voltages higher than the upper limit of this range can destroy the sensor by overvoltage. The energies of the various particles can be detected, as shown in Figure 3.24, by using different methods of analysis [5].

3.3.7 Probes for Biological and Chemical Voltages

There are a wide variety of methods for obtaining electrochemical signals, such as those in polarography, chromatography, voltammetry, electrophoresis, specific ion electrodes, infrared and visible spectroscopy, and many others. It is difficult and very specific to address this issue in an interesting way from a general engineering point of view. Therefore, this section covers only measurements using microelectrodes to verify the electric potential of nerves and muscles, specific ion electrodes used to measure concentrations of some very specific ionic species in solutions, and electrochemical volumetric probes.

TABLE 3.6
Comparison of the Characteristics of Some Sensors

Characteristic	Geiger-Müller Tube	Photovoltaic Cells	Phototransistors	Thermocouples	Photoconductive Cells
Time response	1.0 ms	10 μs	<1 μs	10 ms	10 μs
Output variable	Current	Voltage	Current	Voltage	Resistance
The need of an external push	High-voltage source	Null	Null	Amplifier	Null
Sensitivity	High	Average	Average	Low	Average
Impedance	High	Low	High	Low	Average
Size	Large volume	Small	Small	Small	Small
Wavelength range	From 10^{-4} to 10^{-1} Å	From 350 to 750 nm	From 250 to 1100 nm	From 1 to 40 μm	From 350 to 4000 nm

To observe the voltages of nerves or the interior of cells, microelectrodes with a tip of an atomic thickness of about 1 Å (1 angstrom $= 10^{-10}$ m) are used. Such small magnitude generates signal source impedance in the order of 100 MΩ or more, causing problems such as interferences, easy overloading of the signal source, and displacement of some Hertz of high frequency due to cables' and parasitic capacitances. So the instrumentation amplifier must be precisely designed in order to have a high input impedance, a very low common-mode rejection ratio (CMRR), and a low input noise.

In the case of electrodes for specific ions, they serve to measure the voltage developed between a reference electrode and an electrode of thin glass walls that allows the diffusion of hydrogen ions. There are many measurements made out of these ions (sodium, potassium, mercury, bromine, iodine, fluorine, and calcium). The sensing of these ions is used in pH meters. A pH meter is a scientific instrument that measures the hydrogen-ion activity in water-based solutions, indicating its acidity or alkalinity expressed as pH. The pH meter measures the difference in electrical potential between a pH electrode and a reference electrode, and so the pH meter is sometimes referred to as a "potentiometric pH meter." The difference in electrical potential relates to the acidity or pH of the solution. The pH meter is used in many applications ranging from laboratory experimentation to quality control. The same high-impedance problems encountered in microelectrodes are present in ion-specific electrodes. In order to improve the operation of such high impedance system, it is necessary a voltage range of the reading between 0.0 to 2.0 V, with a 2 mV accuracy to drain a current lower than 100 pA. The process is further complicated by the need for about 1% temperature compensation of the variation in V/°C. With these electrodes, an accuracy of 1% in solutions with concentrations of 1 part per billion (ppb) is possible.

It is also possible to observe specific ion concentrations by measuring the electrochemical current (reaction rate) versus applied voltage in a given solution, such variable voltage may reach a threshold where the reactions occur. These methods are mainly related to cyclic voltammetry, polarography, and voltammetry of bare anodes. The technique can detect the presence of lead and cadmium portions in orders of 1 ppb.

3.3.8 Alcohol Sensors Used in Blood Measurements

Sensors for sensing alcohol in the blood, also known as breath analyzers, can measure the concentration of alcohol in breath air coming from the lungs. They are based on the operation principle of a fuel cell, sketched as a simplified illustration in Figure 3.25. This process of sensing is the reverse of electrolysis, producing a small current from the combination of hydrogen (H_2) and oxygen (O_2). The arrow shown in Figure 3.25 indicates the negative direction of the electron flowing from the anode to the cathode, unlike the conventional current direction.

A blood sensor alcohol can be explained with the chemistry related to fuel cells, where a small voltage will be produced with the alcohol detected by the fuel cell [9]. As the amount of alcohol in the air is directly proportional to the concentration of alcohol in the blood, theoretically, the ethanol could not be fully oxidized into CO_2 and water. Thus, ethanol does not react completely in the body, most likely, and is only partially oxidized to become ethanal.

The anode and cathode reactions of this approach are nearly related by:

$$C_2H_5OH \Leftarrow CH_3CHO + 2H + 2e^- \quad (3.36)$$

With this dissociation, the electron flow through an external circuit results in:

$$\tfrac{1}{2}O + 2H^+ + 2e^- \Leftarrow H_2O \quad (3.37)$$

The fuel cell voltage is proportional to the concentration of reactants due the concentration of alcohol in the air breathed out from the lungs on the cell. Therefore, such fuel cell voltage will provide a measurement of the concentration of alcohol in the blood. This type of fuel cell is the basis of many breath analyzers used worldwide by highway police.

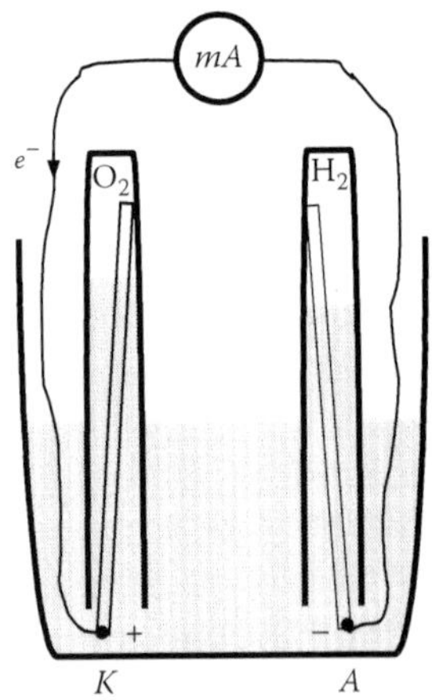

FIGURE 3.25 Fuel cell's working principle.

3.3.9 Other Magnetic Sensors

In order to measure very low magnetic fields, it is possible to use the so-called superconducting quantum interference device (SQUID), which is an arrangement of superconducting junctions capable of measuring up to 0.2 microgauss-cm. A SQUID is very complex to operate and to use since it uses cryogenic material such as liquid helium.

For accurate magnetic field measurements, in the order of kilogauss (1 ppm), it is usual to use a magnetometer of nuclear magnetic resonance (NMR). This device is based on the precession of the nuclear spin (usually hydrogen) in an external magnetic field. The output is a frequency. Therefore, all advantages provided by a precision frequency/time can implemented in the full-fledged system.

Another way to measure magnetic fields is the use of a magnetometer for flux measurement. Its operation is based on the excitation of a ferrite core with an AC field excitation, and the response is observed as proportional to the magnetic field [11–13].

3.4 MECHANICAL SENSORS

Mechanical-based sensors provide a measurement based on their physical propertions or displacement of some material in their construction. It is common to have elastic sensors, pneumatic sensors, differential pressure sensors, and turbine sensors or rotating disks.

3.4.1 Elastic Sensors

Elastic sensors are based on nonpermanent deformation of elastic materials such as springs, diaphragms, bellows, and coils that allow reversible length changes. They follow a proportional deformation law expressed as:

$$F = kx \tag{3.38}$$

3.4.2 Pneumatic Sensors

Pneumatic sensors are based on an expanding gas, such as compressed air, escaping from an injector gun with a variable aperture (Figure 3.26a), such as the flapper-nozzle injector. They usually follow a square law of a nozzle opening covered by a movable lid, as shown in Figure 3.26b. The pressure is expressed by:

$$P_o = \frac{P_i}{1+16\left(d_i^2 x^2 / d_o^4\right)} \tag{3.39}$$

where

d_i is the nozzle diameter
d_o is the diameter of the output aperture
P_i is the input pressure
P_o is the measured pressure
x is the lid displacement with respect to the nozzle

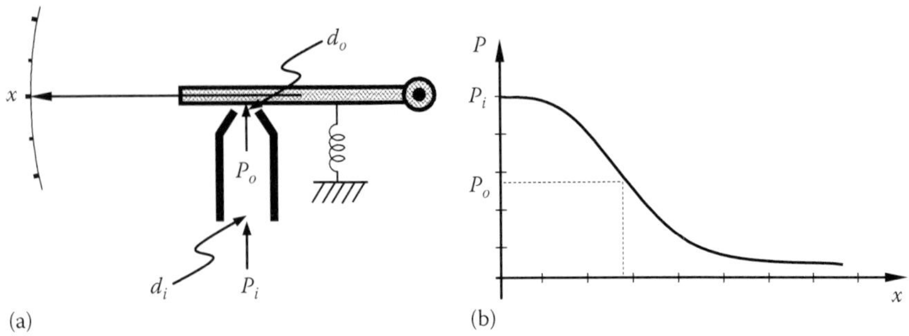

FIGURE 3.26 Flapper-nozzle sensor: (a) simplified mechanical structure, and (b) output characteristic.

3.4.3 Differential Pressure Sensors

Differential pressure sensors are based on the relative difference of the pressure of a fluid moving along its axis, the physical modeling is based on Bernoulli's equations, and the flow (m^3/s) is measured along two cross-sectional areas, indicated in Figure 3.27. Observing the Venturi tube in Figure 3.27 and using the Bernoulli's law, one may establish the following relations:

$$\frac{v_1^2}{2g} + \frac{P_1}{\rho g} = \frac{v_2^2}{2g} + \frac{P_2}{\rho g} \tag{3.40}$$

$$Q = A_1 v_1 = A_2 v_2 \tag{3.41}$$

where

indexes "1" and "2" refer respectively to each cross-sectional area, A_1 and A_2.
v_1 and v_2 are the fluid velocities at the smaller and larger sections of the pipe ("vena contracta"), respectively
P_1 and P_2 are fluid pressures at the smaller and larger sections of the pipe ("vena contracta"), respectively
ρ is the fluid density
g is the gravity constant

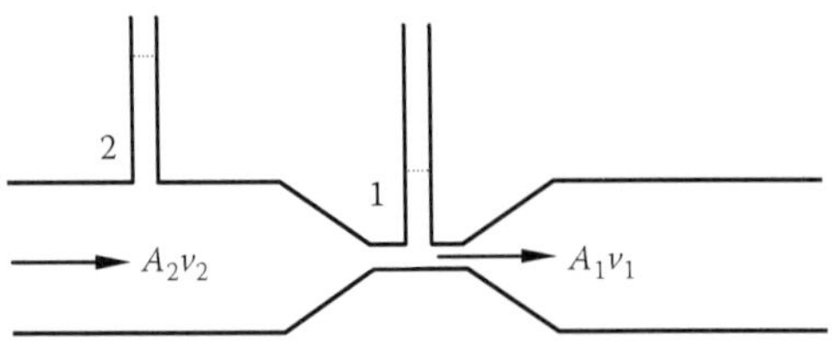

FIGURE 3.27 Flux measured by a differential pressure sensor.

Replacing the values of v_1 and v_2 of Equations 3.28 and 3.27, assuming constant density of the fluid, and rearranging terms, we derive:

$$Q^2\left(\frac{1}{A_1^2}-\frac{1}{A_2^2}\right)=\frac{2}{\rho}\left(P_1-P_2\right) \tag{3.42}$$

which can be explicated regarding Q as follows:

$$Q=\sqrt{\frac{2}{\rho\left(\frac{A_1^2-A_2^2}{A_1^2A_2^2}\right)}\left(P_1-P_2\right)}=C\sqrt{\left(P_1-P_2\right)} \tag{3.43}$$

where

$$C=\sqrt{\frac{2A_1^2A_2^2}{\rho\left(A_1^2-A_2^2\right)}}$$

$$A=\pi d^2/4$$

3.4.4 Turbine Sensors

The turbine sensors are based on measuring the rotation of a small turbine, sometimes known as a turnstile, layered axially inside the duct of the fluid passing through it at a certain speed. The angular speed of the turbine is approximately proportional to the fluid flow. The peculiarities of this sensor are that it offers resistance to the fluid, and it is usually expensive and with an accuracy of about ±0.3%.

Another way of flow measurement is using a magnetic pickup whose reluctance of the magnetic core path varies with the rotation of the turbine in the fluid, as shown in Figure 3.28. Special care should be taken with this sensor since the fluid will have higher speeds in central parts of the pipe whenever the viscosity is not very low.

3.4.5 Rotation Sensors

Rotary shaft sensors can be optical or magnetic. The magnetic type is based on the speed of a gear wheel varying the magnetic path (reluctance) of a coil, which, in turn, produces a pulsating EMF. The optical type interrupts the light beam of an LED of a

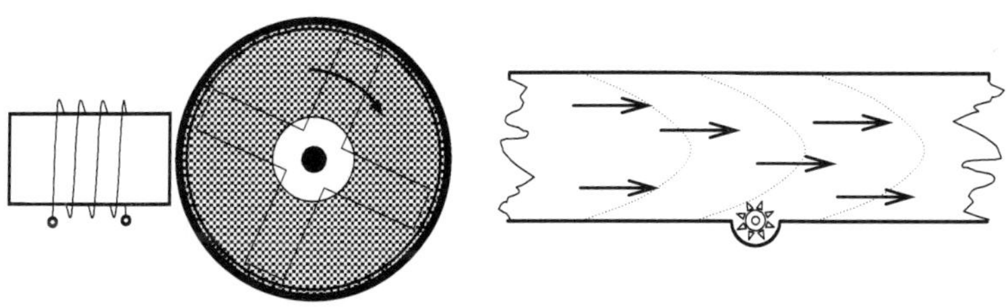

FIGURE 3.28 Turbine sensor (or turnstile).

phototransistor (optical or photocoupler pair). These sensors can be used for velocity measurements (rotation) or position.

3.4.6 Shaft Encoders (Rotation or Position)

The term shaft encoder is applied for the association of a digital code with the angular position (absolute type), or a pulse train associated with a shaft speed (incremental type). The magnetic shaft encoder is shown in Figure 3.29, and its operation is similar to the turbine sensor, except that it is driven directly by the shaft in which position or rotation is under measurement. A second train of pulses can also be generated when one needs to know the rotation direction. This is done by a second layer concentric with the first orifice, just slightly offset by a δ angle (Figure 3.30). The rotational direction of the movement is assessed by a pair of either a LED or photodiode-phototransistor circuits.

There are a wide variety of optical sensors, and two types of them are shown in Figures 3.31 and 3.32. Two particular advantages of optical sensors are that they convert directly the motion information into a digital code, ready for digital data acquisition, and are not affected by electromagnetic interferences. Their resolution can be increased by increasing the number of opaque or transparent strips, or by using a greater number of detector pairs. The Gray code is used to allow variation of 1 bit per count, thus reducing the transition pulse impact on the sensor source. These sensors are cheap and their cost increases with their resolution.

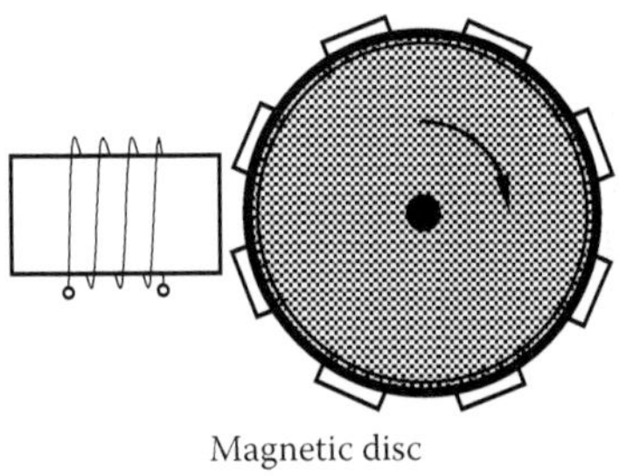

FIGURE 3.29 Magnetic shaft encoder.

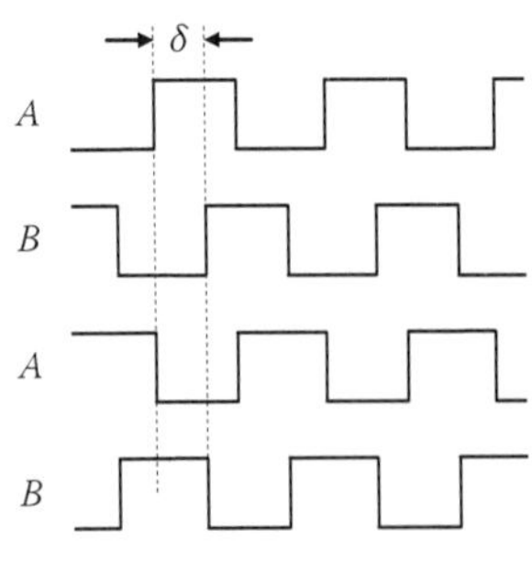

FIGURE 3.30 Output signals of an incremental shaft encoder.

FIGURE 3.31 Optical shaft encoder.

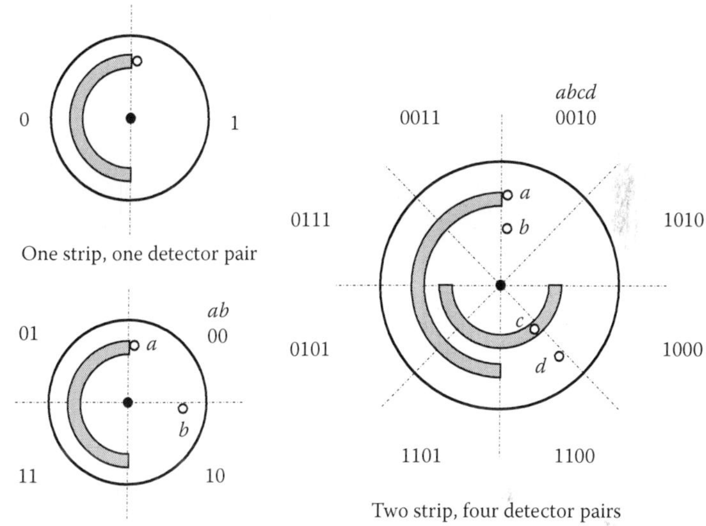

FIGURE 3.32 Resolution with optical shaft encoders (0 = open, 1 = closed).

Resolvers are a rotary type of sensors that act as a synchronous AC servomotor or generator with two delayed fields. Brushless resolvers (brushless DC motor, BLDC) are economical and very accurate [12–14].

Feedback movement sensors can be used to provide information on speed and position to electronic controls in a closed loop as well as to switch BLDC motors (DC motor commutation). They usually contain no optical or electronic internal components and are immune to electrical noise, heat, shock, and vibration. They represent a robust position sensor (control and accuracy), widely used in robotics to detect axis position $x(\theta)$ and convert the output signals ($\sin\theta$ and $\cos\theta$) into digital position data, which are then taken to a microprocessor (Figure 3.33). Resolvers are used in the feedback control loop of brushless DC motors; a nonvolatile memory (such as PROM or another one) will allow a look-up table to provide angular data from the resolver into switching signals used by power electronic transistors controlling the motor.

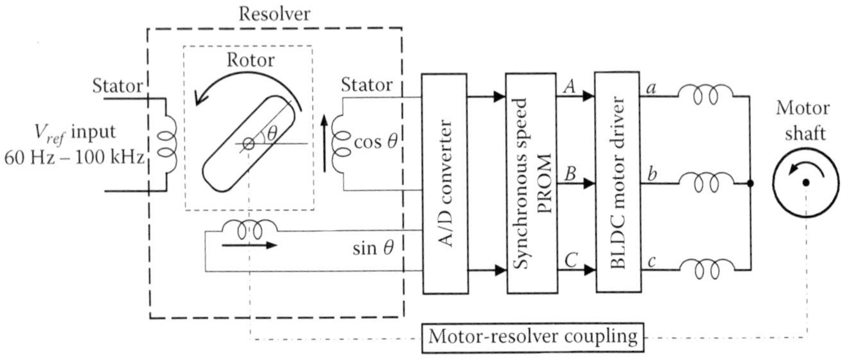

FIGURE 3.33 BLDC resolver.

There are lots of different types of resolvers; however, a typical resolver has three windings: a primary winding and two secondary windings. These windings are made using copper wire and are usually formed on the resolver's stationary element, the stator. Sometimes, the rotor is made from a material such as iron or steel and is arranged such that it will couple varying amounts of energy into the secondary windings, depending on its angle of rotation. When exciting both phases' inputs, voltages A and B will produce two outputs with magnitudes defined in Equations 3.44 and 3.45:

$$v_y = V_a \sin\theta + V_b \cos\theta \tag{3.44}$$

$$v_x = V_a \cos\theta - V_b \sin\theta \tag{3.45}$$

As the reading values of shaft encoders can be continuous, the most important features of resolvers are the resolution (highest precision by the number of bits that can be used to represent an angular position) and the maximum tracking rate (revolutions per minute, rpm).

Resolver-converters calculate digital position instantaneously with signals sin θ and cos θ. Figure 3.34 illustrates how instantaneous position can be indexed by two sine/cosine waveforms. There is a high-frequency modulation to be implemented either inside or outside the resolver. Most of these devices work in a closed-loop

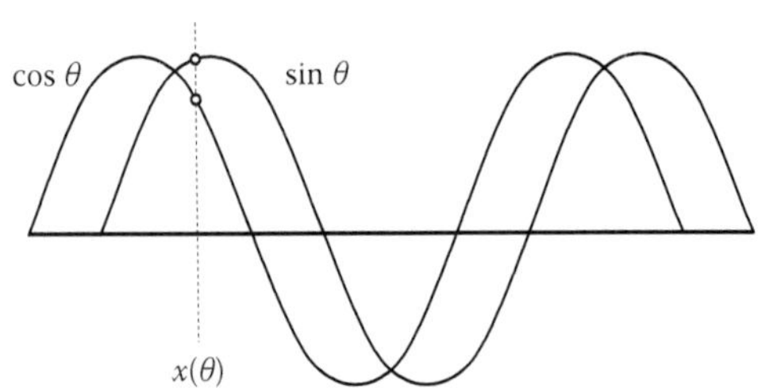

FIGURE 3.34 Output signals of a resolver-to-digital (R/D) converter.

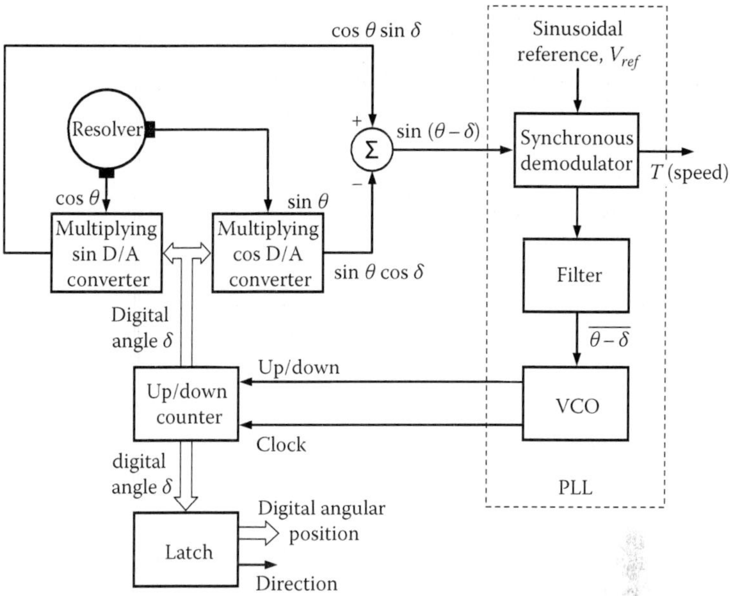

FIGURE 3.35 Processing the output signals of a resolver (R/D converter).

tracking, as in the external angle modulation depicted in Figure 3.35. They can detect both frequency and counting direction, which can increase (up) or decrease (down) when the VCO output is "0" or "1," respectively. Such measurements are established, respectively, by the amplitude error and the positive or negative signal in the phase detector output (synchronous demodulator).

Information about rotor position or speed is helpful to appear as functions of time since that is the way it will be controlled or positioned. Figure 3.36 shows the high-frequency modulation, and it is clear that a better precision is achieved when using the envelope of signals with respect to their rotor position at the excitation frequency. The envelope can be used if the rotor of the resolver is in high frequency rate, so the fundamental envelope frequency is only a small fraction of the excitation frequency. The resolver system will work similar to an AC carrier with reference signal V_{ref} used for its excitation. Brushes and slip rings are avoided in modern resolvers, because a solid rotor will directly vary the coupling between the primary and each secondary, the whole system behaves as a rotating transformer. In this way, defining the rotor mechanical angle by θ, one can resolve this angle into its trigonometric components. Exciting both input phases by a primary voltage V_{ref}, say, voltages $V_{\sin}$ and $V_{\cos}$, and using a standard transformation ratio $a = 0.5$, it is possible to produce two outputs whose magnitudes, and therefore the secondary sine and cosine waves, with respect to the excitation frequency will be expressed, respectively, as:

$$V_{\sin} = a \cdot V_{ref} \sin\theta$$
$$V_{\cos} = a \cdot V_{ref} \cos\theta$$

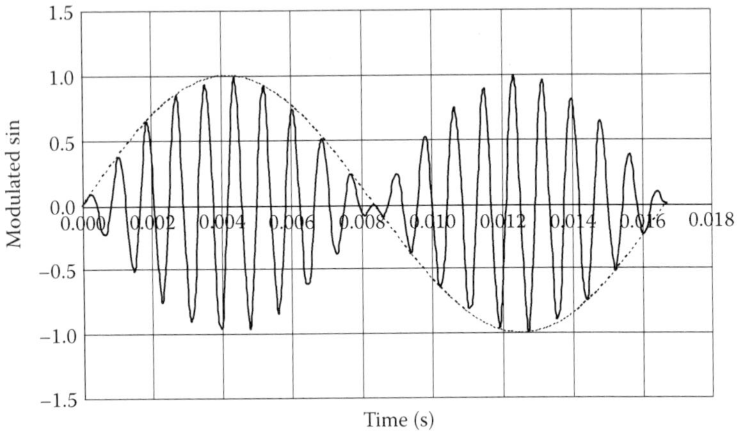

FIGURE 3.36 Modulated sin signal of a resolver (R/D converter).

There is a ratiometric tracking conversion, since the resolver secondary signals represent the sine and cosine of the rotor mechanical angle θ, and the ratio of signal magnitudes will be a tangent rate of sine and cosine given by the above set of equations [13–15]:

$$\theta = arc\tan\left(\frac{\sin\theta}{\cos\theta}\right) = arc\tan\left(\frac{V_{\sin}}{V_{\cos}}\right)$$

The above trigonometric relationship will allow a phase-locked loop (PLL) counter to track the resolver position θ since:

$$\sin(\theta - \delta) = \sin\theta\cos\delta - \sin\delta\cos\theta$$

If the difference between the two angles is relatively small ($\theta - \delta < \pi/6$), we can make:

$$\sin(\theta - \delta) \cong \theta - \delta$$

Assuming a binary word status of the up-down counter to be δ. The two instantaneous resolver output voltages modulated by the sine and cosine functions of the angle can be multiplied

$$\sin\omega t \cdot \sin\theta\cos\delta$$
$$\sin\omega t \cdot \cos\theta\sin\delta$$

With this, an error amplifier subtracts these signals, giving:

$$\sin\omega t(\sin\theta\cos\delta - \cos\theta\sin\delta) = \sin\omega t\sin(\theta - \delta)$$

where $\theta - \delta$ is the instantaneous angular error.

As explained in Section 5.3.2, demodulation or removal of the carrier signal is necessary to obtain the digital primary and secondary signals and the average of the rectified secondary voltage signal performed by the resolver-to-digital converter (R/D). The digital converter has to demodulate the resolver-formatted signals to remove the carrier and to provide the digital rotor angle. Therefore, the digital angle δ is incremented or decremented (direction) by the voltage-controlled oscillator pulses until the average error $\theta - \delta$ is zero, i.e., $\delta = \theta$, tracking in this way the resolver angle θ. As illustrated in Figure 3.35, the speed direction is positive whenever θ is positive. The rpm speed is calculated by $f = pn/120$ or $n = 60/T$ for the two-pole ($p = 2$) and T-period resolver.

EXERCISES

3.1 Discuss the operation and properties of thermocouples used in instrumentation. Exemplify a case using a simple amplifier circuit for its connection to an instrument.

3.2 Make an outline and use it to explain the operation of the differential transformer linearly variable (LVDT) and give a good practical application example.

3.3 The thermal parameters of the material used in a temperature sensor are as follows: $\alpha = 3.8 \times 10^{-3}\ °C^{-1}$, $\beta = -26.0 \times 10^{-7}\ °C^{-2}$, and $\gamma = 1.21 \times 10^{-10}\ °C^{-3}$. What is the error introduced by this nonlinearity for a temperature variation of 50 °C, and how one could determine the maximum linearity fractional error as a function of temperature?

3.4 What are the limits C_{min} and C_{max} that can be obtained from a capacitive sensor as depicted below with dielectrics $\varepsilon_1 = 8.85 \times 10^{-12}$ F/m and $\varepsilon_2 = 13.00 \times 10^{-12}$ F/m, $w = 4$ cm, $d = 0.1$ cm, and $L = 5$ cm? Suggest a measurement scale range to be used with this sensor.

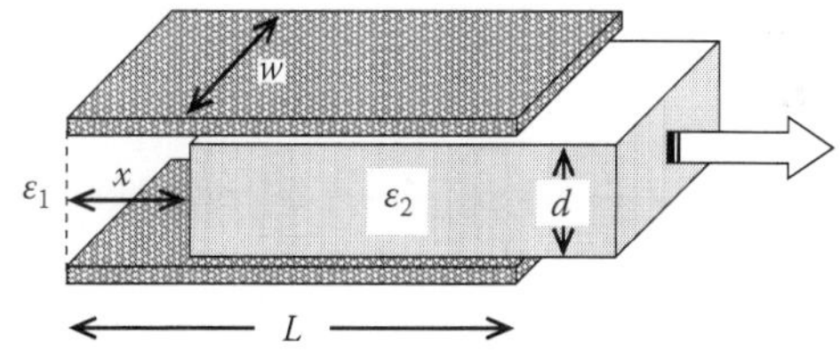

3.5 A resistive divider is used as an interface for an AC multimeter to be connected to the public grid voltage of 220 V/60 Hz. The resistors used are $R_1 = 740\ k\Omega$ and $R_2 = 10\ k\Omega$. Introduce any necessary changes in this divider and establish the limits of the range of voltage such that the losses do not exceed 0.1 W as error margin. What is the fractional error introduced in the measurements if the multimeter input resistance is $R_{instr} = 50\ k\Omega$?

3.6 Using two strain gauges, explain and demonstrate mathematically how they can be efficiently compensated for ambient temperature variations in the case of deformation measurement of elastic membranes. Also, sketch how to locate them on the membrane and suggest the most conventional electric circuit to be used.

3.7 Draw, i.e. sketch, an optical disk for measurement of a rotary shaft position with two opaque strips and two position detection sensors. Use the binary Gray code. Describe the disk operation along a look-up table for the indexing the quadrants.

3.8 Dimension the components represented in Figure 3.9 for a fraction $\delta = 0.001$ of a strain gauge deformation sensor with a resistance $R_2 = 100\ \Omega$ and a $V_{DC} = 12\,V$. What is the value of v_o? Suggest values for the compliance range.

3.9 Develop an approximate water depth sensor to be used in a river using commercial aluminum pipes, one having a diameter of 5 cm and the other one 1 cm, both with a length of 6 m.

REFERENCES

1. Paton, B.E., *Sensors, Transducers and Labview*, Virtual Instrumentation Series, Prentice Hall PTR, Upper Saddle River, NJ, October 1998, ISBN: 13-978-0130811554, ISBN: 10-0130811556.
2. Halvorsen, H.P., *Data Acquisition in LabVIEW*, University College of Southeast Norway, Notodden, Norway, 2016.
3. Silveira, P.R. and Santos, W.E., *Automation and Discrete Control* (*Automação e Controle Discreto*), Editora Érica, São Paulo, Brazil, 229pp., 1993.
4. Diefenderfer, A.J., *Principles of Electronic Instrumentation*, Editora W.B. Saunders Co., Philadelphia, PA, 667pp., 1972.
5. Zelenovsky, R. and Mendonça, A., *PC: A Practical Guide to Hardware and Software* (*Um Guia Prático de Hardware and Software*), 2nd edn., Editora MZ, Rio de Janeiro, Brazil, 760pp., 1999.
6. Eveleigh, V.W., *Introduction to Control Systems Design*, TMH Edition-Tata McGraw-Hill Publishing Company, New York, 624pp., 1972.
7. Bowron, P. and Stephenson, F.W., *Active Filters for Communications and Instrumentation*, McGraw-Hill Book Company, Berkshire, U.K., 285pp., 1979.
8. Auslander, D.M. and Agues, P., *Microprocessors for Measurement and Control*, Osborne/McGraw-Hill, Berkely, CA, 310pp., 1981.
9. Bolton, W., *Industrial Control and Measurement*, Longman Scientific and Technical, Essex, England, 203pp., 1991.
10. Horowitz, P. and Hill, W., *The Art of Electronics*, Cambridge University Press, London, England, 716pp., 2000.
11. Larminie, J. and Dicks, A., *Fuel Cell Systems Explained*, John Wiley & Sons, West Sussex, England, 308pp., 2000.
12. Helfrick, A.D. and Cooper, W.D., *Modern Electronic Instrumentation and Measurement Techniques*, Prentice-Hall-PH, Englewood Cliffs, NJ, 324pp., 2009, ISBN: 978-81-317-0888-0.
13. Bakhoum, E.G., *Micro- and Nano-Scale Sensors and Transducers*, CRC Press, 2015, ISBN: 9781482250909.
14. Yurish, S. Ed., *Modern Sensors, Transducers and Sensor Networks, Advances in Sensors*, Vol. 1, International Frequency Sensor Association Publishing (IFSA), 2012, ISBN: 978-84-615-9613-3.
15. Reddy, S.C.M. and Raju, K.N., Inverse tangent based resolver to digital converter—A software approach, *International of Advances in Engineering & Technology*, 4(2), 228–235, September 2012.

4 Electronic Instruments for Electrical Engineering

4.1 INTRODUCTION

Chapter 3, discussed how physical quantities could be converted into electrical signals using sensors and transducers. As discussed, data acquisition (from an instrument or computer) will be analyzed by measuring physical variables through transducers. The physical phenomenon is converted on a corresponding electrical signal variation. Then, a computer program will have a code to either control equipment or to provide graphical user information for humans who are using the equipment. The quantity usually handled in measurements of physical variables is voltage. However, other physical quantities can be considered, such as power or current. Those variables can be converted for computer-based data processing. Chapter 5 will discuss how to obtain quality indices based on the technical evaluation of pulse characteristics with their power, as function of time.

A typical challenge in instrumentation is to decouple the electrical signal variation related to the physical phenomenon from the variation caused by its display. For example, the maximum deflection of the DC moving coil to drive D'Arsonval type mechanism requires a minimum current of 50 μA through a coil resistance of 200 Ω, which results in 10 mV. Therefore, measurement of high impedances is seriously affected by the current deviation. Even small portions of current being drained from the measured quantity to move the pointer instrument will compromise the accuracy of the measurement. This effect can be minimized by electronic amplifiers. However, currently, in the vast majority of dials, digital indicators, or modern electronic voltmeters, current deviation can be virtually disregarded.

4.2 INSTRUMENT IN DIRECT CURRENT WITH AMPLIFIERS

Modern instruments are driven by voltage in direct current. Commonly, mathematical relationships are used to convert these DC values into other variables, and then processed and interpreted as frequency, alternating voltage, pressure levels from transducers, speed, etc. This section addresses the basic DC meter and its transformations to adapt it to other types of measurements that are not fundamentally DC voltage values.

4.2.1 Standard DC Electronic Voltmeter

Figure 4.1 illustrates an electronic DC voltmeter using a field-effect transistor (FET) amplifier in a Darlington configuration with a bipolar transistor (BJT). The input impedance of the instrument is constant and equal to about 10^7 Ω, which is

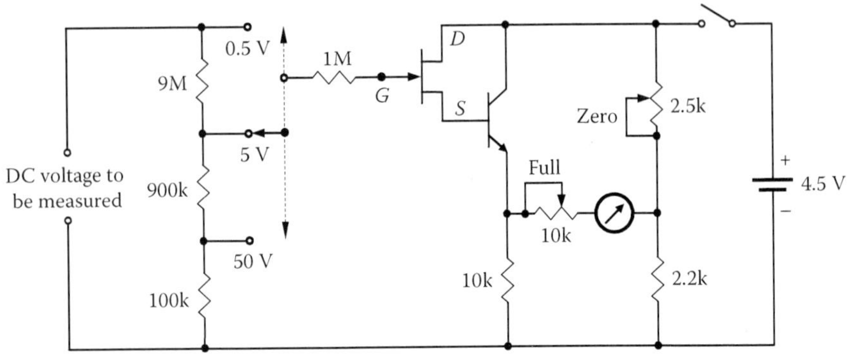

FIGURE 4.1 Electronic voltmeter with field-effect transistor.

virtually unaffected by the typical input impedance of a FET transistor, somewhere between 10^{10} and 10^{14} Ω. The BJT-FET Darlington transistor constitutes one arm of a balanced bridge made of series resistance, which sets the zero indication of the instrument. The full scale is made by a potentiometer, which sets the maximum deflection of the milliammeter.

4.2.2 Electronic Voltmeter-Ammeter

Figure 4.2 illustrates the electronic voltmeter-ammeter using an operational amplifier. The instrument input impedance is constant and equal to 1 MΩ. This value is not affected by the typical input impedance of operational amplifiers, which can range from 10^8 to 10^{10} Ω. The problems presented by these configuration settings are the bias currents (drift) and the offset current and voltage. Instrumentation amplifiers have low offset, i.e. low deviation current; in addition their low drift current features are appropriate for operation in DC measurements, since they have outstanding performance for signal conditioning purposes.

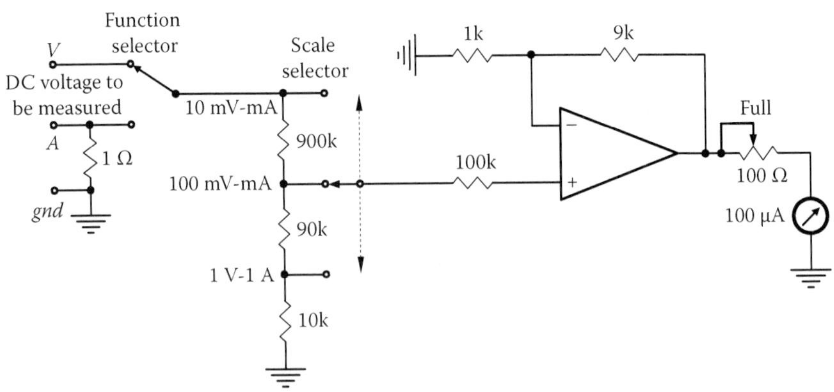

FIGURE 4.2 Electronic voltmeter-ammeter.

4.3 COMMON CIRCUITS USED IN INSTRUMENTATION

When the voltage to be measured is not direct current, or when there is a need to make other types of physical measurements, such as resistance, power, energy, and temperature, the DC voltmeter has to be adapted so that it can correctly interpret these signals. The AC signal may also contain residual DC current or other distortions that have to be considered, and in each case a suitable solution must be properly adopted.

4.3.1 Operating Principle of the Chopper Amplifier

The chopper amplifier meter is used in measurements of DC or low-frequency AC currents. The chopper amplifier minimizes the voltage and current deviations (offsets and drifts) when high gains are required. Its operating principle is shown in Figure 4.3.

A practical implementation of this circuit is shown in Figure 4.4. The numbers indicated in each of the metal-oxide-semiconductor FETs (MOSFETs) transistor

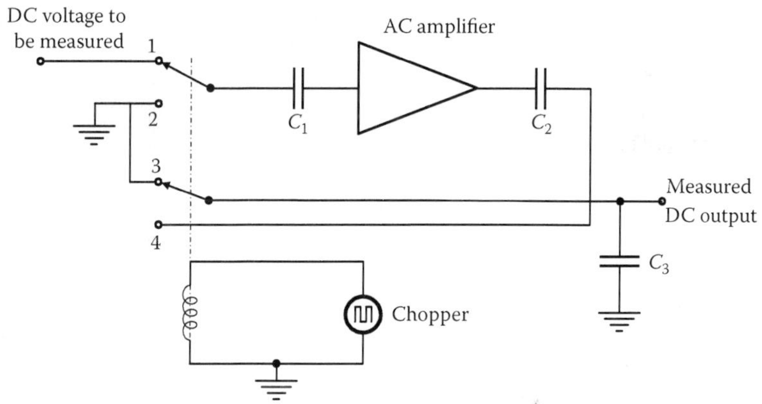

FIGURE 4.3 Operating principle of the chopper amplifier.

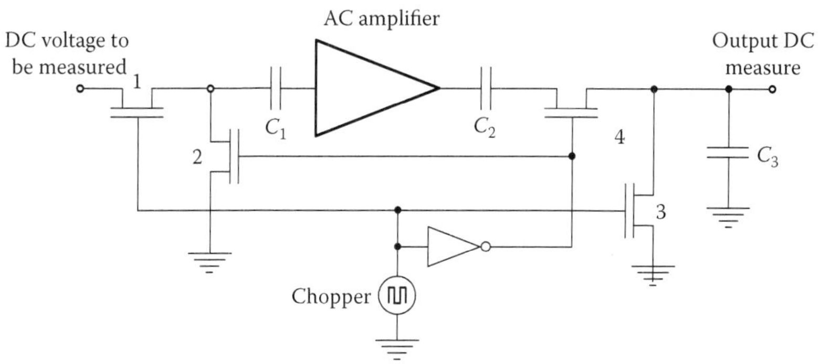

FIGURE 4.4 Implementation of the chopper amplifier with MOSFETs.

correspond to those indicated on the switch contacts in Figure 4.3. The switches in positions 1 and 3 allow the charge of capacitor C_1 through the virtual short circuit across the operational amplifier inputs, and to discharge capacitor C_2 in the meter output. When the switches move to positions 2 and 4, capacitor C_1 discharges itself and recharges capacitor C_3 from capacitor C_2, the charge voltage of which appears across the voltmeter output.

4.3.2 Alternating Current Voltmeters with Rectifier

Alternating current voltmeters are based on a standard DC voltmeter. A rectifier is used to convert the low-harmonic content of the AC voltage (usually less than 10%) into the correspondent DC value, which is eventually output to the display. The applied amplifier in AC voltmeters should be stable in environmental conditions, and have a unity gain. The signal to be measured can be rectified before or after amplification, which may cause distinct effects on the output signal, as discussed below.

Some configurations of signal rectification are shown in Figure 4.5. Each one of that configuration should be applied based on the requirement of its particular application and the designer criterion, such as cost, simplicity, voltage levels, admissible ripple, loading the measuring circuit, and measuring speed.

If the measured AC signal is first rectified and then amplified, its amplitude should loosely exceed the voltage level of the diode knee before indicating any significant output value. Therefore, for very low values of voltage, about 1.0 V or so, the error becomes unacceptable. For a negligible reading error, this setting is recommend but only for high values of measured voltage, say above 100 V. Some advantages of rectifying before the amplification are simplicity, fewer number of components, and lower cost. Figure 4.6 shows a circuit with this arrangement where one can observe a voltage-to-current conversion as part of the process.

Owing to large differences in the input impedances of operational amplifiers, it is common to use a high-value input resistance R_{in}. The value of this resistance must be large enough to serve as the instrument input, and it must be, simultaneously, small enough compared with the input impedance of the internal amplifier. The instrument input current will depend on R_{in}. The trimmer R limits the current through the

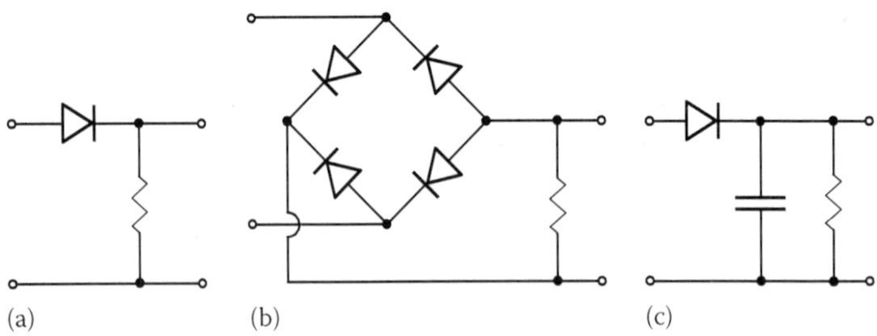

FIGURE 4.5 Rectifiers for instrumentation: (a) half wave, (b) full wave, and (c) peak detector.

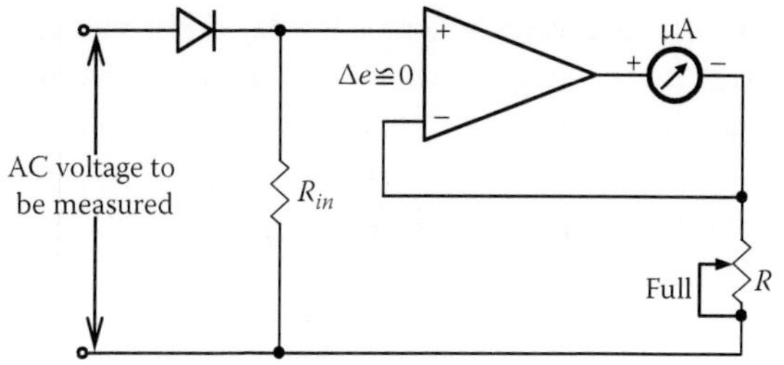

FIGURE 4.6 AC-meter with rectification before signal amplification.

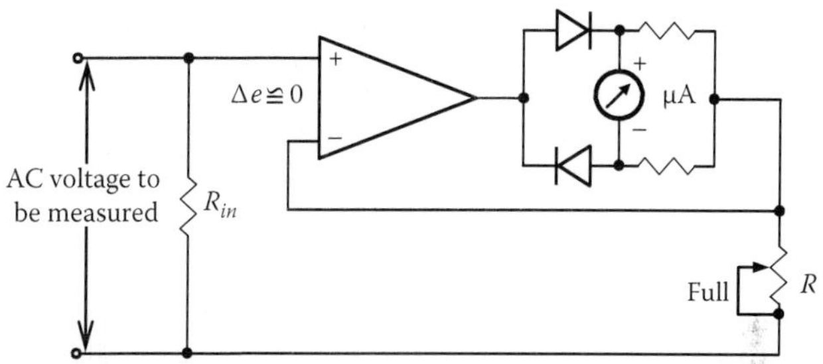

FIGURE 4.7 AC-meter with rectification after signal amplification.

microammeter, and it serves to adjust the instrument full scale. The zero of the scale can be adjusted in the offset trimmer of the amplifier.

If the AC signal to be measured is first amplified and then rectified, the input impedance of the instrument can approximate the high impedance input of the operational amplifier (Figure 4.7). This configuration also solves the problem of voltage-to-current conversion and the nonlinearity of diode rectifiers. It can also be applied in the case of very small signals.

In general, voltmeters measure well DC levels, they can also be adapted to measure average or effective sinusoidal signals. Therefore, they are calibrated according to their specific needs; however, measurement errors are introduced in nonsinusoidal waveforms. If the form factor of the nonsinusoidal wave is known, it can still be measured by a DC voltmeter, introducing the corresponding correction. Usually, this distinction is made by using terms like effective amount for pure sine waves and true effective value (true rms) when the waveforms have more than 10% of harmonic content. For a pure DC signal, this factor is unity. In fact, the form factor expresses the harmonic content of the signal, and it is defined as:

$$k = \frac{\text{Effective value (rms)}}{\text{Average value (DC)}} = \frac{\sqrt{\frac{1}{T}\int_0^T e^2 dt}}{\frac{1}{T}\int_0^T e\,dt} \tag{4.1}$$

For a sinusoidal full-wave voltage across the AC meter output in Figure 4.7, the form factor will be:

$$k = \frac{E_m / \sqrt{2}}{2E_m / \pi} = 1.11 \tag{4.2}$$

4.3.3 Voltmeters of True Effective Value ($V_{true-rms}$)

The rms voltmeter is used to measure complex or nonsinusoidal wave shapes. The rms voltage value in the display is proportional to the active power contained in the waveform (generated heat), which, in turn, is proportional to squared voltage V_{rms} including all distortions. The generated heat is then sensed by a thermocouple sensor, which generates a corresponding voltage proportional to the true-rms voltage.

The most challenging problem for this methodology is the inherent nonlinearity of the thermocouple sensor. Therefore, a compensation for such nonlinearity must be provided. This can be done by using two thermocouples, one for measurement and another for nonlinearity compensation, as in the schematic shown in Figure 4.8. In this scheme, the display reading (V_o) can be done within a common thermal environment. After both thermocouples reach the equilibrium temperature, the indication on the display corresponds to the measured true rms.

A more elegant, modern, simple, comprehensive, and compact way to measure true-rms values is using Fourier transform. This method includes acquisition of the instantaneous voltage or current signal used to determine the Fourier coefficients, as described in Chapter 6. These coefficients correspond to the amplitudes of each harmonic component presented in the instantaneous signal. Limitations of this method include slower time response for the full acquisition of the signals, data processing, and result display.

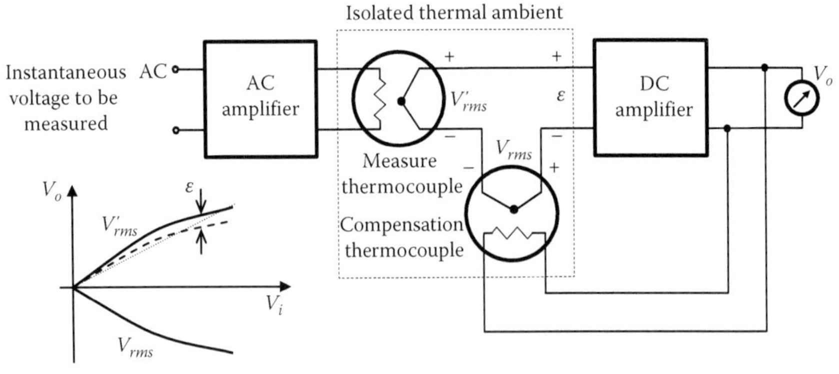

FIGURE 4.8 Voltmeter to measure true rms.

4.4 ELECTRONIC MULTIMETERS

An electronic multimeter should be able to read at least voltage and current, direct or alternating, and resistance. More recently, the facilities of electronic schemes present, as other possibilities, the addition of other quantities such as capacitance, inductance, temperature, frequency, diode, and transistors testing. Sometimes, an audible indicator (beep) is available for continuity tests.

A simple schematic of the basic core of a self-biased electronic multimeter is shown in Figure 4.9, which consists of the following modules:

1. DC bridged amplifier with the display
2. Scale selector with attenuator and filter
3. Rectifier for AC signals (not shown in this figure)
4. Battery and internal circuits as a resistance reference and filter for resistance measurements
5. Selector of functions (voltage, current, AC, DC, R, etc. not shown)

4.4.1 General Characteristics of the Analog Voltmeters

The analog voltmeter, with or without the aid of an amplifier, should be characterized by the following general features:

1. Input impedance (Z_i)
   a. The output impedance must be at least ten times greater than the input impedance.
   b. The sensitivity to parasitic capacitances must be optimized at higher frequencies.
   c. Probes can reduce the capacitance, with the typical loss of about 20 dB efficiency.
2. Voltage ranges
   1-3-10 with 10 dB separation or similar

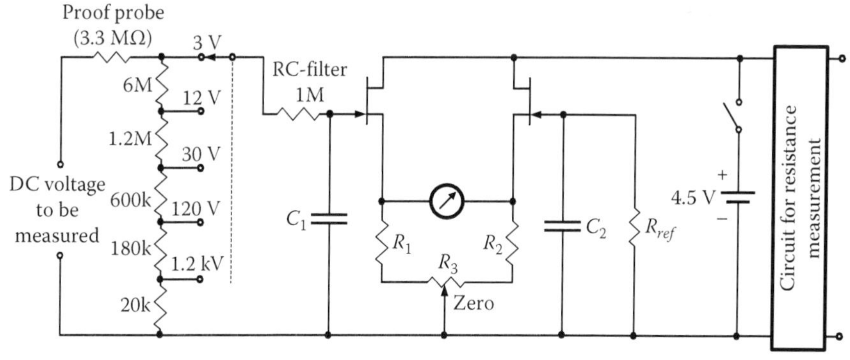

FIGURE 4.9 Basic core of an electronic multimeter.

3. Decibels

   When a wide range of values with small differences along their variation is evaluated, it is better to express such range in decibels, assuming a given reference. Therefore, it is possible to establish a reference range in decibels with the equation for power:

$$P_{ref} = \frac{V_{ref}^2}{R} \tag{4.3}$$

   Taking the power reference value as $P_{ref} = 1$ mW, the following references can be established:

   0 dB_m for $R = 600\ \Omega$ resulting in 0.7746 V

   0 dB_m for $R = 50\ \Omega$ resulting in 0.2236 V

   Also $V_{ref} = 1$ V can be used as voltage reference and thus has a relative measurement V/V_{ref} as:

   0 dB_v for any standard impedance system

4. Resolution and band-pass frequency ranges

   The noise depends on the frequency range Δf for which the instrument has been designed. The broader the range of frequencies, the greater will be the noise. For example, for an instrument with a range between 10 Hz and 10 MHz, the resolution is 1.0 mV; for an instrument that measures 5 MHz only, the resolution is 1.0 μV.

5. Battery operation

   To build a portable instrument for field works, it is essential to use batteries. When there are ground loops in the area, it is also preferable to use equipment with battery, which also has the advantage of not introducing noise from the power supply.

6. Measurements of AC currents

   For AC currents, ammeter pliers are used in which the primary coil has only one turn and the secondary coil has several wound around a ferrite core. An alternative is a Hall effect device (HED), which can be used to measure either alternating or direct current.

7. General guidelines

   a. There are no problems in DC measurements beyond the measurement isolation.

   b. For alternating current measurements with distortions less than 10%, with or without harmonics, the average value voltmeter (DC) gives better accuracy and higher sensitivity per unit of capital invested.

   c. For frequencies higher than 10 MHz, the peak detector voltmeter is the most economical and acceptable choice for a waveform that is reasonably sinusoidal.

   d. For nonsinusoidal waveforms, it is recommended the true-rms voltmeter or an algorithm based on instantaneous data acquisition be analyzed by Fourier series.

4.5 GENERAL CLASSIFICATION OF DVMs

Depending on the cost and the desired characteristics of the measurements, digital voltmeters (DVMs) can be classified into four types:

1. Ramp or single integration (single slope)
2. Double integration (dual slope)
3. Continuous balance
4. Successive approximations

4.5.1 Standard Single-Slope or Ramp Voltmeter

The ramp voltmeter measures time by decade counting units (DCUs) making a linear ramp voltage either change from zero to the DC level of the input signal, or decrease the DC voltage input level to zero. The time measurement is performed by a pulse counter (time-voltage converter), depicted in Figure 4.10, where the counter is set at point 1 and reset at point 2, and the ramp is equal to the reference value. Figure 4.11 shows the implementation, where a logic gate performs the comparison, allowing the counter display to indicate the output. This system is called as a "single slope meter".

The single-slope meter can reduce the errors introduced by analog integrators depending on the resistor and capacitor precisions used for the comparison ramp. The accuracy of the digital ramp will depend on the crystal (clock) used to drive the digital processor controlling and monitoring the ramp type voltmeter. Note that the ramp voltage can also be achieved with a progressive digital counting in a microprocessor or computer.

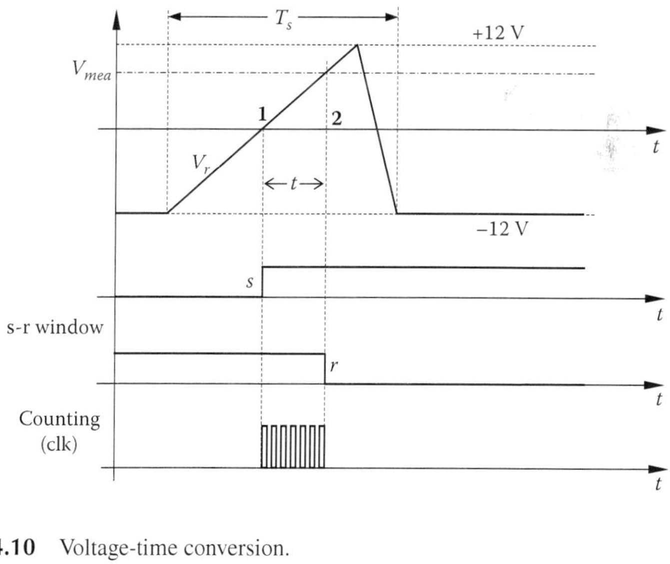

FIGURE 4.10 Voltage-time conversion.

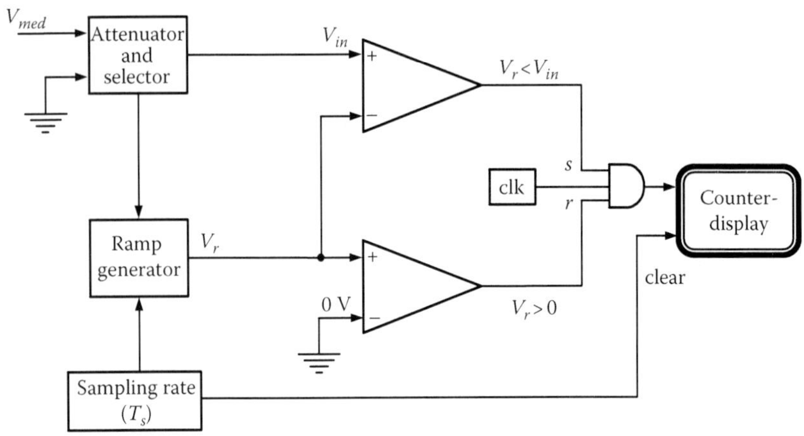

FIGURE 4.11 Schematic operating function of the single-slope voltmeter.

Typically, the single-slope voltmeter has the following characteristics:

- 10 MΩ attenuator to measure between 100 and 1000 V
- DC voltage amplifier with a fixed gain of 1000 providing input for up to 10 V in an independent comparator in the selected range
- 4.5 kHz oscillator for the pulse counter
- Three or four DCUs of the form [1888] or [18888]
- Sampling rate controlled by a relaxation oscillator (set/reset) with two samples per second
- Digital-to-analog converter with an integrator circuit to generate a voltage slope that depends on an accurate and stable RC circuit

4.5.2 Dual-Slope Voltmeter

The dual-slope voltmeter has a higher performance when compared to the single-slope voltmeter. The single-slope method requires a very precise comparison, whereas the dual-slope has one integrator that takes the positive slope within a given fixed time. When the slope is reversed, the same integrator will return to zero conditions, and the time is measured, making the dual-slope scheme independent of an accurate comparison. The integration order is not relevant. Figure 4.12 shows a schematic diagram for the dual-slope voltmeter.

The voltage V_{mea} is the one to be measured. The first integration occurs within a predetermined fixed time, T_i, where the internal integration voltage ramps from zero to V_i, as shown in Figure 4.13:

$$V_i = \frac{1}{RC}\int_{-T_i}^{0} V_{med}dt = \frac{T_i}{RC}V_{med} \tag{4.4}$$

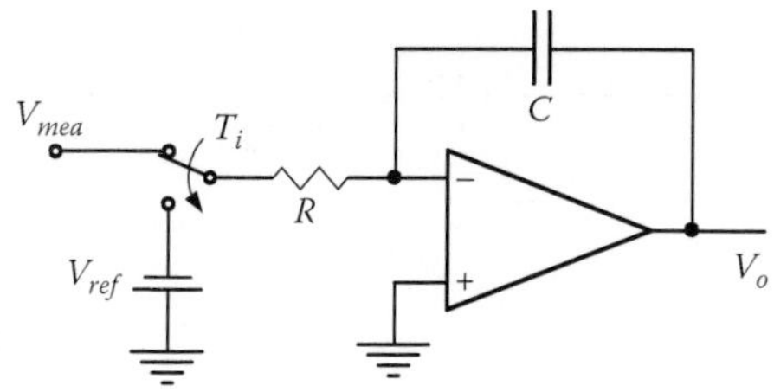

FIGURE 4.12 Principle of the dual-slope voltmeter.

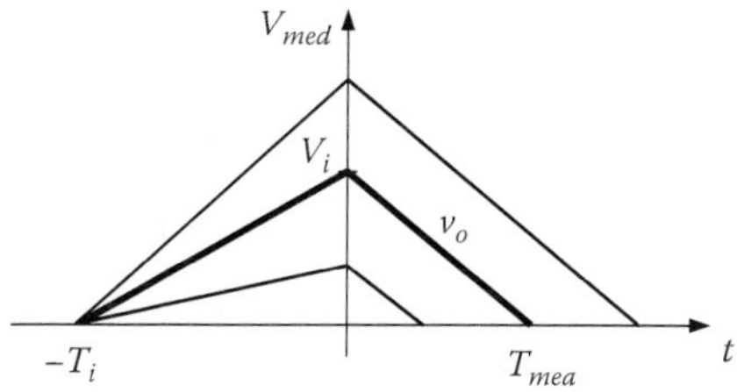

FIGURE 4.13 Diagram of the dual-slope voltmeter.

After the voltage reaches V_i (within the pre-determined time T_i) there is an inverse integration time T_{med} that will take time as required to bring the integrated voltage back to zero.

The second integration is conducted after reversing the switch position, indicated in Figure 4.12, to the negative terminal of the voltage source V_{ref}. The capacitor initial charge voltage V_i is brought to zero at a time "t". The system measures the number of pulses. The voltage variation across the operational amplifier output is then:

$$v_o = V_i - \frac{1}{RC}\int_0^t V_{ref}dt = V_i - \frac{t}{RC}V_{ref} \tag{4.5}$$

Replacing Equation 4.4 in (4.5) when $v_o = 0$, after $t = T_{med}$, we get:

$$V_{med} = \frac{T_{med}}{T_i}V_{ref} \tag{4.6}$$

Note that Equation 4.6 does not rely on R or C, but on the device constants V_{mea} and T_i. Moreover, the capacitor charging current is proportional to V_{mea}, and the discharge current is proportional to V_{ref}.

It is not required any sample-and-hold circuit because the integrator calculates the average value. The meter using such a dual-slope technique is usually slow and not suitable for computer data acquisition and real-time control. However, the dual-slope technique is often used in portable instrumentation, typically used by technicians working in the field.

To obtain a significant improvement in these sorts of converters, it is advisable to introduce auto-scaling (auto-ranging), and an automatic correction of zero

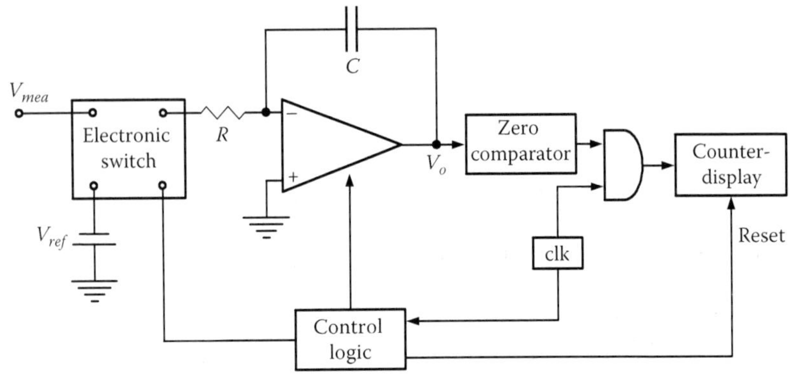

FIGURE 4.14 Digital voltmeter with dual slope.

(auto-zeroing) to compensate offsets, drifts, and capacitor leakage currents. In the auto-zeroing technique, when a conversion is made, the offset voltage is stored in a capacitor to compensate for the input voltage offsets.

A practical sketch of the dual-slope technique is shown in Figure 4.14, in which the capacitor C is fully charged to the level of the measured DC voltage. The time starts at zero at the same instant the capacitor begins to discharge, driven by the internal reference voltage V_{ref}. The full discharge time is proportional to the measured voltage V_{mea}.

4.5.3 Continuous Balance Voltmeter

Another method that eliminates the analog-digital conversion is shown in Figure 4.15, which makes use of an instrument that tracks the capacitor voltage and compares it with the reference voltage V_{dig} generated by a digital counter with the measured voltage value V_{mea}. The capacitor current will be null when these two values are equal, and the display can be carried out in a digital counter. This technique has mainly two advantages: (1) it uses the same reference capacitor to measure the input quantities, and cancels out the dependence on the stability and accuracy of the capacitor; and (2) the AC line network can be chosen for the digital integration period (1/60 s), and the instrument becomes insensitive to the AC network buzzes and harmonics. As for

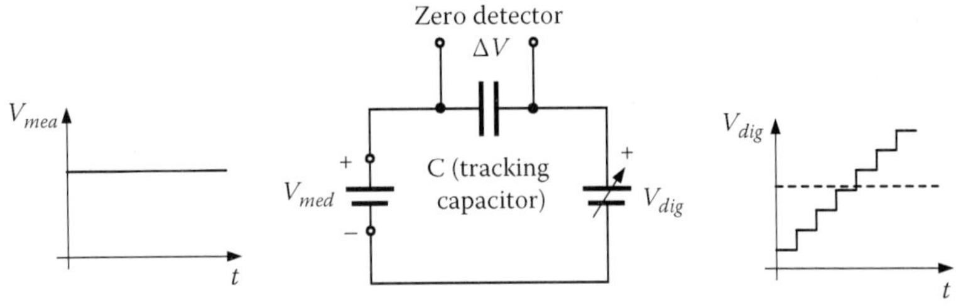

FIGURE 4.15 Measuring technique by charge balance.

disadvantages, the continuous balance voltmeter is inherently slow poor accuracy, when compared to the successive approximations voltmeter.

4.5.4 Voltmeter of Successive Approximations

The voltmeter of successive approximation is efficient and relatively inexpensive. It uses the binary regression (bisection) to match the measured quantity wherein successive estimates are obtained from the most significant bit (MSB) to the least significant (LSB). The procedure of this sort of voltmeter is postulated as follows:

1. A successive approximation method, considers a procedure where for each reading estimation, a binary word will have its most significant bit tested initially if the binary number is greater or lesser than the one to be measured. When that is greater, the bit is set to "0," otherwise, i.e. smaller, it is set to "1." The steps are repeated until all bits have been tested.
2. The range of maximum and minimum values should be known in advance for the amount of the desired number.
3. The required number of estimates is exactly equal to the number of bits required for conversion of the largest decimal number to fit in the range of the binary numbers.

Figure 4.16 shows the block diagram for a successive approximation voltmeter using a register for displacement called as successive-approximation-register (SAR). This architecture allows sample rates under 5 megasamples per second (Msps), and the resolution for SAR ADCs most commonly ranges from 8 to 16 bits. These features make this SAR voltmeter ideal on data/signal acquisition for industrial controls. The operational amplifier performs the comparison between the analog signal level being digitized V_{dig} and the measured value V_{mea} and defines which is higher or lower.

Example	
address	data
0000	0000
0001	0101
0010	0010
0011	0111
0100	0100
0101	0111
0110	0110
0111	0101
1000	1000
1001	1001
1010	1010
1011	1001
1100	1100
1101	1101
1110	0110
1111	0011

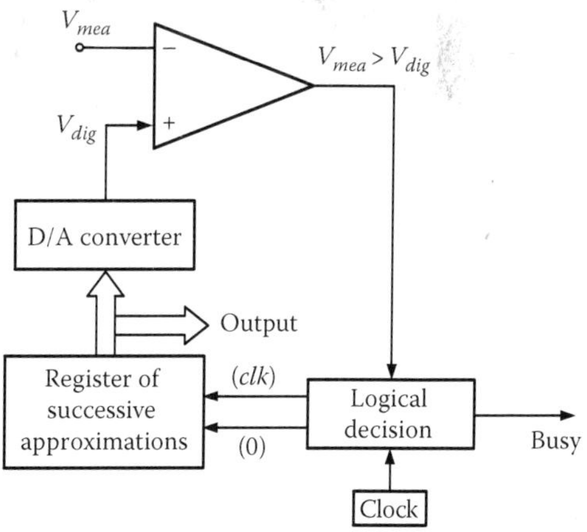

FIGURE 4.16 Voltmeter of the successive approximation type.

The logic decision block is responsible for checking whether the estimate is higher or lower than the read value while maintaining or not the logic "1." This type of voltmeter has a good trade-off between time response and accuracy.

4.5.5 General Characteristics of a DVM

The general characteristics of digital voltmeters are as follows:

- Use discrete values for the measurements, thereby avoiding interpolation errors of reading and parallax errors; DVM is more suitable for interfacing with computers since it uses a digital treatment for the signals.
- Unlike microammeters that are currently being used to trigger a mechanism, the DVMs use their own instrument voltage source to excite an electronic digital display, thereby preventing overloading the measurement circuit.
- Measuring range between 1 and 1000 V.
- Absolute accuracy: ±0.005% of the indication in a four-and-a-half-digit display (±1/1888).
- Stability:
  Short-term up to 0.002% in 24-hour time
  Long-term up to 0.008% in 6-month time
- Resolution: 1.106 (1 ppm), for example, 1 Ω/1 MΩ.
- Typical input characteristics:
  $R_i = 10$ MΩ
  $C_i = 40$ pF
- Calibration: internal standard with a stabilized voltage.
- Can easily provide output signals to the computer.
- Additional measures can result in a ratio between two voltages that may represent voltage, current, and resistance, or any other pair of physical variables.

4.6 OHMMETERS AND MEGOMETERS

Resistance may be measured by a relationship between the voltage across the instrument terminals and the current through it. The internal resistance of the instrument, a voltmeter or an ammeter, is inevitably included in the measure. Consider, initially, the measurement configuration shown in Figure 4.17, where the voltage drop on the ammeter is included in the measured voltage. Although the current through the

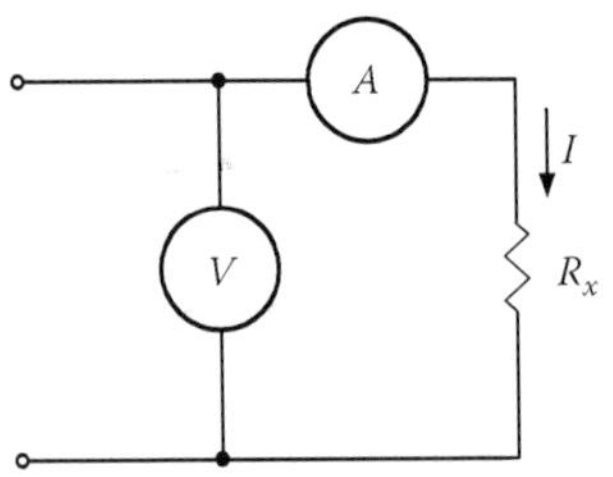

FIGURE 4.17 Current and voltage measurements across a resistor.

unknown resistance R_x will be measured correctly, the voltage will not. The voltage will be increased by the ammeter series resistance. Quantitatively, the voltage value measured is expressed as:

$$V = I\left(R_a + R_x\right) \tag{4.7}$$

where

I is the measured current

R_a is the ammeter resistance

If $R_x \gg R_a$, R_a will have only little effect on the measurement, and V will be a reasonable value. Thus, for large values of R_x, an instrument becomes reasonable in this condition, as ordinary megometer.

One possible configuration for voltage and current measurements of resistance is shown in Figure 4.18. In this case, although the voltage across the unknown resistance R_x will be correctly measured, the current will not. The current is increased by the largest parallel resistance of the voltmeter. Quantitatively, the current value measured is expressed as:

$$I = \frac{V}{R_v} + \frac{V}{R_x} \tag{4.8}$$

where R_v is the voltmeter resistance.

If $R_x \ll R_v$, R_v will have little effect on the current, and the value of I will be reasonable. Thus, for small values of R_x, an instrument such as an ohmmeter becomes reasonable in this configuration. Quantitatively, the measured current value can be expressed as:

$$I = \frac{V}{R_a + R_{sc} + R_x} \tag{4.9}$$

where R_{sc} is the ohmmeter scaling resistance.

The ohmmeter itself is the most direct way of measuring resistance, though it is less precise. It relies solely on a calibrated current measurement on a fixed scale across a series resistance, as shown in Figure 4.19.

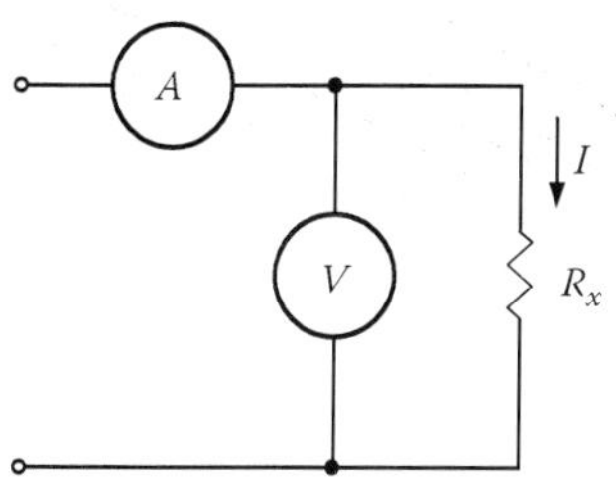

FIGURE 4.18 Voltage and current measurements across a resistor.

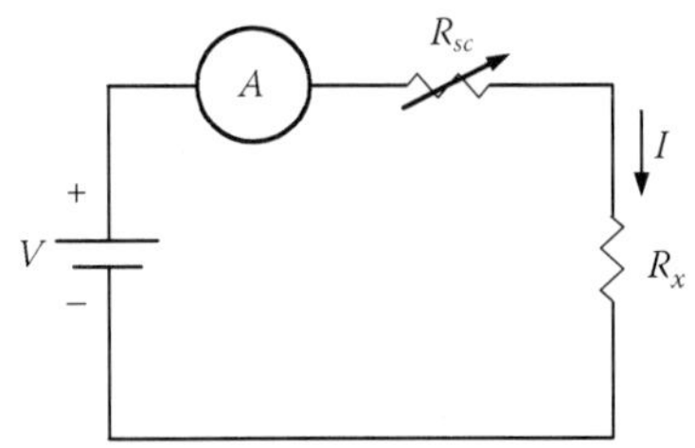

FIGURE 4.19 Standard ohmmeter.

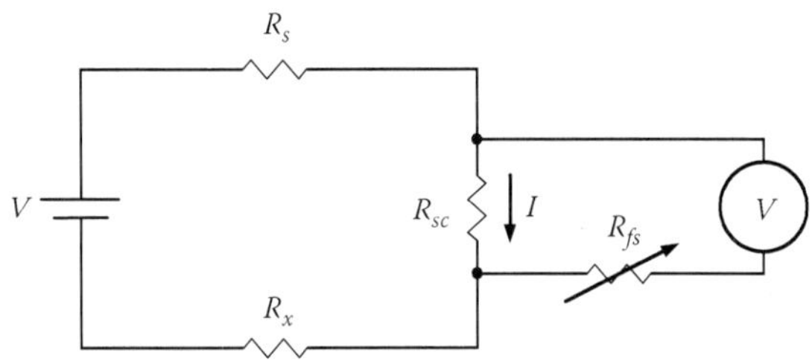

FIGURE 4.20 Improved version of ohmmeter.

Any value can be assigned for the full-scale adjustment R_{sc}, as that scale indicates the correct value of a standard resistance, usually two-thirds of the full-scale value. It serves as measurement and reference readings. Once the scale is adjusted, the sum $R_a + R_{sc}$ should not change. This is seen, therefore, as a nonlinear scale (hyperbolic).

The main limitations of the ohmmeter are as follows: a limited range for acceptable accuracy, which is a hyperbolic function varying greatly with R_x, and which may increase beyond the knee point; and the need to reset the instrument before each measurement.

An improved configuration of this simplified ohmmeter version is shown in Figure 4.20, where a series-parallel combination of resistors allows for a choice of the most suitable scales to track the measured values. In this configuration, a voltmeter is used instead of an ammeter, with a resistance R_s in series with R_x and R_{sc}.

The resistance R_E imposes the scale of the ohmmeter. It is modified by resistance R_{fs} (full scale) in order to adjust the full scale, that depends on the voltmeter resistance R_v. The total value of this association is:

$$R_E = \frac{R_{sc}\left(R_{fs} + R_v\right)}{R_{sc} + R_{fs} + R_v} \tag{4.10}$$

The value of full scale (short circuit or no resistance) is obtained by shorting up R_x and then adjusting the deflection of the meter pointer or digital meter (DVM) to its full-scale value. Thus, removing the short circuit across R_x, the resistance value is read by the voltmeter, which modifies the scale to show directly resistance values.

By changing resistance R_{fs}, one can vary the measurement ranges in decades, keeping substantially the same current through the voltmeter. This configuration can also be used to read either voltage or currents, after which this instrument becomes known as VOM (volt-ohm-milliamp).

4.7 DIGITAL NETWORK ANALYZERS

Digital network analyzers, wattmeters, varmeters, and energy meters are devices that interpret the interaction between voltage and current to express, respectively, active power, reactive power, and the amount of energy in an electrical circuit. Particularly, digital network analyzers are based on the voltage and current values obtained in the instrument terminal measurements [1,2].

4.7.1 Analog-to-Digital Network Analyzers

Integrated circuit boards or data acquisition circuits with signal processors can be used as a core for the development of digital network analyzers requiring only a little information from the network signal, such as the instantaneous voltage $v(t)$ and the instantaneous current $i(t)$.

Assuming, initially, quantities under periodic ($T = 1/60$ s) and sinusoidal condition, and on the basis of the rms digital detection of the maximum values of those quantities, V_m and I_m, one can derive the following parameters.

1. $$V_1 = \frac{V_m}{\sqrt{2}} \quad \text{(effective amount to be recorded directly on the display)} \tag{4.11}$$

2. $$I_1 = \frac{I_m}{\sqrt{2}} \quad \text{(effective amount to be recorded directly on the display)} \tag{4.12}$$

3. $$f = \frac{1}{T} \tag{4.13}$$

   The alternating voltage period can be obtained by detecting the average period, or half-period, using a PLL (phase-lock-loop) circuit, the p-q theory, or, exceptionally, using cycles of the 50 or 60 Hz source. Under distortion in the sine waveform, the wave period T is not really constant along all the measurements. In the specific case of U.S. power grids, this period can be considered fairly constant, since the country's power network is fully interlinked. However, in the case of independent generators (e.g., microgrids), as with asynchronous generators, this value may be extremely variable. Therefore, to determine a value close to the true value, a filter must be implemented (analog or digital) in the data acquisition system with the function of approximating the fundamental waveform periods.

   If a wave analyzer is used only for a fairly harmonic-free public network, it is advisable to use $T = 1/60$ s as the fundamental period. The following relationships are possible:

4. $$|Z| = \frac{V_1}{I_1} = \frac{V_{m1}}{I_{m1}} \tag{4.14}$$

5. $|S| = V_1 I_1$ (assuming absence of harmonics) or $V_{rms} I_{rms}$ (4.15)

6. $P = \frac{1}{T}\int_0^T p(t)\,dt = \frac{1}{T}\int_0^T v(t)i(t)\,dt$ (value on the display) (4.16)

The complete solution of that integral is simple, and it results in the average power obtained by numerical methods, such as trapezoidal, Newton-Raphson, Euler direct or inverse, or Runge-Kutta methods. However, the basic theory of circuits for sinusoidal waveform voltage and current, where the subindex "1" indicates the fundamental period, defines the instantaneous active power as $p(t) = v(t) \cdot i(t)$. It can be obtained from the average value of that expression within a period between zero and $T = 1/60$ s, or multiple periods, to improve the average period, as:

$$\begin{aligned} v(t) &= V_{m1}\sin(\omega_1 t) \\ i(t) &= I_{m1}\sin(\omega_1 t + \theta_1) \\ p(t) &= v(t)\cdot i(t) = V_{m1} I_{m1}\sin(\omega_1 t)\sin(\omega_1 t + \theta_1) \end{aligned} \tag{4.17}$$

From the trigonometry identities:

$$\sin B \cdot \sin A = \frac{1}{2}\left[\cos(A - B) - \cos(A + B)\right]$$

$$p(t) = \frac{1}{2}V_{m1} I_{m1}\left[\cos(\omega_1 t + \theta_1 - \omega_1 t) - \cos(\omega_1 t + \theta_1 + \omega_1 t)\right] \tag{4.19}$$

$$p(t) = \frac{1}{2}V_{m1} I_{m1}\left[\cos(\theta_1) - \cos(2\omega_1 t + \theta_1\right]$$

where

- $v(t)$ and $i(t)$ are, respectively, the instantaneous voltage and current
- V_{m1} and I_{m1} are, respectively, the fundamental peak amplitude of voltage and current
- ω_1 is the fundamental angular frequency

As $V_1 = V_{m1}/\sqrt{2}$ and $I_1 = I_{m1}/\sqrt{2}$, then simplifying and integrating the above equation to obtain the average of $p(t)$, between 0 and T, we get:

$$P = \frac{1}{T}\int_0^T V_1 I_1\left[\cos(\theta_1) - \cos(2\omega_1 t + \theta_1)\right]dt = V_1 I_1\cos(\theta_1) \tag{4.20}$$

where

- V_1 and I_1 are, respectively, the fundamental values of the effective voltage and current
- Subindex "1" means the fundamental order (or harmonic of the first order), subindex "2" means second harmonic order, and so on

7. Once the average power is obtained, by integrating Equation 4.19 or 4.20, the power factor can be then defined as:

$$\cos\theta_1 = \frac{P}{|S|} \tag{4.21}$$

where θ_1 is the lag or displacement angle between the fundamental current and voltage.

8. $\bar{W} = \int_0^{T_{mea}} p(t)\,dt = PT$ is the average power measured in a period between 0 and T_{mea}, which can be integrated to totalize the energy values within this same period:

$$W = W_1 + W_2 + \cdots + W_n \tag{4.22}$$

The reactive power, resistance, and reactance can be calculated as:

9. $Q = |S| \cdot \sin\theta_1$ (4.23)
10. $R = |Z| \cdot \cos\theta_1$ (4.24)
11. $X = |Z| \cdot \sin\theta_1$ (4.25)

The harmonic content strongly affects the parameter measurement of electrical networks, and it is discussed in detail in Chapter 6.

4.8 POWER SOURCES AND ADAPTERS

The development of power sources and adapters feeding measurement instruments is critical because they are responsible for reading errors and many unwanted noises. Such noise can be caused by cumulative errors in pointer indicators, electric fields, magnetic fields, electromagnetic fields, or in the inaccuracy of digital displays and in difficulty in stability of the two last displayed digits. Figure 4.21a and b shows the Thévenin equivalent circuits, for batteries and power adapters, respectively, based on the network. Note that in Figure 4.21, the total ripple amount originated in the power source, or in parallel circuits (conducted distortion) connected to the same source, or

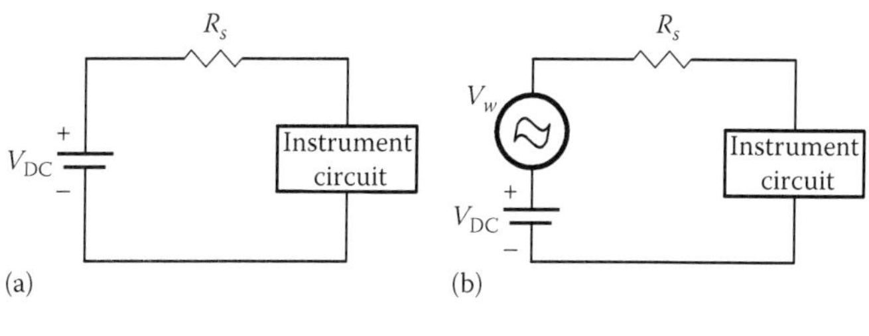

FIGURE 4.21 Equivalent Thévenin sources of instruments: (a) DC equivalent circuit, (b) AC equivalent circuit with decoupling by-pass capacitor..

even induced (radiated distortion) by close circuit represented by V_w. In many cases, it is recommended to use decoupling capacitors as close as possible to the integrated circuits, bypass capacitors, symmetrical circuits with respect to zero voltage (ground), or configurations such as in antiphase (push-pull), as discussed in Chapter 1.

The advantage of using batteries is that they are independent of the ground loop circuit to be monitored, and they do not introduce instrument noise or typical ripples from rectified sources. Furthermore, the instrument becomes portable. In terms of limitations, the supply voltages of these sources decrease rapidly (battery discharge) with the prolonged use of the instrument, and precautions should be taken at the instrument design stage so that the measurements are only minimally affected. In the case of rectified sources, the noise problem is much more serious, in that all measurements can be affected by ripples and intense pulse might originate in the current supply. If it is not introduced, special care such as filtering, use of integer multiple of the network frequency, and special compensation measures, the instrument accuracy may be seriously affected.

Note that in the present state of technology, switched power supplies should be avoided in precision instruments for good-quality measurements and data acquisition readings.

EXERCISES

4.1 Differentiate the internal operation of the actual rms voltage meters and the ordinary AC voltage meters.

4.2 What is the approximate percentage improvement in an instrument using a conventional peak detector factor with the ideal operational amplifier, and $C = 0.01$ μF (no discharge resistor), compared with an unfiltered half-wave rectifier and ideal diodes, both powered by a voltage $v = 10.0 \sin \omega t$ at a frequency of 10 kHz.

4.3 Explain the working principle of the dual-slope digital voltmeter, indicating its strengths and weaknesses.

4.4 In a dual-slope DVM, $V_{ref} = 1.2$ V, $C = 0.1$ μF, $R = 10$ kΩF with a fixed integration time $T_i = 0.01$ s, and reverse integration time $T_r = 0.1$ s. What is the reading of the voltage value that is expected in the display? Discuss using this example to emphasize the advantages of this reading technique.

4.5 Give the general characteristics of DVMs with a brief explanation about each one.

4.6 In the electronic voltmeter-ammeter in Figure 4.2, determine input impedance and ratio output current/input current to make the instrument operate under the conditions given by the switch position, shown in the figure, using a resistive divider and an indication of 60 μA in the microammeter. What is the maximum voltage readable with this instrument? Assume that the microammeter resistance is equal to 200 Ω and the maximum current indication is equal to 100 μA.

4.7 Discuss the most common criteria to choose the best analog voltmeter for a required performance for a particular purpose.

4.8 In a fundamental current of amplitude, $I_1 = 10$ A, containing harmonic currents of the second and third orders, with amplitudes, respectively, of $I_2 = 5$ A and $I_3 = 3.333$ A, and displacement factor $\cos\theta_1 = 0.92$, calculate what is the effective current, total harmonic current distortion, and power factor if these are considered harmonic distortions?

4.9 Justify the presence of each circuit component in Figure 4.1.

4.10 In order to measure the coil self-inductance and leakage capacitance, such a coil is placed in parallel with a 120-pF capacitor. It was observed that the resonance frequency was at 10 MHz. After the capacitor is replaced by another one of 40 pF, the resonance frequency measured again was 15 MHz. What are the values found for the self-inductance and leakage capacitance of this coil?

REFERENCES

1. Buso, S. and Mattavelli, P., *Digital Control in Power Electronics*, 2nd edn., Synthesis Lectures on Power Electronics, Hudgins, J., Series Editor, Morgan & Claypool Publishers, 2016, ISSN: 1931-9525.
2. Silveira, P.R. and Santos, W.E., *Automation and Discrete Control (Automação e Controle Discreto)*, Érica, São Paulo, Brazil, 229pp., 1998 (in Portuguese).

BIBLIOGRAPHY

Auslander, D.M. and Agues, P., *Microprocessors for Measurement and Control*, Osborne/McGraw-Hill, Berkely, CA, 310pp., 1981.

Bakshi, A.V. and Bakshi, U.A., *Electronic Measurements and Instrumentation*, Technical Publications, New Delhi, India, 507pp., 2007.

Bowron, P. and Stephenson, F.W., *Active Filters for Communications and Instrumentation*, McGraw-Hill Book Company, Berkshire, U.K., 285pp., 1979.

Clarkson, P.M., *Optimal and Adaptive Signal Processing*, CRC Press, Boca Raton, FL, 1993, ISBN: 9780849386091.

Diefenderfer, A.J., *Principles of Electronic Instrumentation*, W.B. Saunders Co., Philadelphia, PA, 667pp., 1972.

Graeme, J.G., Tobey, G.E., and Huelsman, L.P., *Operational Amplifiers: Design and Applications*, BURR-BROWN Editors, Tokyo, Japan, 473pp., 1973.

Helfrick, A.D. and Cooper, W.D., *Modern Electronic Instrumentation and Measurement Techniques*, Prentice-Hall-PH, Englewood Cliffs, NJ, 324pp., 2009, ISBN: 978-81-317-0888-0.

National Instruments, *Instrumentation Catalogue Measurement and Automation*, tutorial edition, National Instruments, Austin, TX, 2012.

Peled, A. and Bede, L., *Digital Signal Processing, Theory, Design, and Implementation*, John Wiley & Sons, New York, 1978.

Peng, F.Z., Akagi, H., and Nabae, A., A new approach to harmonic compensation in power systems, *Proceedings of the IEEE/IAS Annual Meeting Conference*, Pittsburgh, PA, pp. 874–880, 1988.

Taub, H., *Digital Circuits and Microprocessors*, University of Michigan, McGraw-Hill and Sons, New York, 1982, ISBN: 0070629455, 9780070629455.

5 Signal Simulators and Emulators

5.1 INTRODUCTION

Simulators and signal generators are tools serving to artificially create signals in various ways so that they can behave like true signals generated in processes, equipment, or instruments. These signals are characterized by the envelope, frequency contents, phase, and levels of direct current and value ranges. In most cases, they do not supply driving power. The signal generators have to be adjustable by the user for a given purpose to make the output signals consistent with the instrument's capabilities in terms of magnitude, phase, and frequency. To this end, attenuators, filters, processors, and amplifiers are selected.

An emulator is a tool made to function as a piece of hardware capable of driving power to circuits or machines, with a built-in design and management operating system that makes such an emulator perform, as the specific real system is either expensive, or not available to be off-line, or may not be switched off, or too dangerous to be controlled. For example, a nuclear power plant is unsafe to turn off, dangerous to control, and not available to be off-line, so a piece of hardware could be used to simulate that plant for real-world applications, that is, to have an emulator. Another example would be an expensive microprocessor, which is fragile and difficult to stop the internal operations; an emulator could be built to make a very robust equipment to be programmed, with step-by-step debugging tools for improving the performance of the code compiled and assembled to work on that particular microprocessor. Instrumentation could also be designed based on emulators, for example, when blasting and smelting metallurgical materials, it is very hard to develop instrumentation for such industrial hot temperature processes, and an emulator could be used to make designers capable of studying the best instrumentation analysis of those processes.

5.1.1 The Attenuator

The attenuator can reduce the signal amplitude to a known value through a previous calibration to match the source impedance or load. Impedance matching is important because if the load impedance is less than that of the source output, then there is an overloading on the source. Conversely, if the load impedance is greater than the source output impedance, there will be an underused power source. Regardless of the attenuation level, the output signal of the attenuator is based on a fixed impedance. The power reduction of the input signal after the attenuator is constant:

$$A(dB) = 10\log\frac{P}{P_{ref}} = 20\log\frac{V}{V_{ref}} = 20\log\frac{I}{I_{ref}} \quad (5.1)$$

For example, for proper power measurements, it is used as a standard, one dB_m where $P_{ref} = 1$ mW through a 50 Ω resistor [1,2].

5.1.2 Types of Attenuators

Signal attenuators are widely used in interfaces between instrumentation circuits to match compatible connections and to maximize the signal energy transfer between them. The most common types of attenuators are the resistive divider for low frequencies and the attenuator type π for higher frequencies. To optimize the signal energy transfer, attenuators can be of a symmetrical type having the same impedances Z_i and Z_o as seen from the input and output terminals, respectively; or otherwise, they are of asymmetric type.

The resistive divider type of attenuator has the configuration shown in Figure 5.1a. Resistor R_1 is part of the source impedance, while resistor R_2 is part of load impedance. The relationship between the output and input voltage signals is given by:

$$\frac{v_o}{v_i} = \frac{R_2}{R_1 + R_2} \tag{5.2}$$

A resistive divider attenuator can match voltage levels, but not load and source impedances, in order to obtain a maximum transfer of signal levels.

A π-type attenuator has the configuration shown in Figure 5.1b. Its input and output impedances are, respectively, given by:

$$Z_i = R_1 // \left(R_2 // Z_o + R_3 \right) \tag{5.3}$$

$$Z_o = R_2 // \left(R_1 // Z_i + R_3 \right) \tag{5.4}$$

The relationship between the output and input voltage signals of a π-type attenuator is given by:

$$\frac{v_o}{v_i} = \frac{R_2}{R_2 + R_3} \tag{5.5}$$

A type-π configuration is essential for high-frequency signals because the attenuator must take into account the reflected waves, i.e., to verify that the wavelengths are close

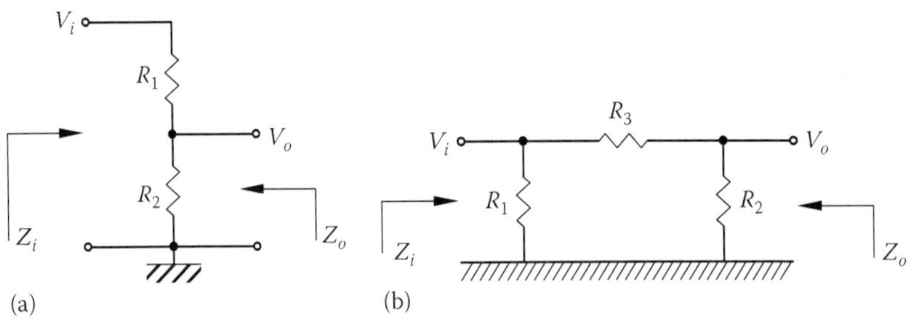

FIGURE 5.1 Signal attenuator types: (a) resistive divider and (b) π.

to the physical lengths of each used component. The input and output impedances Z_i and Z_o, and the attenuation v_o/v_i can be independently selected for this type of attenuator.

The signal energy transfer from the attenuator input to the load (output) is maximum when the impedances Z_i and Z_o (analog values) match. Thus, the resistors in parallel R_1 and R_2 can be used, respectively, for matching the input and output impedances. The reflection ρ and transmission τ wave coefficients are generally expressed, respectively, by [1–3]:

$$\rho = \frac{V_-}{V_+} = \frac{Z_L - Z_c}{Z_L + Z_c} \tag{5.6}$$

$$\tau = \frac{V_L}{V_+} = \frac{2Z_L}{Z_L + Z_c} \tag{5.7}$$

where

V_- is the negative traveling wave (traveling toward the load source)
V_+ is the positive traveling wave (traveling toward source load)
V_L is the voltage across the line load terminals
$Z_c = \sqrt{X_L X_C} = \sqrt{L/C}$ characteristic impedance of the line
Z_L is the load impedance

On the basis of these relationships, there is no wave reflection when the load impedance is exactly equal to the characteristic impedance of the line, $Z_c = Z_L$. That is, all the energy of the input wave is transferred to the load, and then this finite length line cannot be distinguished from a line of infinite length (without reflection). These relationships cannot be obtained with an attenuator of the resistive divider type.

The π-type attenuator is the most common and versatile. It is manufactured with standard resistors with up to 20 dB attenuation for frequencies up to 100 MHz, which can go up to the gigahertz order. As in any combination of resistors, the attenuator has spurious shunt and series capacitances and inductances at higher frequencies. The reactance formed by them becomes considerable, particularly the capacitive one, having a heavy influence on the attenuation factor. To minimize this problem, attenuators are made with lower power factors and connected in cascade. The resistors used in attenuators are specially made of rods and discs for frequencies on orders up to gigahertz. Figure 5.2 is a block diagram of an example of sinusoidal wave generator using resistive attenuators, frequency counter, and automatic control of the attenuation level.

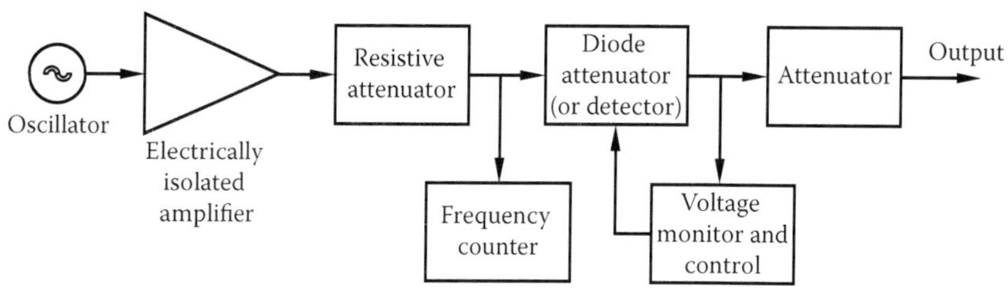

FIGURE 5.2 Sinusoidal waveform generator.

5.2 WAVEFORMS FOR ELECTRONIC INSTRUMENTS

The electronic instrumentation serves as an aid in the data assessment from processes or products. For this, the user makes use of signals generated in the equipment itself or of externally injected signals from signal generators. To be most useful, the injected signals must have general characteristics such as the following [4,5]:

- Signals with adjustable amplitude
- Well-defined and stable frequency
- Signals without undesired distortions

One way to measure the quality of a pulse or square wave generator is by a technical evaluation of the generated pulse characteristics. A widely used pulse pattern is given in Figure 5.3, whose standard terminologies are as listed in Table 5.1.

Generally speaking, the origin of electrical pulses used in instrumentation can be passive (molding pulses) or active (generating pulses) using a sinusoidal or relaxation oscillator. The output impedance of the pulse generator must match the impedance of the system being tested to prevent wave reflections and therefore unwanted distortions imposed on the original signal.

Many types of signal generators are commercially available, programmable or not, which should be selected according to the purpose of the tests. The most common ones are as follows:

- Function simulator
- Function generator
- Pulse and sweeping generator

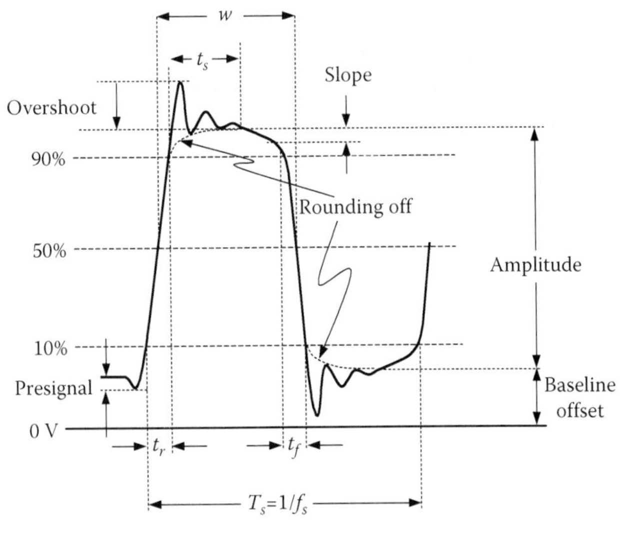

FIGURE 5.3 Terminology of a time-domain pulse.

TABLE 5.1
Definition of Details of a Signal Pulse

Parameter	Definition
Ringing	It is the transient oscillation during the signal settling period until it reaches a stable level
t_s	Settling time of the signal transition
t_r	Rise time that the pulse takes to go from 10% to 90% of its amplitude
t_f	Fall time that the pulse takes to go from 90% to 10% of its amplitude
Rounding time	It is the overdamped transition of the pulse incursion soon after the rise time, or shortly before the fall time, until the beginning of the stable low level of the pulse
Amplitude	It is the difference between the minimum stable level and maximum stable level of the signal
PRR	Pulse repetition rate or switching rate
Overshoot	It is the maximum signal excursion beyond the range defined by the ringing oscillation end
Slope	It is the difference between the maximum stable level and the minimum stable level of the pulse
Offset	It is the DC level of the baseline (offset) with respect to the zero-voltage line at the beginning of the stable lower level of the pulse
Switching frequency (f_s)	It is the repetition frequency of the pulses with a period T_s (switching time)

- Audio generator
- RF generator
- Video and sound generator
- Systems emulator

All these devices have a high-quality oscillator as a central core, whose waveforms can commonly be sine, square, and sawtooth with an adjustable DC offset.

5.3 SIGNAL GENERATORS AND SIMULATORS WITH FREQUENCY SYNTHESIZER

An extension of the frequency counter instruments with the electronic display is the electronic adjustment of the output frequency, called frequency synthesizer. It performs corrections on the content and frequency variations of the simulated signal at each time interval to ensure minimum distortion of amplitude and phase; such corrections are made by direct or indirect methods.

5.3.1 Indirect Method of Frequency Synthesis

The indirect method is based on a PLL (phase-locked loop) performing a frequency correction using a phase correction between the output and the internally generated frequency in a local oscillator with a standard high stability and quality.

Generally, the internal frequency reference is provided by a quartz crystal (from 1 to 10 MHz).

The problems associated with this method are as follows:

- Not all the modulation frequency can be removed with filters.
- The output of the filter affects the frequency (in phase and amplitude) of the center frequency of the voltage-controlled oscillator (VCO).
- To minimize distortion of undesired harmonics, it is necessary to use a large time constant, which then makes time for substantive changes in the VCO frequency too long.
- When the reference frequency is low, there are problems with the low resolution of the synthesizer caused by ripples in the BP filter output; as a result, the above problems are aggravated. High resolution is obviously a desirable feature in synthesizers.

One way to overcome the problems of the indirect method discussed here is to use synthesizers with multiple or complex loops, the discussion of which would be out of context in an introductory book like this.

5.3.2 Phase-Locked Loop

Phase-locked loops (PLLs) are circuits for tracking the phase of a local signal generator with the phase of an external signal oscillator, regardless of the size or format of both. That is, the local signal and internal signal generated are based on the information contained in the external signal: zero crossings or frequency. The quality of the local oscillator is responsible for the quality of the signal generated by instrumentation and other applications. There are many examples of the uses of these circuits in the literature, such as to control the harmonic content of signals, control the reception of radio signals, and control and drive the power devices. The following text discusses the most common types of PLLs [5–9].

The operating principle of the PLL is shown in Figure 5.4. The phase angle between two slightly different frequencies, f_1 and f_2, is detected as a square wave with a phase difference, θ. This square wave is filtered to obtain an average value, which drives a VOC. Therefore, it generates the frequency f_2. Thus, the speed rate of f_2 is

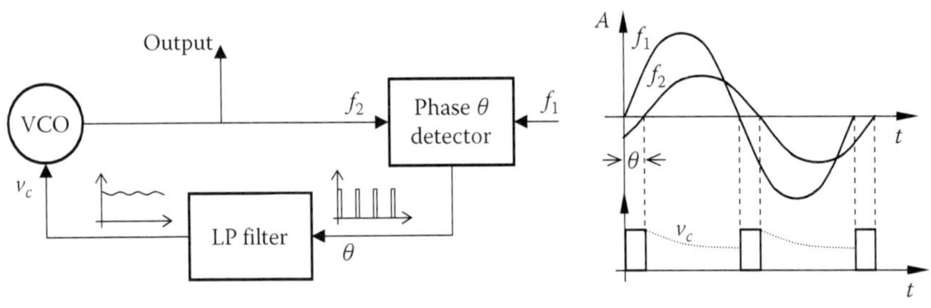

FIGURE 5.4 PLL locking of two distinct frequencies.

subjected to the phase difference between f_1 and f_2, causing the oscillator frequency f_2 to be locked in phase with frequency f_1, sometimes called synchronous demodulator.

In practice, there are many alternatives to implement a PLL according to the desired purpose: filtering, phase locking, shape conversion of the signal, etc. Note that the frequencies f_1 and f_2 may have any repetitive amplitude or format, not necessarily of the same type:

1. Standard PLL, or locking to an internal VCO with the captured external frequency [12,13].
2. PLL locking an internal frequency (x-tal) to a submultiple of the output frequency. Figure 5.5 shows the block diagram of a PLL where VOC is used at low frequencies, or a current-controlled oscillator (ICO) is used at higher frequencies [10–14].
3. Locking an internal frequency (ramp generator) to an external frequency. The beginning and end of the ramp voltage define the sweep oscillation period of the internally generated sweep frequency.
4. PLL for digital locking of an internal frequency generated in a look-up table containing the variations of the external frequency, or only by instants marked by their voltage zero crossings (VZC), as shown in Figure 5.6.

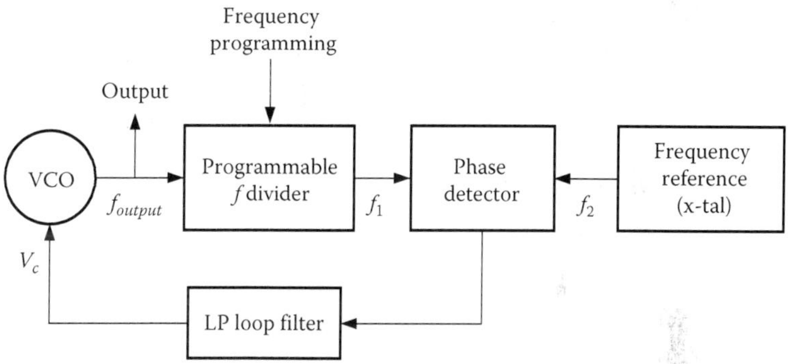

FIGURE 5.5 Phase-locked loop with frequency divider.

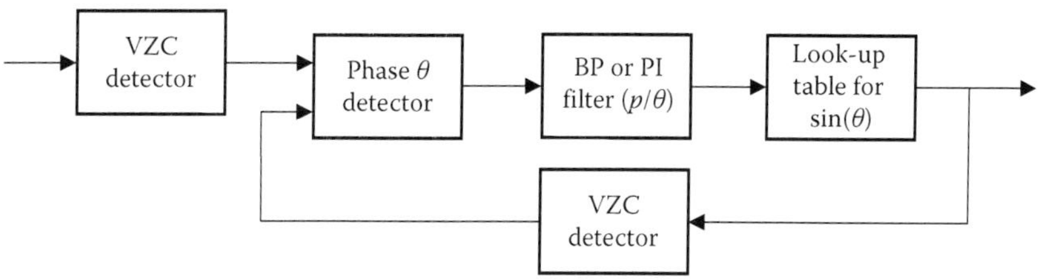

FIGURE 5.6 PLL with look-up table and VZC detector.

5. Using a synchronous PLL with a synchronous detector [10–13]. This PLL is based on the following trigonometric identity:

$$\sin(a)\cos(b) = \frac{1}{2}\left[\sin(a-b)+\sin(a+b)\right] \quad (5.8)$$

where $a=\omega_1 t+\theta$ and $b=\omega_2 t$.
Therefore:

$$\sin(\omega_1 t+\theta)\cos(\omega_2 t) = \frac{1}{2}\left\{\sin\left[(\omega_1-\omega_2)t+\theta\right]+\sin\left[(\omega_1+\omega_2)t+\theta)\right]\right\} \quad (5.9)$$

In real circuits, it may appear that many derived harmonics should be filtered out. In steady state, $\omega_1 \to \omega_2$, and therefore, the product of sine times cosine tends to:

$$\sin(\omega_1 t+\theta)\cdot\cos(\omega_2 t) \cong \frac{1}{2}\left\{\sin\left[(\omega_1+\omega_2)t+\theta\right]+\sin\theta\right\} \quad (5.10)$$

However, when the frequency $\omega_1-\omega_2$ is very low compared with $\omega_1+\omega_2$, it can be separated from the high frequency by a low-pass filter to eliminate the sinusoidal term or an RC circuit of average value, as shown in Figure 5.7 [11–16].

6. PLL for conversion of *abc* into *αβ0* systems suitable for three-phase systems. In this case, the PLL attempts to minimize the sum of errors $\Delta\alpha+\Delta\beta$ by comparing it to the zero reference, as shown in Figure 5.8. The result of the conversion *abc* into *αβ0* is $\alpha\cdot\cos\theta+\beta\cdot\sin\theta$ as discussed in reference.

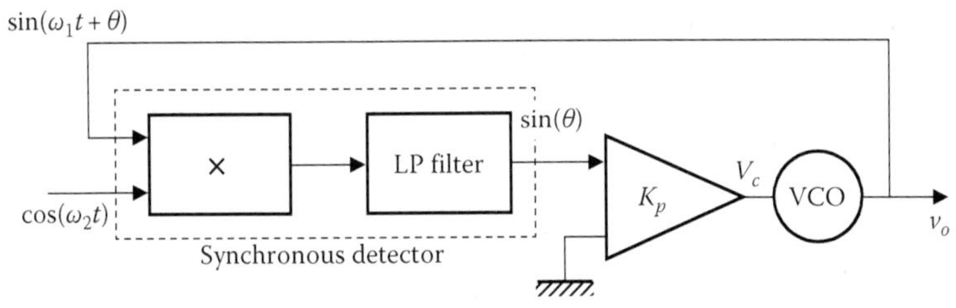

FIGURE 5.7 PLL using product of functions.

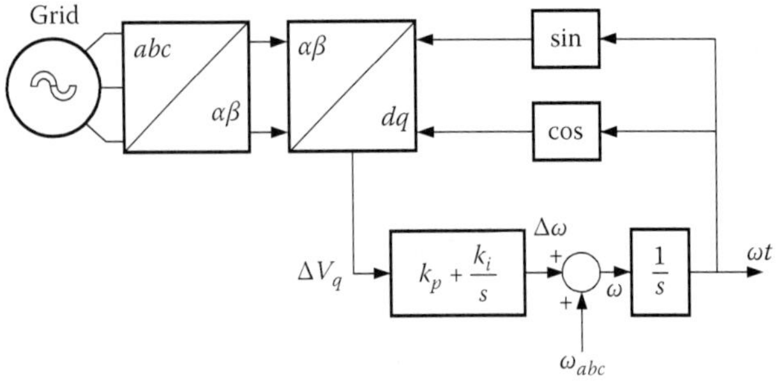

FIGURE 5.8 PLL with coordinate conversion *abc-αβ0*.

5.3.3 Direct Method of Frequency Synthesis

In the direct method of frequency synthesis, rather than generating a frequency for the phase difference between internal and external signals, the frequency is generated from a reference, while another frequency is derived from this one. Figure 5.9 shows the case of a synthesizer with an 18.0 MHz reference that is multiplied, divided, and mixed to generate 10 output steps of 0.1 MHz from 2.0 to 2.9 MHz using auxiliary oscillators. Also, an intermediate reference of 16 MHz is used to rebuild the original frequency. Each synthesizer module constitutes a standard frequency decade, and these are all equal to each other. In the output of each decade, the accuracy gains of decimal frequency are considered, as represented in Table 5.2, using an auxiliary oscillator with a highest frequency of 2.9 MHz, which is suitable for three decades. Each one is connected in cascade soon after the band-pass filters, as shown in Figure 5.9.

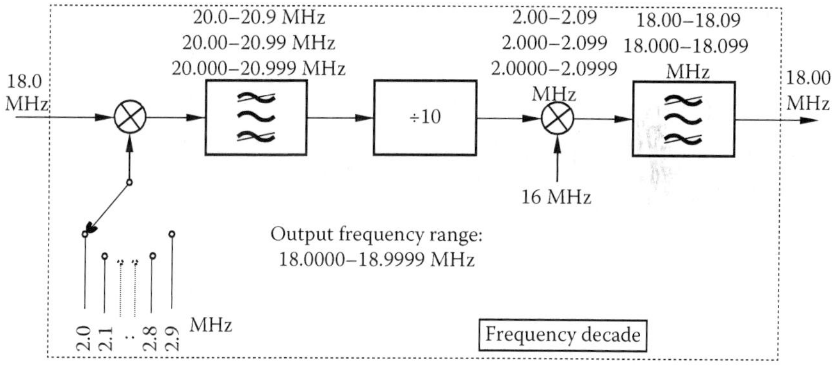

FIGURE 5.9 Decade of a frequency synthesizer.

TABLE 5.2
Decades of Direct Frequency Synthesis (Refer to Figure 5.9)

Stage	Input (MHz)	Processing	Output (MHz)
Decade 1	18.0	18 + 2.9 = 20.9	18.09
		20.9 ÷ 10 = 2.09	
		2.09 + 16 = 18.09	
Decade 2	18.09	18.09 + 2.9 = 20.99	18.099
		20.99 ÷ 10 = 2.099	
		2.099 + 16 = 18.099	
Decade 3	18.099	18.099 + 2.9 = 20.999	18.0999
		20.999 ÷ 10 = 2.0999	
		2.0999 + 16 = 18.0999	

5.4 SIGNAL GENERATORS BY FREQUENCY DIVISION

Signal generators by frequency division offer some advantages over the previous ones. The basic element is a high Q oscillator operating in the range of 256–512 MHz. Factor $Q=f/\Delta f$ is here defined as the bandwidth Δf with respect to the tuned frequency f.

The frequencies are divided by flip-flops whose outputs are square waves producing odd harmonics only, which must be removed by filters. The output can be adjusted manually and mechanically. It uses a PLL correction only for small frequency bands, and therefore it may appear small phase shifts are generating few lateral side frequencies. The signal generator by frequency division can cover only a certain range, which requires an inductor to select each frequency band and a variable capacitor for tuning adjustments.

If the flip-flop output is perfectly symmetrical, only odd harmonics are presented. However, harmonic orders may also appear at high frequencies; the most problematic of these are those orders nearer to the fundamental, especially the second order. To solve this, two filters can be used, one low-pass (LP) to reject the lower side of the spectrum and one high-pass (HP) to reject the higher side, with a separation occurring in the geometric average of the output frequencies, as shown in Figure 5.10. The geometric average is calculated by:

$$f = \sqrt{f_{low} f_{high}} \tag{5.11}$$

To switch from HP to LP filter, a mechanical device is used. To have a clean spectrum with no modulation, every time the frequency, as well as the modulation, is divided by 2. But if the modulation frequency is to be added to the generator output signal, then it is necessary to include a correction circuit controlled by a frequency band selector switch.

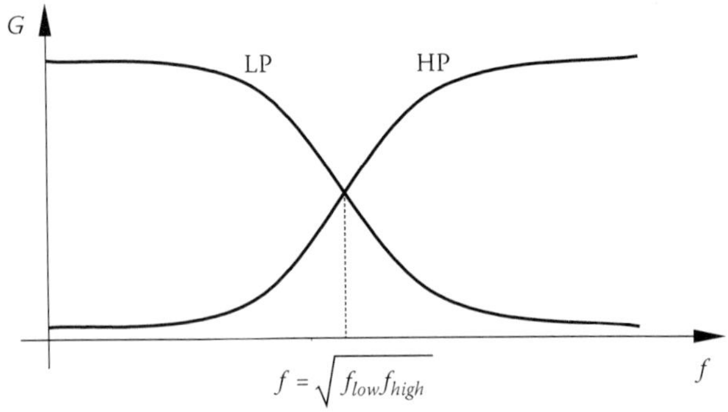

FIGURE 5.10 Frequency division.

5.5 SIGNAL GENERATOR WITH MODULATION

The signal generator can be modulated by frequency (MF) or amplitude (MA). Most generators provide an amplitude modulation close to, but never exactly, 100%. A varactor diode in the tuned circuit can produce this modulation frequency, as shown in Figure 5.11.

5.6 FREQUENCY SWEEPING GENERATOR

The sweep generator generates signals with variable frequency in a linear fashion, repeated automatically and within a given range of values. This instrument is used to analyze the frequency response of circuits. A typical scheme of this instrument is shown in Figure 5.12.

Figure 5.13 shows a circuit for compensating the nonlinearity between the sweeping control voltage V_c and the oscillator frequency. Every time the sweeping ramp voltage exceeds each of the levels V_1, V_2, and V_3, the corresponding diode comes into conduction, and the resistance in series with the diode becomes parallel with R_i, thereby smoothly altering the gain of the amplifier. One can theoretically achieve many waveforms with this circuit, whose outputs are smoothed by the exponential characteristics of the diodes. If the generator is of the digital type, this correction is made in the device programming.

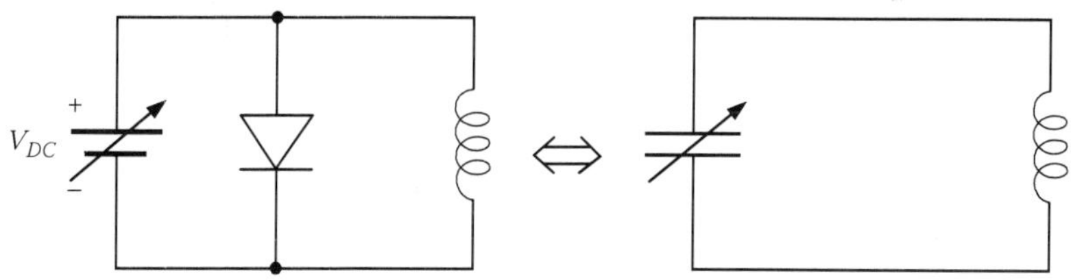

FIGURE 5.11 Signal modulation.

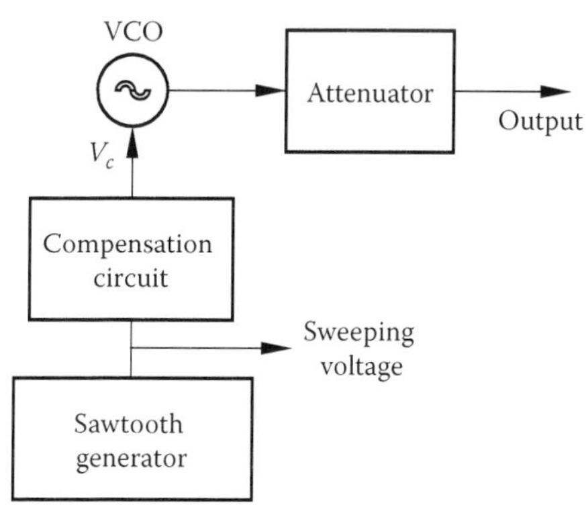

FIGURE 5.12 Diagram of a sweep generator.

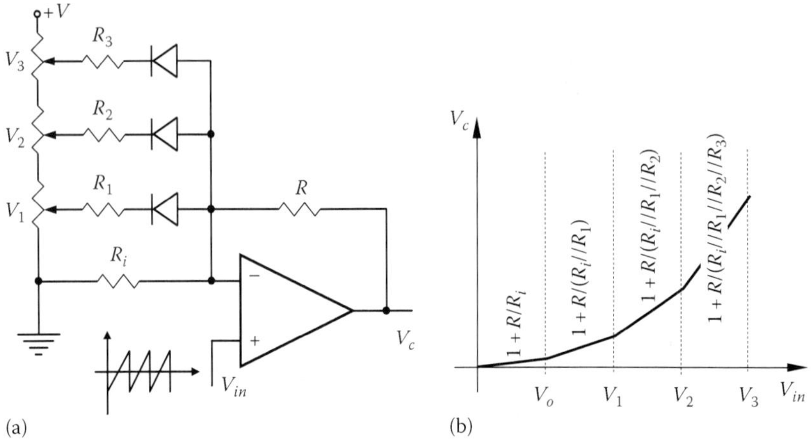

FIGURE 5.13 Circuit for nonlinearity compensations: (a) compensation circuit and (b) diagram of the voltage gains.

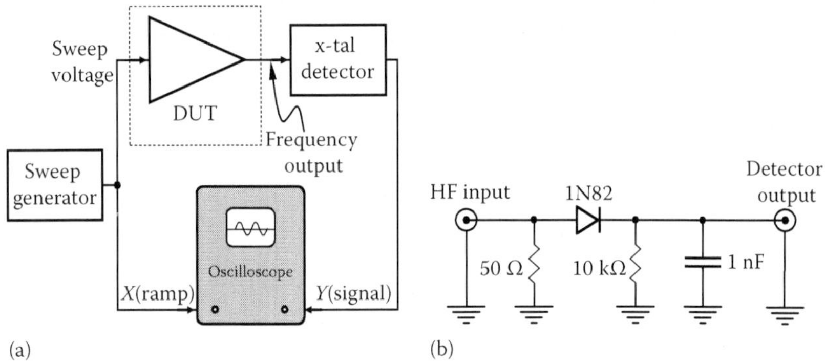

FIGURE 5.14 Amplifier frequency response frequency by experimental evaluation: (a) connection diagram and (b) detector circuit.

Figure 5.14 illustrates an application of the sweep generator to determine the frequency response of a radio frequency amplifier. A signal is generated with constant amplitude and variable frequency by applying a sawtooth voltage in a VCO. The obtained signal is injected into the device under test (DUT) represented in Figure 5.14 by a linear amplifier. It should respond to the input signal across its passband between the lower and upper cutoff frequency. The detection of the DUT frequency response to the input signal is done by an oscilloscope, whose input X is activated by the same scan applied to the VCO, and input Y is activated by the signal obtained from the amplifier output. This test results in the response band of the DUT.

5.7 PULSE GENERATORS AND RECTANGULAR WAVES

Pulse generators and rectangular waves are mostly used in power electronics and distributed simulators and emulators. The fundamental difference between pulse and rectangular wave is the duty cycle defined in a period T, i.e., it is the ratio between the average (V_{DC}) value and the peak (V_{do}) value of the function expressed as [17–20]:

$$D(\text{duty cycle}) = \frac{V_{DC}}{V_{do}} = \frac{W}{T} \quad (5.12)$$

where W is the pulse width.

The square wave generator necessarily uses the relationship $D=0.5$. In this regard, the duty cycle of the pulse generator is adjustable from 0.0 to 1.0, controlling the average power supplied to the circuit under test.

5.8 FUNCTION GENERATORS, SIMULATORS, AND AUDIO GENERATORS

Function generators, simulators, and audio generators are commonly used in laboratories for the analysis of low-frequency circuits. The most common forms of output pulses are sine, triangle, square, and sawtooth. The frequency scales range from fractions of Hz to several hundred kHz.

In particular, the square wave generator is used to measure linearity in audio circuitry producing another simultaneous output with a triangular wave or sawtooth, which is able to excite the horizontal deflection of an oscilloscope.

Audio generators often use the Wien oscillator to produce a sine oscillation, which is taken as a stable industrial standard with little distortion. The square wave oscillators are used as multivibrators.

EXERCISES

5.1 What are the input and output impedances of the π-attenuator represented in the figure below, and what is the attenuation factor in dB?

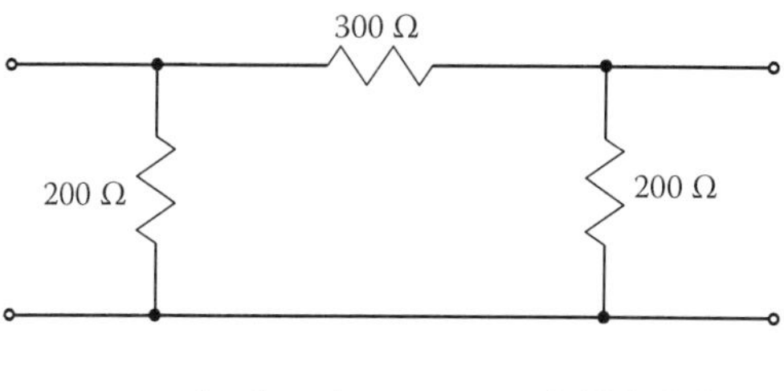

5.2 Design a π-attenuator to feed an instrument of 300 Ω input impedance for receiving electrical signals from a sensor rated at 50 Ω and 750 V. The maximum excursion of the input signal across the instrument input should not exceed 5 V.

5.3 Distinguish between direct and indirect methods of frequency synthesizer readings giving a brief explanation about the operational diagram of each.

5.4 A 30-watt amplifier has a 32 Ω output impedance, and it is going to be connected to a 10 W and 4 Ω loudspeaker. Calculate the attenuator π so that there is proper matching impedance between the amplifier and the loudspeaker. What is the resulting voltage attenuation?

REFERENCES

1. Ramo, S., Whinnery, J.R., and Van Duzer, T., *Fields and Waves in Communication Electronics*, 3rd edn., John Wiley & Sons, Tokyo, Japan, p. 28, 1994, ISBN: 13-978-0471585510, ISBN: 10-0471585513.
2. Boylestad, R.L. and Nashelsky, L., *Electronic Devices and Circuit Theory*, Editor Pearson Education, Upper Saddle River, NJ, 2009, ISBN: 8131725294, ISBN: 9788131725290.
3. Horowitz, P. and Hill, W., *The Art of Electronics*, Cambridge University Press, Cambridge, England, 1989.
4. Nordin, N.S., Omar, R., Sulaiman, M., Krismadinata, and Elias, M.F.M., Comparative study of cascaded H-Bridge multilevel inverter model based on power electronic simulator (PSIM), *Third IET International Conference on Clean Energy and Technology (CEAT)*, Kuching, Malaysia, pp. 1–6, 2014, DOI: 10.1049/cp.2014.1498.
5. Michel, L., Chériti, A., and Sicard, P., Physical and system power electronics simulator based on a SPICE kernel, *2010 IEEE 12th Workshop on Control and Modeling for Power Electronics (COMPEL)*, Boulder, CO, pp. 1–6, 2010, DOI: 10.1109/COMPEL.2010.5562422.
6. Aguirre, L.A., *Fundamentals of Instrumentation (Fundamentos de Instrumentação)*, Editor Pearson Education, São Paulo, Brazil, 2013, ISBN: 978-85-8143-183-3.
7. Pavan, M., Bailapudi, K., and Sinha, N., Fuzzy logic controlled wind turbine emulator (WTE), *International Conference on Information Communication and Embedded Systems (ICICES)*, Chennai, India, pp. 1–8, 2016, DOI: 10.1109/ICICES.2016.7518843.
8. Himani, G. and Dahiya, R., Development of wind turbine emulator for standalone wind energy conversion system, *IEEE Sixth International Conference on Power Systems (ICPS)*, New Delhi, India, pp. 1–6, 2016, DOI: 10.1109/ICPES.2016.7583999.
9. Bowron, P. and Stephenson, F.W., *Active Filters for Communications and Instrumentation*, Editor McGraw-Hill Book Company, Berkshire, U.K., 285pp., 1979.
10. Helfrick, A.D. and Cooper, W.D., *Modern Electronic Instrumentation and Measurement Techniques*, Prentice-Hall-PH, Englewood Cliffs, NJ, 324pp., 2009, ISBN: 978-81-317-0888-0.
11. Diefenderfer, A.J., *Principles of Electronic Instrumentation*, 3rd edn., Editor W.B. Saunders Co., Philadelphia, PA, 667pp., 1994, ISBN: 10-0030747090, ISBN: 13-978-0030747090.
12. Graeme, J.G., Tobey, G.E., and Huelsman, L.P., *Operational Amplifiers: Design and Applications*, Editor Burr-Brown, Tokyo, Japan, 473pp., 1971.
13. Eveleigh, V.W., *Introduction to Control Systems Design*, TMH Edition-Tata McGraw-Hill Publishing Company, New York, 624pp., 1972.
14. Vijay, V., Giridhar Kini, P., Viswanatha, C., and Adhikari, N., Regenerative load emulator with battery charging for evaluation of energy management in microgrid with distributed renewable sources, *Modern Electric Power Systems* (*MEPS*), IEEE, Wroclaw, Poland, pp. 1–6, 2015, DOI: 10.1109/MEPS.2015.7477157.

15. Erkaya, Y., Moses, P., Flory, I., and Marsillac, S., Steady-state performance optimization of a 500 kHz photovoltaic module emulator, *IEEE 43rd Photovoltaic Specialists Conference (PVSC)*, Portland, OR, pp. 3205–3208, 2016, DOI: 10.1109/PVSC.2016.7750257.
16. Man, E.A., Sera, D., Mathe, L., Schaltz, E., and Rosendahl, L., Thermoelectric generator emulator for MPPT testing, *International Aegean Conference on Electrical Machines & Power Electronics (ACEMP), International Conference on Optimization of Electrical & Electronic Equipment (OPTIM) & 2015 International Symposium on Advanced Electromechanical Motion Systems (ELECTROMOTION)*, Side, Turkey, pp. 774–778, 2015, DOI: 10.1109/OPTIM.2015.7427051.
17. Tien Lang, T., *Électronique des Systéme de Measures*, Editor Masson, Paris, France, 1997.
18. Regtien, P.P.L., *Electronic Instrumentation*, 2nd edn., Editor VSSD, Delft, Belgium, 2009.
19. Auslander, D.M. and Agues, P., *Microprocessors for Measurement and Control*, Editor Osborne/McGraw-Hill, Berkely, CA, 310pp., 1981.
20. Gonzatti, F. and Farret, F.A., Mathematical and experimental basis to model energy storage systems composed of electrolyzer, metal hydrides and fuel cells, *Energy Conversion and Management*, 132(15), 241–250, January 2017, https://doi.org/10.1016/j.enconman.2016.11.035.

6 Advanced Harmonic Analysis for Power Systems

6.1 INTRODUCTION

Chapter 3 discussed sensors and transducers, which are devices capable of measuring and converting physical, biological, or chemical changes into electrical signals, usually represented by voltage signals. Chapter 5 presented signal generators whose waveforms can be reconstructed from the analysis of their components by simulating the signals contained in the most common phenomena found in current technologies and sciences. This chapter addresses such signals that must be, somehow, captured by an instrument, and recorded and converted into electrical signals to be conveniently studied, analyzed, or applied. After the electrical signal has been acquired and stored by the measuring instrument, it can be processed and converted into the most appropriate and useful form for users. Such devices are known as signal analyzers, and they can assume many forms according to the convenience of the signal under application.

There has been a worldwide concern with the quality of the electrical energy supplied to consumers by utilities, which is normally related to transient variations and distortions in the alternating voltages. Such a concern is perfectly acceptable considering the degree of sophistication of modern electronic devices that may have their operation hampered by iron losses and by radiated or conducted noise. The determination of the harmonic content of waves by engineers means a mathematical analysis of what is essential to identify in a signal, and what is just noise, distortion, or interference. The signal analysis may also be related to other fields of engineering like communications, computing, imaging, aerospace travel, power systems, power electronics, large structures, road traffic, railway, aviation, maritime, and inland waterway, among others. Some domestic examples include personal computer, multiple features of a mobile phone, DVD drive, and TV functions, which are easily affected by distortions in the supply line and radiations from the environment. Signal analysis are useful for evaluation of vibration and noise, harmonic voltage and current in equipment in industrial and commercial installations, various phenomena concerning electrical resonance frequencies, mechanical examination of speech and sounds, medical applications, and pattern recognition of signal signatures in civil and mechanical structures correlated to possible fatigue or fault of materials. It is also useful for the study of pathologies reflected in biological signals (electromyography, electrocardiography, electroencephalography, epidemics, etc.).

A technological evolution in the signal analysis field started with personal computers, data loggers, spectrum analyzers, and universal measurement devices (UMDs), among others, all of which can be easily adapted to the desired analysis, both digital and analog, via ready-made data acquisition boards. The digital facilities allow data of voltage and current waveforms to be analyzed with Fourier series and the wavelet transforms on amplitude, frequency, and phase of each harmonic component. The spectrum results can be recorded in memories or printers for further processing, analysis, evaluation, and decision-making.

6.2 HARMONIC ANALYSIS

Harmonic analysis means determining the three most interesting characteristics of each harmonic content: amplitude, harmonic order (or frequency), and phase. With these characteristics, the original signal can be fully reconstructed and distinguished from noise.

From mathematical analyses, any repetitive wave shape with a period T can be decomposed into components of the Fourier series. The components of the Fourier series may be used to reconstruct the original waveform into their instantaneous values. In the case of a generic function, let's say current i_s, with a period 2π and phase α, the Fourier series representation is:

$$i_s(\omega t) = I_{av} + \sum_{n=1}^{\infty}\left[a_n \cos(n\omega t) + b_n \sin(n\omega t)\right] \tag{6.1}$$

where a_n and b_n are the amplitudes of the nth harmonic order determined by:

$$I_{av} = \frac{1}{2\pi}\int_{\alpha}^{2\pi+\alpha} i_s(\omega t)\,d\omega t$$

$$a_n = \frac{1}{\pi}\int_{\alpha}^{2\pi+\alpha} i_s(\omega t)\cos(n\omega t)\,d\omega t \tag{6.2}$$

$$b_n = \frac{1}{\pi}\int_{\alpha}^{2\pi+\alpha} i_s(\omega t)\sin(n\omega t)\,d\omega t \tag{6.3}$$

Note that the general equation for $i_s(\omega t)$ from Equation 6.1 can also be represented, considering $a_n = I_n \sin\phi_n$ and $b_n = I_n \cos\phi_n$ to obtain:

$$I_n = \sqrt{a_n^2 + b_n^2} \quad \text{and} \quad \phi_{fn} = \tan^{-1}\frac{b_n}{a_n} = -n\phi_n$$

where

$\phi_{fn} = -n\phi_n$ is the fundamental angle of the nth harmonic with phase ϕ_n

I_n is the amplitude of the phase current of the nth harmonic order

Therefore, taking into account that $\sin(a+b)=\sin a\cdot\cos b+\sin b\cdot\cos a$, the current represented by Equation 6.1 will be given by:

$$i_s(\omega t)=I_{av}+\sum_{n=1,2,3,\ldots}^{\infty} I_n \sin\left[n(\omega t+\phi_n)\right] \tag{6.4}$$

such that the rms value of the nth harmonic current is given by:

$$I_{sn}=\frac{1}{\sqrt{2}}\sqrt{a_n^2+b_n^2} \tag{6.5}$$

For simplicity and generality, let's establish the equations of the Fourier components for a generic pulse block reference with the following parameters: width, w, amplitude, I_a, and phase, θ. The lagging angle θ is the pulse phase shift set by the ordinate origin and the beginning of the line current block in the AC side of the equipment, as shown in Figure 6.1. The integration limits of the equations expressing the Fourier coefficients are the smallest period that the waveform repeats itself.

To simplify the harmonic analysis, a reference axis of time t passing exactly by the middle of the generic block with a delay θ about the beginning of time axis is used, as shown in Figure 6.1. Such generic block is particularly useful due to its waveform similarity with the pulsed load current i_s commonly used in electronics. Such waveforms are frequently seen in switched power supplies, thyristor power converters, natural switching, and transmission of signals by any electronic switch whose load current or voltage is likely to be represented by a series of blocks as the instantaneous current values shown in Figure 6.1. With this time shift from t' to t, $i_s(\omega t')$ becomes an even function defined as $i_s(\omega t)$, and therefore the coefficient b_n is zero for any n, i.e., the series has only co-sinusoidal terms. After that, the amplitudes of the pulse Fourier components in Figure 6.1 are obtained with the aid of Equation 6.2:

$$I_{av}=\frac{I_a}{2\pi}\left[\int_{-w/2}^{+w/2} d\omega t\right]=\frac{wI_a}{2\pi} \tag{6.6}$$

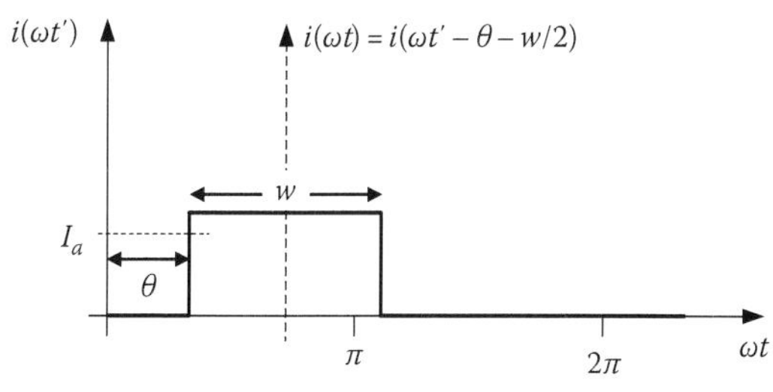

FIGURE 6.1 Typical pulse of a switched current signal (current block).

$$a_n = \frac{I_a}{\pi}\left[\int_{-\pi}^{\pi} \cos(n\omega t)\,d\omega t\right] = \frac{I_a}{\pi}\left[\int_{-w/2}^{+w/2} \cos(n\omega t)\,d\omega t\right]$$

or:

$$a_n = \frac{I_a}{n\pi}\left[\sin\left(\frac{nw}{2}\right) - \sin\left(-\frac{nw}{2}\right)\right] = \frac{2I_a}{n\pi}\sin\frac{nw}{2} \tag{6.7}$$

$$b_n = 0 \quad \text{for any “}n\text{”}$$

Then, the rms value of the line current in the generic train of positive blocks shown in Figure 6.1 can be calculated as:

$$I_s = \sqrt{I_{\text{DC}}^2 + \sum_{n=1}^{\infty} I_n^2} \quad \text{or generically } I_s = \sqrt{\frac{1}{2\pi}\int_{\theta}^{\theta+w} I_a^2\,d\omega t} \tag{6.8}$$

Therefore, Equations 6.6 and 6.7 substituted in Equation 6.1 can represent a series of positive blocks as:

$$F_+(\theta) = \frac{2I_a}{\pi}\left\{\frac{w}{4} + \sum_{n=1}^{\infty}\left[\frac{1}{n}\sin\frac{nw}{2}\cos n\theta\right]\right\} \tag{6.9}$$

If the train of blocks is equal to that of Figure 6.1, but with negative amplitudes interleaved with the positive blocks as shown in Figure 6.2, Equation 6.9 may change in two aspects: (1) inversion of the $F_+(\theta)$ signal from plus to minus and (2) introduction of a delay by 180° in the reference axis, i.e., substitute θ with $180° - \theta$ to result in:

$$F_-(\theta) = \frac{2I_a}{\pi}\left\{-\frac{w}{4} - \sum_{n=1}^{\infty}\left[(-1)^n\frac{1}{n}\sin\frac{nw}{2}\cos n\theta\right]\right\} \tag{6.10}$$

Furthermore, any repetitive shape can be set up using many blocks, like the one represented in Figure 6.1, that can be associated typically with a current waveform.

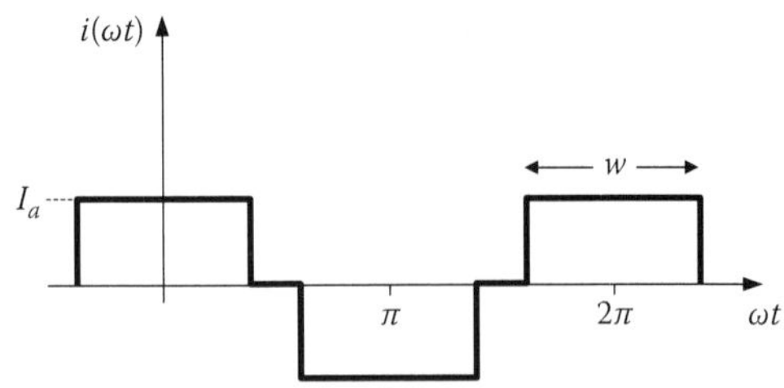

FIGURE 6.2 Typical alternating pulses of a switched signal.

TABLE 6.1
Phase Sequence of Harmonic Order "n" in Three-Phase Systems

Harmonic Order n	$-120n$	Sequence
1	0°, −120°, −240°	+
2	0, −240°, −120°	−
3	0°, 0°, 0°	0
4	0°, −120°, −240°	+
5	0°, −240°, −120°	−
6	0°, 0°, 0°	0

Example:

In the case of a single-phase half-wave square block, symmetric on the zero angle and pulse width $w = 180°$ or $w = \pi$, Equation 6.7 becomes:

$$a_n = \frac{2I_a}{n\pi}\sin\frac{n\pi}{2} \quad \text{and} \quad b_n = 0 \tag{6.11}$$

For $n = 2$, 4, 6,... (all even numbers) in the conditions of Equation 6.11, the sine is zero, and therefore even harmonics amplitudes are null. For odd orders $n = 1, 3, 5, 7, \ldots$, the amplitudes of the harmonics are alternately positive and negative in this order and generically expressed as:

$$a_n = \pm\frac{2I_a}{n\pi} \tag{6.12}$$

It should be noted from Equation 6.12 that the harmonics generated with the positive phase sequence of three-phase systems also have phase sequence as shown in Table 6.1. Furthermore, from Equation 6.12, it can be established $\pm a_n/a_1$:

$$a_n = \pm\frac{a_1}{n} \tag{6.13}$$

6.2.1 Effects of Harmonic Distortions on Network Parameters

Several methods of analysis can be used to calculate harmonic distortion of voltage and current in the definition of distributed generation parameters, such as FFT, wavelets, PLL, or p-q theory [1,2]. Such methods separate the fundamental harmonic component (voltage V_1 or current I_1) from the other harmonic components. The total harmonic distortion (THD) of voltage or current is defined as:

$$THD(V) = \frac{\sqrt{V_{rms}^2 - V_1^2}}{V_1} = \sqrt{\left(\frac{V_{rms}}{V_1}\right)^2 - 1} \tag{6.14}$$

$$THD(I) = \frac{\sqrt{I_{rms}^2 - I_1^2}}{I_1} = \sqrt{\left(\frac{I_{rms}}{I_1}\right)^2 - 1} \tag{6.15}$$

In the case of very small distortions, a data processor is most likely needed for precise solutions of these equations. On the other hand, the effects of THD on network parameters can be taken into account by simple mathematical manipulation. Consider the case of the power factor definition, *PF*, in terms of a purely sinusoidal voltage V_1 that contributes to the total apparent power, *S*, of a machine turned into useful active power expressed by: $P_1 = S \cdot PF$. Therefore, from these definitions:

$$P_1 = V_1 I_1 \cos\theta_1 \quad \text{and} \quad S = V_{rms} I_{rms} \tag{6.16}$$

In the case of a purely sinusoidal voltage, the conventional displacement factor is equal to the power factor. However, under voltage and/or current harmonic condition, the power factor may be defined as:

$$PF = \frac{P_1}{S} = \frac{V_1 I_1 \cos\theta_1}{V_{rms} I_{rms}} \tag{6.17}$$

With the addition of voltage harmonics and DC component, the above settings are affected, as shown by the equations below:

$$V_{rms} = \sqrt{V_{\text{AC}}^2 + V_{\text{DC}}^2} \tag{6.18}$$

where

$$V_{\text{AC}} = \sqrt{V_1^2 + V_2^2 + \cdots + V_n^2} = \sqrt{V_1^2 + \sum_{k=2}^{n} V_k^2} \tag{6.19}$$

Thus, the *THD* definition without the DC component becomes:

$$THD(V) = \sqrt{\frac{V_{\text{AC}}^2 - V_1^2}{V_1^2}} = \sqrt{\left(\frac{V_{\text{AC}}}{V_1}\right)^2 - 1} \tag{6.20}$$

or, including the DC current:

$$THD(V) = \sqrt{\frac{V_{rms}^2 - V_1^2}{V_1^2}} = \sqrt{\frac{V_{rms}^2}{V_1^2} - 1} \tag{6.21}$$

The definition of apparent power changes with the inclusion of the effective harmonic values [6]:

$$S = V_{rms} I_{rms} = \left(\sqrt{V_{\text{DC}}^2 + \sum_{1}^{\infty} V_k^2}\right)\left(\sqrt{I_{\text{DC}}^2 + \sum_{1}^{\infty} I_k^2}\right) \tag{6.22}$$

The active and reactive powers can be obtained by calculating the instantaneous power step by step [3] as:

$$
\begin{aligned}
p(t) = v(t)i(t) &= \left(V_{\text{DC}} + \sum_{m=1}^{\infty} V_m \sin(m\omega t + \theta_m) \right) \cdot \left(I_{\text{DC}} + \sum_{n=1}^{\infty} I_n \sin(n\omega t + \phi_n) \right) \\
&= V_{\text{DC}} I_{\text{DC}} + V_{\text{DC}} \sum_{n=1}^{\infty} I_n \sin(n\omega t + \phi_n) + I_{\text{DC}} \sum_{m=1}^{\infty} V_m \sin(m\omega t + \theta_m) \\
&+ \sum_{m=1}^{\infty} V_m \sin(m\omega t + \theta_m) \cdot \sum_{n=1}^{\infty} I_n \sin(n\omega t + \phi_n)
\end{aligned}
\tag{6.23}
$$

The phase difference between voltage and current of a particular kth harmonic order is set as:

$$
\delta_k = \phi_k - \theta_k \tag{6.24}
$$

In this case, the last sinusoidal term of Equation 6.23 for $\phi_k = \theta_k + \delta_k$ becomes:

$$
\sin(n\omega t + \phi_n) = \sin(n\omega t + \theta_n + \delta_n) = \sin(n\omega t + \theta_n)\cos\delta_n + \cos(n\omega t + \theta_n)\sin\delta_n \tag{6.25}
$$

and therefore, Equation 6.23 becomes:

$$
\begin{aligned}
p(t) &= V_{\text{DC}} I_{\text{DC}} + V_{\text{DC}} \sum_{n=1}^{\infty} I_n \sin(n\omega t + \phi_n) + I_{\text{DC}} \sum_{m=1}^{\infty} V_m \sin(m\omega t + \theta_m) \\
&+ \sum_{m=1}^{\infty} V_m \sin(m\omega t + \theta_m) \cdot \sum_{n=1}^{\infty} I_n \left[\sin(n\omega t + \theta_n)\cos\delta_n + \cos(n\omega t + \theta_n)\sin\delta_n \right]
\end{aligned}
$$

or yet:

$$
\begin{aligned}
p(t) &= V_{\text{DC}} I_{\text{DC}} + V_{\text{DC}} \sum_{n=1}^{\infty} I_n \sin(n\omega t + \phi_n) + I_{\text{DC}} \sum_{m=1}^{\infty} V_m \sin(m\omega t + \theta_m) \\
&+ \sum_{m=1}^{\infty} V_m \sin(m\omega t + \theta_m) \cdot \sum_{n=1}^{\infty} I_n \left[\sin(n\omega t + \theta_n)\cos\delta_n \right] \\
&+ \sum_{m=1}^{\infty} V_m \sin(m\omega t + \theta_m) \cdot \sum_{n=1}^{\infty} I_n \left[\cos(n\omega t + \theta_n)\sin\delta_n \right]
\end{aligned}
\tag{6.26}
$$

where the arguments of the trigonometric identity $\sin(a+b)=\sin(a)\cos(b)+\cos(a)\sin(b)$ in Equation 6.26 are defined as:

$$\begin{aligned} a &= n\omega t + \theta_n \\ b &= \delta_n \end{aligned} \tag{6.27}$$

The average value of $p(t)$ is the average term of Equation 6.26. The sum of the purely sinusoidal terms of Equation 6.26 is null, and the instantaneous power is the sum of the active and nonactive powers. Then:

$$p = p_a + p_{na} \tag{6.28}$$

6.2.1.1 Instantaneous Active Power, p_a

Considering the phase shift in Equation 6.26 as zero, $\phi_k = \theta_k$, the instantaneous active power of order "k" ($\delta_k = 0$ for $m = n$) can be defined depending on the index; if $\delta_k \neq 0$, it sets the instantaneous non-active, passive, or reactive power of order "k" ($\delta_k \neq 0$ and $m = n$). In this case, Equation 6.26 becomes:

$$\begin{aligned} p_a &= V_{\text{DC}} I_{\text{DC}} + V_{\text{DC}} \sum_{m=1}^{\infty} I_m \sin(m\omega t + \phi_n) + I_{\text{DC}} \sum_{m=1}^{\infty} V_m \sin(m\omega t + \theta_m) \\ &\quad + \sum_{m=1}^{\infty} V_m \sin(m\omega t + \theta_m) \cdot \sum_{m=1}^{\infty} I_m \sin(m\omega t + \theta_m) \\ &= V_{\text{DC}} I_{\text{DC}} + V_{\text{DC}} \sum_{m=1}^{\infty} I_m \sin(m\omega t + \phi_n) + I_{\text{DC}} \sum_{m=1}^{\infty} V_m \sin(m\omega t + \theta_m) \\ &\quad + \sum_{m=1}^{\infty} V_m I_m \sin^2(m\omega t + \theta_m) \end{aligned} \tag{6.29}$$

As the active power is based on the average values, the sinusoidal terms of Equation 6.29 are canceled out, resulting in:

$$p_a = V_{\text{DC}} I_{\text{DC}} + \sum_{m=1}^{\infty} V_m I_m \sin^2(m\omega t + \theta_m) \tag{6.30}$$

and so, as $\sin^2 a = \frac{1}{2}(1 - \cos 2a)$, Equation 6.30 becomes:

$$\begin{aligned} p_a &= V_{\text{DC}} I_{\text{DC}} + \sum_{m=1}^{\infty} \frac{1}{2} V_m I_m \left[1 - \cos 2(m\omega t + \theta_m)\right] \\ &= V_{\text{DC}} I_{\text{DC}} + \sum_{m=1}^{\infty} \left\{ \frac{V_m}{\sqrt{2}} \frac{I_m}{\sqrt{2}} - \frac{V_m}{\sqrt{2}} \frac{I_m}{\sqrt{2}} \left[\cos 2(m\omega t + \theta_m)\right] \right\} \end{aligned} \tag{6.31}$$

where the first product within brackets is the active harmonic power of order m and the second term is the intrinsic harmonic power that does not contribute to the net transfer of energy or to losses in conductors [6].

6.2.1.2 Instantaneous Nonactive Power, p_{na}

If $\delta_k \neq 0$ for $m=n$, then Equation 6.26 loses all these terms. If $\delta_k=0$ for $m=n$, the nonactive power equation (Equation 6.31) becomes:

$$
\begin{aligned}
p_{na} &= V_{DC}\sum_{m=1} I_m \sin\left(m\omega t+\phi_m \mp \delta_m\right)+I_{DC}\sum_{n=1} V_n \sin\left(n\omega t+\phi_n\right) \\
&\mp \frac{1}{2}\sum_{n=1} V_n I_n \sin\left(\delta_n\right)\sin 2\left(n\omega t+\phi_n\right)+\sum_{m}\sum_{n\neq m} V_n I_m \sin\left(n\omega t+\phi_m\right)\sin\left(m\omega t+\phi_m\right)
\end{aligned} \tag{6.32}
$$

There is not a transfer of useful energy associated with the nonactive instantaneous power; even so, these currents contribute to additional losses in the power conductors [6].

6.2.1.3 Active or Average Power, P

The active, effective, real, or useful power is the average of the instantaneous power [6] given by:

$$
P=\frac{1}{kT}\int_t^{t+kT} p(t)\,dt=\frac{1}{kT}\int_t^{t+kT} v(t)\cdot i(t)\,dt \tag{6.33}
$$

Equation 6.33 can be split into the active power, P_1 (calculated by the fundamental of the Fourier expansion), and the harmonic active power, P_h, including inter- and subharmonics:

$$
P=P_1+P_h \tag{6.34}
$$

The active fundamental power is given by:

$$
P_1=\frac{1}{kT}\int_t^{t+kT} p(t)\,dt=V_1 I_1 \cos\theta_1 \tag{6.35}
$$

while the active harmonic power is given by the other terms of the total power as:

$$
P_h=V_{DC}I_{DC}+\sum_{k>1} V_k I_k \cos\theta_k=P-P_1 \tag{6.36}
$$

6.2.2 Effects of Harmonic Distortion on the Power Factor

As suggested in Section 6.2.1, the effects of the harmonic distortion on the power factor are observed in the average active power, including harmonic components, in:

$$P = \frac{1}{T}\int_0^T vidt = \frac{1}{T}\int_0^T \sum_{i,k}\left[V_i \sin\left(i\omega t + \phi_i\right)\cdot I_k \sin\left(k\omega t + \theta_k\right)\right]dt \tag{6.37}$$

where

$v_i(t) = V_i \sin(i\omega t + \phi_i)$ is the instantaneous harmonic voltage of order "*i*"
$i_k(t) = I_k \sin(k\omega t + \theta_k)$ is the instantaneous harmonic current of order "*k*"

From the trigonometry: $\sin(\text{a})\cdot\sin(\text{b}) = \frac{1}{2}\left[\cos(a-b) - \cos(a+b)\right]$ which by replacing in Equation 6.37 results in:

$$\sin\left(i\omega t + \phi_i\right)\cdot\sin\left(k\omega t + \theta_k\right)$$
$$= \frac{1}{2}\left\{\cos\left[\left(i\omega t + \phi_i\right) - \left(k\omega t + \theta_k\right)\right] - \cos\left[\left(i\omega t + \phi_i\right) + \left(k\omega t + \theta_k\right)\right]\right\}$$
$$\sin\left(i\omega t + \phi_i\right)\cdot\sin\left(k\omega t + \theta_k\right)$$
$$= \frac{1}{2}\left\{\cos\left[\left(i-k\right)\omega t + \left(\phi_i - \theta_k\right)\right] - \cos\left[\left(i+k\right)\omega t + \left(\phi_i + \theta_k\right)\right]\right\} \tag{6.38}$$

So, replacing Equation 6.38 in Equation 6.37 yields:

$$P = \frac{1}{T}\int_0^T \frac{1}{2}\sum_{i,k} V_i I_k \left\{\cos\left[\left(i-k\right)\omega t + \left(\phi_i - \theta_k\right)\right] - \cos\left[\left(i+k\right)\omega t + \left(\phi_i + \theta_k\right)\right]\right\} dt \tag{6.39}$$

As a sum of two cosine integral terms exists over a full cycle *T*, it raises two possibilities:

1. If $i \neq k$, the resulting integral is zero and $\overline{P} = 0$.
2. If $i = k$, Equation 6.39 becomes:

$$P = \frac{1}{T}\int_0^T \frac{1}{2}\sum_{k} V_k I_k \left[\cos\left(\phi_k - \theta_k\right) - \cos\left(2k\omega t + \phi_k + \theta_k\right)\right]dt \tag{6.40}$$

Again, the sum of integrals of the cosine terms over a full cycle *T* is zero, and the average active power becomes:

$$P = \frac{1}{2}\sum_{k} V_k I_k \left[\cos\left(\phi_k - \theta_k\right)\right] = \sum_{k} V_{sk} I_{sk} \left[\cos\left(\phi_k - \theta_k\right)\right] \tag{6.41}$$

where V_{sk} and I_{sk} are the effective values of voltage and current components, respectively.

The definition of average active power including the DC component is:

$$P = \frac{1}{T}\int_0^T v \cdot i dt = V_{DC}I_{DC} + \sum_{k=1}^{n} V_{sk}I_{sk}\left[\cos\left(\phi_k - \theta_k\right)\right] \tag{6.42}$$

The graphical interpretation of the kth harmonic components is shown in Figure 6.3. All vectors in this figure are rotating for each order k with angular velocity ω_k, and the overall vector summation constitutes the representative vector of the apparent power, S, including the harmonic components.

In a system with distorted power supply, the displacement factor (power factor, including the harmonics) shall be defined as:

$$PF = \frac{P}{S} = \frac{V_{DC}I_{DC} + \sum_1^n V_{sk}I_{sk}\cos\left(\phi_k - \theta_k\right)}{V_{rms}I_{rms}} \tag{6.43}$$

From what is presented in this section, the power factor of waveforms without direct current may be calculated as:

$$PF = \frac{P}{S} = \frac{\sum V_{sk}I_{sk}\cos\left(\phi_k - \theta_k\right)}{V_{rms}I_{rms}} = \frac{V_{s1}I_{s1}\cos\left(\phi_1 - \theta_1\right)}{V_{rms}I_{rms}} + \frac{\sum_2^n V_{sk}I_{sk}\cos\left(\phi_k - \theta_k\right)}{V_{rms}I_{rms}} \tag{6.44}$$

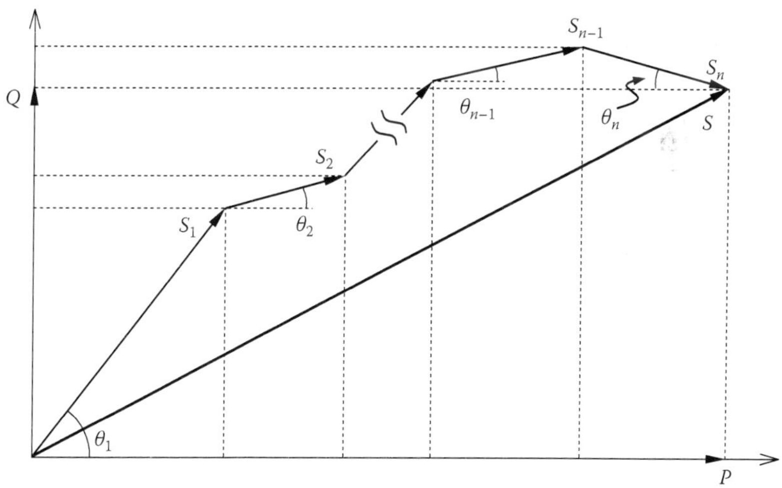

FIGURE 6.3 Graphical interpretation of the sum of "n" harmonic components.

The power factor is defined by Equation (6.45), considering the fundamental terms on the numerator, and the effective terms on the denominator [4,5].

$$PF = \frac{P}{S} = \frac{V_{s1} I_{s1} \cos\theta_1}{V_{rms} I_{rms}} \tag{6.45}$$

If the DC level is taken into consideration for the active power formulation, the average active power becomes:

$$P = \frac{1}{T}\int_0^T p\,dt = V_{\mathrm{DC}} I_{\mathrm{DC}} + \frac{1}{2}\sum V_{mk} I_{mk} \cos(\phi_k - \theta_k) = V_{\mathrm{DC}} I_{\mathrm{DC}} + \sum V_{sk} I_{sk} \cos(\phi_k - \theta_k) \tag{6.46}$$

In this case, the definition of power factor including all harmonics is:

$$PF = \frac{P}{S} = \frac{V_{\mathrm{DC}} I_{\mathrm{DC}} + \sum V_{sk} I_{sk} \cos(\phi_k - \theta_k)}{V_{rms} I_{rms}} = \frac{V_{s1} I_{s1} \cos(\phi_k - \theta_k)}{V_{rms} I_{rms}} + \frac{V_{\mathrm{DC}} I_{\mathrm{DC}} + \sum_2^n V_{sk} I_{sk} \cos(\phi_k - \theta_k)}{V_{rms} I_{rms}} \tag{6.47}$$

In actual systems, due to constructive and practical considerations of the generator, and the so-considered infinite bus, a good approximation represents the generated voltage by regarding only its fundamental component V_1, without any DC component. There are filtering effects of transmission lines, line chokes, transformers, and power equipment in general, especially at higher frequencies, and the large equivalent inertia constants of the interconnected systems. Therefore, $V_{rms} \approx V_1$ when substituted into Equation 6.17 results in:

$$PF = \frac{P}{S} = \frac{I_{s1}}{I_{rms}} \cos\theta_1 \tag{6.48}$$

In such cases, the definitions of $THD(V)$ and $THD(I)$ can be given, respectively, by:

$$\frac{V_{rms}}{V_1} = \sqrt{THD^2(V)+1} = 1 \quad \text{and} \quad \frac{I_{rms}}{I_1} = \sqrt{THD^2(I)+1} \tag{6.49}$$

Therefore, replacing Equation 6.17 with the inverse of the previous equations, the power factor definition given by Equation 6.48 becomes:

$$PF = \frac{P}{S} = \frac{\cos\theta_1}{\sqrt{THD^2(I)+1}} \tag{6.50}$$

The kth order harmonic distortion, D_k, is defined as the relationship between the amplitude component of the kth order frequency, and the signal amplitude of the fundamental frequency, $D_k = V_k/V_1$ [4,5]. Then, the THD is given by:

$$THD = \sqrt{\sum_{2}^{n}\left(\frac{V_k}{V_1}\right)^2} = \sqrt{\frac{V_{rms}^2 - V_1^2}{V_1^2}} = \sqrt{\frac{V_{rms}^2}{V_1^2} - 1} = \sqrt{\sum_{2}^{n} D_k^2} \tag{6.51}$$

6.3 HARMONIC DISTORTION ANALYZERS

Signal distortion analyzers serve to decompose electrical signals into their most significant aspects, i.e., related to amplitude, frequency, and phase. To this purpose, there are different instruments, such as wave analyzer, distortion analyzer, spectrum analyzer, audio analyzers, and modulation analyzers, which give a relatively precise information about the frequency content of a physical phenomenon.

Analog frequency analyzers consist of filters associated with a voltmeter and a frequency selector used to measure amplitude and frequency of the electrical components of a particular wave shape. These types of analyzers, usually, cannot detect signal phase directly. However, if the signals are converted into digital values through data acquisition boards, and computer mathematical analysis, the phases of the signal harmonic components can be easily captured, as discussed in Chapter 5. If the signals are analogically processed, they may be detected by means of analog filters using polystyrene capacitors of low tolerance. The ordinary types of filters are Butterworth (maximization of the flat region), Chebyshev (maximization of the gain edge sharpness), and Bessel (minimizing of input-output phase delay), among others digitally or analogically processed.

A spectrum analyzer sweeps the signal frequency band of a physical phenomenon (usually a few hertz or fractions of hertz), audio frequency (order of kHz), radio frequencies (order of MHz), and cellular and satellite frequencies (order of gigahertz). They can be analog or digital analyzers and display a plot of amplitude versus frequency covering operating ranges from approximately 0.01 Hz up to 500 GHz with the technology SiGe (silicon-germanium).

The RF analyzers are of the heterodyne type usually with an intermediate frequency (IF) accurately generated by a local oscillator, i.e., tuning of the local oscillator shifts the various signal frequency components to the passband of a standard commercial IF amplifier. The amplitude of the given sinusoidal frequency is determined by rectifying the output signal of the IF amplifier and brought to the display as a filtered DC component. Figure 6.4 shows the block diagram of heterodyne type analyzer.

The existing methods for harmonic content evaluation are very diverse, but mostly can be classified into methods by filter tuning, heterodynes, and spectral analysis.

6.3.1 Harmonic Analyzer with a Tuned Filter

This type of harmonic analyzer is old-fashioned since it requires individual tuning of each harmonic filter in such a way to isolate amplitude and frequency of

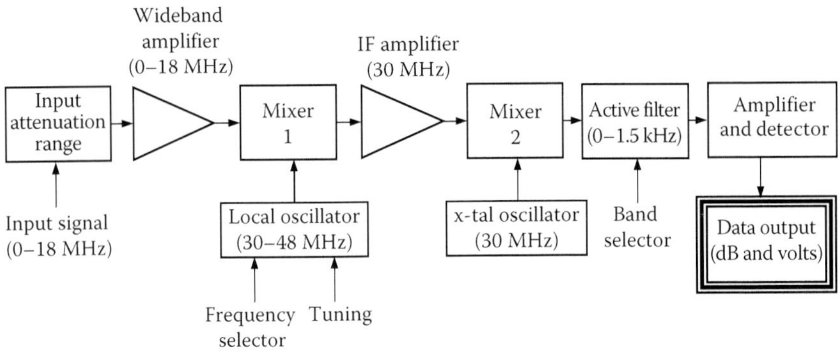

FIGURE 6.4 Heterodyne wave shape analyzer.

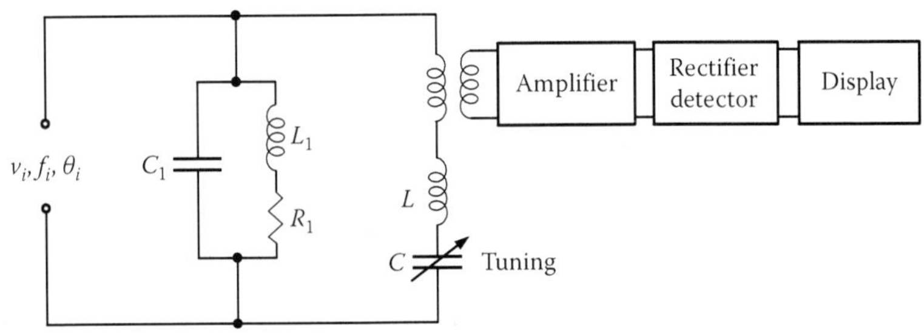

FIGURE 6.5 Tuned circuit.

the desired component of a given waveform. The operating principle is shown in Figure 6.5. In the circuit, there is a parallel tuned circuit with a tank circuit compensating variations in the alternating current resistance of the series resonant circuit (skin and proximity effects), and also the amplifier gain in the instrument frequency range.

There are drawbacks using the tuned harmonic analyzer: (1) at the low frequencies, the L and C values are very high and therefore are bulky and heavy components; (2) it is difficult to detect harmonic orders very close; (3) it is very labor-intensive for tuning and reading the harmonic components; (4) the phase reading is complicated.

6.3.2 Heterodyne Harmonic Analyzer or Wave Distortion Meter

The heterodyne harmonic analyzer works by suppression of the fundamental harmonic to measure the THD. The heterodyne harmonic analyzer allows a negligible harmonic distortion with a less constrained active filter selectivity for suppressing the fundamental of the signal. Figure 6.6 shows this analyzer block diagram.

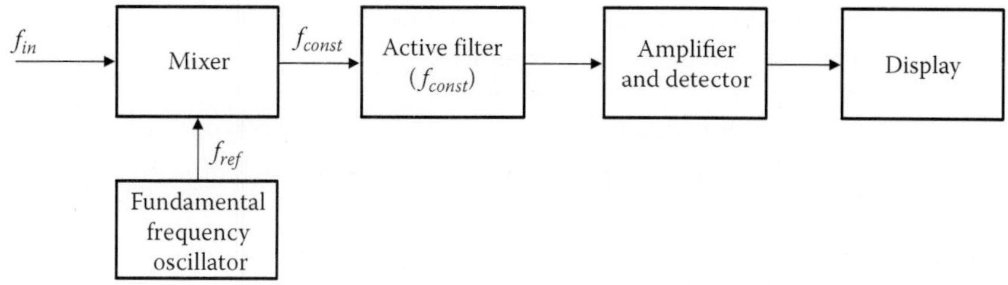

FIGURE 6.6 Heterodyne circuit.

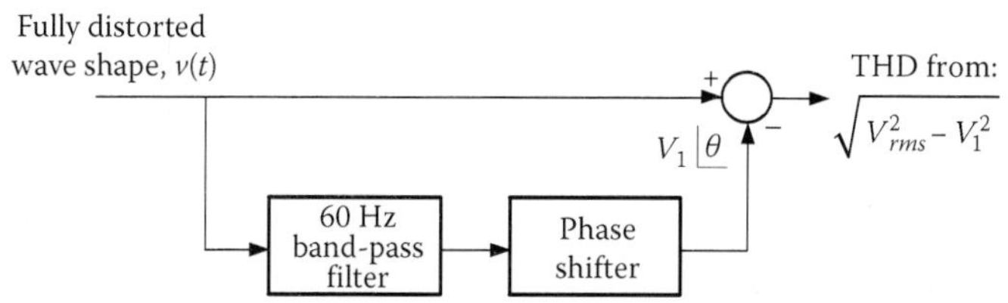

FIGURE 6.7 Feed-forward filter to obtain the THD.

6.3.3 Determination of THD

The THD of a network voltage or current can be an isolated circuit using a feed-forward circuit as shown in Figure 6.7. The effective value of the voltage network $v(t)$ (Equation 6.14), including all harmonics, is placed in the feed-forward filter input. In steady state, a band-pass filter isolates the fundamental signal of 60 Hz to enter a 180° phase shifter, as discussed in Chapter 2. This phase shifter adds the modulated amplitude and phase of the fundamental input signal, then a subtracting circuit takes this from the fundamental in the feed-forward circuit depicted in Figure 6.7. As a result, the output has just the total harmonic content with all components of frequency, amplitude, and phase identical to the input. The output result is the net harmonic distortion given by $\sqrt{V_{rms}^2 - V_1^2}$.

6.4 SPECTRUM ANALYZERS

Spectral analysis is useful to determine the amplitude, frequency, and phase of each harmonic component for the observed signal. This study is related mainly to the emission of electromagnetic noise in the radio frequency range of the equipment (conducted) and from the equipment (radiated). The usefulness of this analyzer depends on its dynamic range defined by the relationship between the smallest signal that can be detected above the noise level and the largest signal that does not generate spurious signals larger than the smallest signal that can be separated. The spectral analysis of these signals can be implemented on the basis of the Fourier transform. The characteristics of this method are as follows:

1. It uses a computational method to obtain the Fourier component, in which the most common ones are the fast Fourier transform (FFT) and the wavelet transform.
2. It uses the output data of a digital-analog converter (ADC) with all its limitations: its speed and available memory.
3. It causes delays in data interpretation due to the process of the calculation algorithm and the need to wait for the acquisition of all data to start the calculations.

Another possible methodology for harmonic measurement in power systems is the use of wavelets. A wavelet transform is a mathematical function that can divide in time a given function, or a continuous signal in different parts of a scale. Usually, it sets up a frequency range for each component range. Each scale component can then be studied with a resolution that matches its scale. A wavelet transform is the representation of a function by wavelets. The wavelets are copies in scale (known as daughter wavelet) and translated from a waveform length defined from an oscillation that decreases rapidly (known as mother wavelet). The wavelet transform is advantageous with respect to the FFT, since it can represent functions containing discontinuities and sharp peaks making it able to reconstruct finite, nonperiodic, and nonstationary signals.

The wavelet theory applies to many subjects. All wavelet transforms may be considered forms of time-frequency representation for continuous signals (analog), being so related to harmonic analysis. Practically, almost all discrete wavelet transforms use banks of filters in time. These filter banks are called scaling coefficients in the wavelet nomenclature and may contain filters either with a finite impulse response (FIR) or infinite impulse response (IIR). The wavelets forming a continuous wavelet transform (CWT) are subject to the principle of Fourier analysis of the uncertainty regarding the theory of sampling; i.e., given a signal with an event on it, one cannot simultaneously set a response scale in time and frequency for this event. The time response scale of all uncertainty, with regards to this particular event, has lower bandwidth. Thus, the scalogram of a continuous wavelet transform of this signal, such as an event, marks an entire region in the time-scale plan rather than of a single point. Similarly, the basis of the discrete wavelet can be considered in the context of other forms of the uncertainty principle.

For all purposes, the wavelet transform is separated into three base classes: continuous, discrete, and multiresolution.

Spectrum analyzers generate x–y signals on an oscilloscope screen, where y represents the signal intensity on a dB scale, usually logarithmic, and x is the frequency, generally speaking, a linear scale. Figure 6.8 shows the result of a frequency domain representation of the waveform Fourier components (f_o and its multiples or harmonic orders) of the input or output of a device under test (DUT).

6.4.1 Spectrum Analyzers by Frequency Scanning

Spectrum analyzers can operate in real or virtual time (or equivalent time), such as those used in digital oscilloscopes. The virtual time is based on the fact that there is

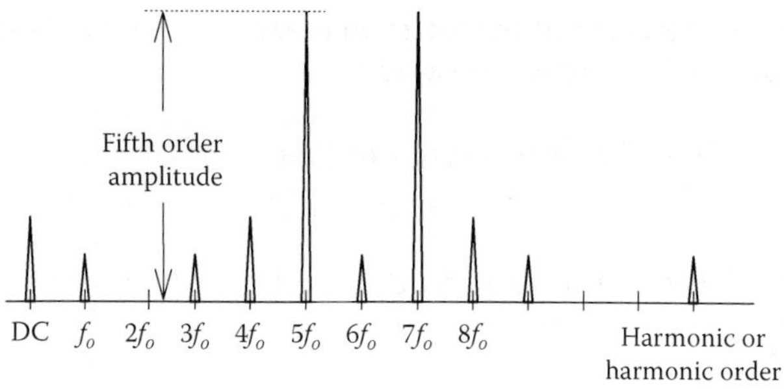

FIGURE 6.8 Frequency spectrum.

a cycle repetition T_v of the signal under study, as shown in Figure 6.9. Thus, one can virtually extend the reading time for a long period, i.e., it comprises a high-frequency signal read by virtual points taken sequentially in a given number of cycles.

Figure 6.10 shows a typical circuit for spectral analysis where the output of a first mixer (400 MHz) appears only when the scanning circuit detects the existence of a harmonic component through its input. This analyzer must have an intermediate frequency above the maximum frequency to be read at the input (300 MHz).

The Nyquist-Shannon theorem proves that the output signal is subject to aliasing and that it was not properly grasped. Frequency portions for which the sampling rate is insufficient according to this theorem need to be removed before filtering,

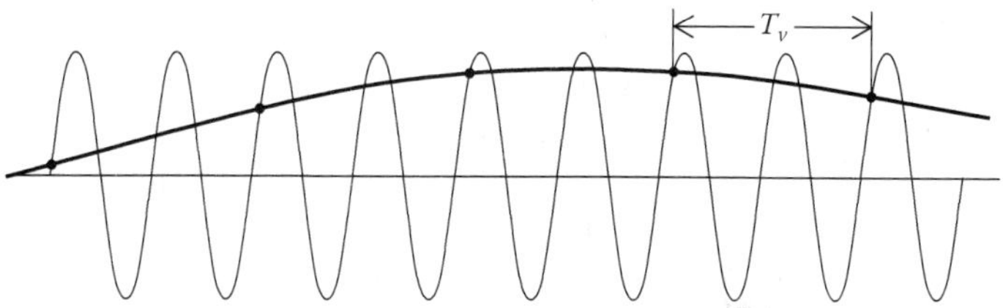

FIGURE 6.9 Real and virtual sampling time (T_v).

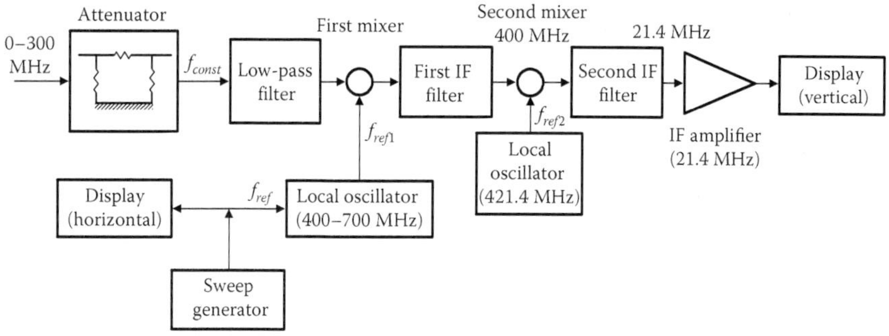

FIGURE 6.10 Spectrum analyzer.

as they can no longer be identified after sampling. For example, a signal set at a 60 Hz sampling rate would show up only at 60 Hz after sampling. This signal, which has been virtually created by undersampling, is indistinguishable from the original signal since part of the original information has been wiped out due to the insufficient sampling rate.

Good-quality spectrum analyzers offer many facilities such as scan ranges, central frequency, filter bandwidth, and various scale displays, among others. The read frequencies can go from fractions of hertz to hundreds of gigahertz gathered in various scale ranges. More advanced models allow the spectrum to avoid storage flickering, absolute amplitude calibration, additional storage for comparing the acquired information of the spectra details on the screen itself, and many other facilities. Note that these spectrum analyzers check one frequency at a time and display it in a spectrum scan as a whole. By checking one at the time, it is possible to miss transient variations, and in the case of narrow frequency bands, the scan rate must be kept very slow. Moreover, the scanning period T_v should be carefully chosen, because if it is a multiple of the measured frequency period, the interpretation of the result is a DC value between zero and the signal amplitude under analysis.

The scanning problem of spectrum analyzers can be circumvented using in their place the real-time analyzers. There is a very wide range of principles, from the mathematical data analysis spectrum to a spectrum construction by Z-transform using a dispersive filter. In this filter, the time delay is proportional to the frequency in such a way so as to replace the band-pass filter in the analysis using the local oscillator sweep. Most modern real-time analyzers have stopped using frequency-selective tuned filters and are using Fourier analysis, more particularly fast Fourier transform (FFT) or wavelets. These instruments convert analog signals into digital codes using extremely fast analog-to-digital conversion to mount the full frequency spectrum with a digital program. Since all frequencies are obtained from a single set of samples within a single time window, the apparatus has high sensitivity and high speed to be used for both steady and transient states. The lower the frequency analyzed, the better is the quality of the spectral data.

Resolution limitations imposed when using FFT or wavelet refer to the window of time that infers the data out of it. The best possible resolution frequency, that is, when all samples are in one signal cycle (range), is limited by:

$$f_r = \frac{1}{T} \tag{6.52}$$

Modern digital techniques allow very high sampling frequencies. The maximum sampling frequency is restricted to low-order harmonics only by the ADC response limitations.

The Nyquist-Shannon theorem predicts that the highest frequency of a complex signal that can be analyzed is half the sampling frequency, which means two samples per cycle of the highest frequency, at least:

$$f_{signal} = \frac{f_{sampling}}{2} \quad \text{or} \quad T_{sampling} = \frac{T_{signal}}{2} \tag{6.53}$$

With respect to the signal amplitude, the dynamic range R_d of an A/D converter sets its limit levels between 1 and 2^N, and it is given by:

$$R_d = 20\log 2^N \tag{6.54}$$

where

R_d is the ratio between the highest and the lowest signal amplitude that can be measured without overload (or distortion)

N is the number of bits

A good application example for either the Fourier or wavelet analysis is the search for signals of intelligent manifestations coming from outer space, where the signals are in a narrow frequency range of 35 dB below the receiver noise level that can be detected in time of 1 min. This corresponds to the radio flow of less than 1 μW over the entire terrestrial disc.

6.4.2 Mathematical Analysis of Power Signals

Any rms (or effective) current or function value can be obtained from:

$$I_{rms} = \sqrt{\frac{1}{T}\int_0^T i^2 dt} \tag{6.55}$$

In the case of the waveform shown in Figure 6.2, the effective value is:

$$I_{rms} = \sqrt{\frac{1}{T}\int_{t_1}^{t_1+w} I^2\, dt} = \sqrt{\frac{w}{T} I^2} = I\sqrt{D} \tag{6.56}$$

where $D = w/T = \omega w/\omega T$.

The following are the three solved problems:

1. Calculate the rms and average values of the waveform shown in Figure 6.11. For single-phase waveforms, like that given in Figure 6.11, under the same conditions, $\omega w = 180°$ and $\omega T = 360°$. Therefore:

$$D = \frac{1}{2} \quad \therefore \quad I_{rms} = \frac{I}{\sqrt{2}} \tag{6.57}$$

For three-phase waveforms, with $\omega w = 120°$ and $\omega T = 360°$, it comes:

$$D = \frac{1}{3} \quad \therefore \quad I_{rms} = \frac{I}{\sqrt{3}} \tag{6.58}$$

The average value is then given by:

$$I_{av} = \frac{1}{T}\int_{t_1}^{t_1+w} I dt = \frac{w}{T}\cdot I = I \cdot D \tag{6.59}$$

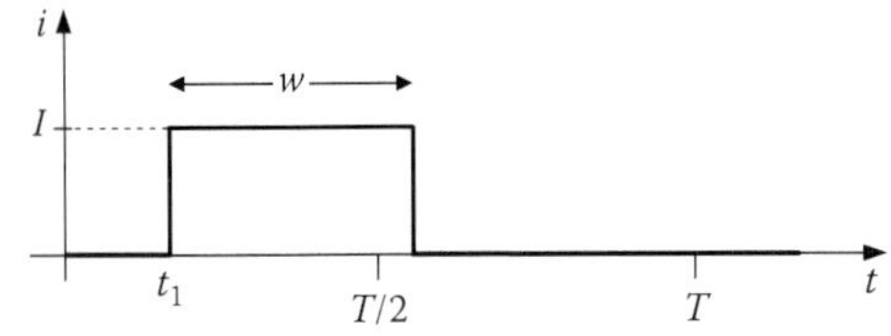

FIGURE 6.11 Pulse of a current signal with level "*I*."

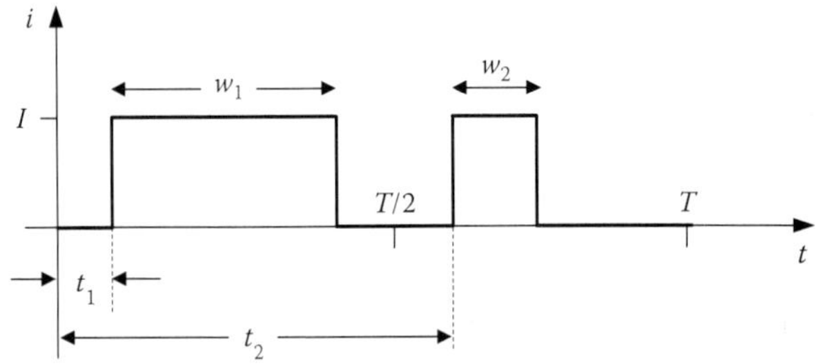

FIGURE 6.12 Multiple pulses of current signal with level "*I*."

2. Compute the rms current value of the waveform shown in Figure 6.12. In the general case of train blocks with different widths and same height, the rms value may be calculated by the expression:

$$I_{rms} = \sqrt{\frac{1}{T}\left[\int_{t_1}^{t_1+w_1} I^2 dt + \int_{t_2}^{t_2+w_2} I^2 dt + \cdots\right]} = \sqrt{\frac{I^2}{T}\left[w_1 + w_2 + \cdots\right]}$$
$$= I\sqrt{\sum_i D_i} \tag{6.60}$$

Example 1:

$$\Sigma w_i = T \quad I_{rms} = I \tag{6.61}$$

and if:

$$\Sigma w_i = T/2 \quad I_{rms} = I/\sqrt{2} \tag{6.62}$$

3. Compute the rms value of the multipulse waveform shown in Figure 6.13. In general, the effective value of the current waveform, like the one shown in Figure 6.13, may be expressed by:

$$I_{rms} = \sqrt{\frac{1}{T}\left[\int_{t_1}^{t_1+w_1} I_1^2\, dt + \int_{t_2}^{t_2+w_2} I_2^2 dt + \cdots\right]} = \sqrt{\frac{1}{T}\left[w_1 I_1^2 + w_2 I_2^2 + \cdots\right]}$$
$$= \sqrt{\sum_i D_i I_i^2} \tag{6.63}$$

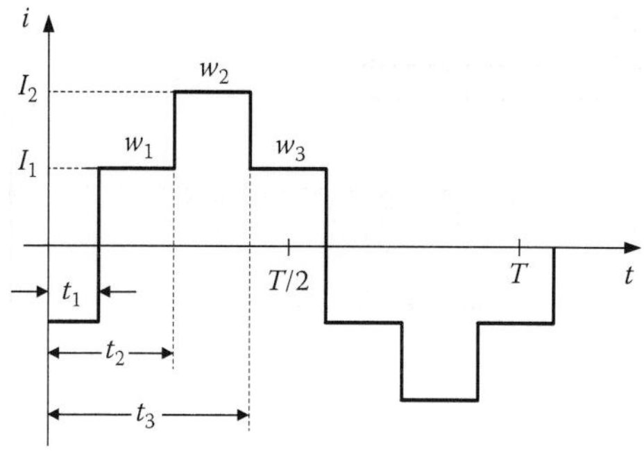

FIGURE 6.13 Signal with multiple pulses of alternating current.

Example 2:

$$\text{If}: w_1 = w_2 = w_3 = 60°, \quad I_1 = I \quad \text{and} \quad I_2 = 2I \tag{6.64}$$

In the case of Figure 6.13, the pulses can be combined in positive and negative cycles to give:

$$I_{rms} = \sqrt{2 \cdot \frac{60I^2 + 60(2I)^2 + 60I^2}{360}} = \sqrt{\frac{I^2 + 4I^2 + I^2}{3}} = \sqrt{2}I \tag{6.65}$$

EXERCISES

6.1 Explain the operation of a harmonic distortion analyzer based on Fourier's theorem. What are the reasons that make them slow? What is the main theoretical limitation for measurement of higher frequencies?

6.2 Calculate the harmonic components up to the fifth order in Figure 6.10, if $I_a = 10$ A and $w = 8.333$ ms.

6.3 Dimension the train blocks in the figure below so as to minimize their harmonic content.

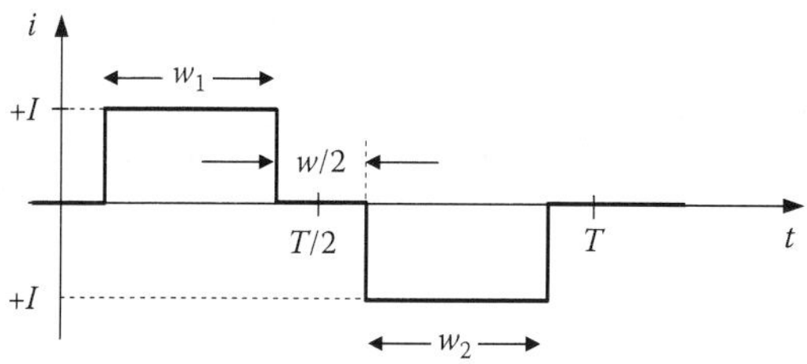

6.4 Prove, analytically, that the train of block currents in the figure of Problem 6.3 has odd harmonics of orders $n = 5, 11, \ldots, 6p - 1$ with positive amplitudes and odd orders $n=7, 13, \ldots, 6p+1$ with negative amplitudes generally expressed by:

$$a_n = \pm \frac{2I_a}{n\pi}$$

REFERENCES

1. Clarkson, P.T., *Optimal and Adaptive Signal Processing*, CRC Press, Boca Raton, FL, 529pp., 1993.
2. Peled, A. and Bede, L., *Digital Signal Processing, Theory, Design, and Implementation*, John Wiley & Sons, New York, 1978.
3. Leão, R.P.S., Sampaio, R.F., and Antunes, F.L.M., *Harmonics in Electrical Systems (Harmônicos em Sistemas Elétricos)*, Elsevier Editor Ltd, Rio de Janeiro, Brazil, 2014, ISBN: 978-85-352-7439-4.
4. Rashid, M.H., *Power Electronics: Circuits, Devices & Applications*, 4th edn., Pearson Editors, London, England, 2013, ISBN: 13-978-0133125900, ISBN: 10-0133125904.
5. Mohan, N., Undeland, T.M., and Robbins, W.P., *Power Electronics: Converters, Applications, and Design*, John Wiley Inc., Hoboken, NJ, 2002, ISBN: 10-047122 69392002, ISBN: 13-978-0471226932.
6. IEEE Standard Definitions for the Measurement of Electric Power Quantities Under Sinusoidal, Nonsinusoidal, Balanced, or Unbalanced Conditions, IEEE Std 1459-2010 (Revision of IEEE Std 1459-2000), vol., no., pp. 1–50, March 19, 2010, DOI: 10.1109/IEEESTD.2010.5439063.

BIBLIOGRAPHY

Auslander, D.M. and Sagues, P., *Microprocessors for Measurement and Control*, Osborne/McGraw-Hill, Berkeley, CA, 1981.

Bastos, A., *Analog and Digital Instrumentation for Telecommunications (Instrumentação Eletrônica Analógica e Digital para Telecomunicações)*, MZ Editora, Rio de Janeiro, Brazil, 2002.

Bega, E.A., *Industrial Instrumentation (Instrumentação Industrial)*, 2nd edn., Interciência, São Paulo, Brazil, 2003.

Bose, B.K., *Power Electronics and Variable Frequency Drives*, IEEE Press, New York, 1997.

Bowron, P. and Stephenson, F.W., *Active Filters for Communications and Instrumentation*, McGraw-Hill Book Co. Ltd, London, U.K., 1979.

Eveleigh, V.W., *Introduction to Control Systems Design*, TMH Edition-Tata McGraw-Hill Publishing Company, New York, 624pp., 1972.

Horowitz, P. and Hill, W., *The Art of Electronics*, Cambridge University Press, London, U.K., 716pp., 2000.

Stefani, R.T., Savant, C.J. Jr., Shahian, B., and Hostetter, G.H., *Design of Feedback Control Systems*, Saunders College Publishing, Boston, MA, 1994.

7 Instrumentation and Monitoring for Distributed Generation Systems

7.1 INTRODUCTION

The consistent increase in energy needs and the relevant concerns with environmental impacts are propelling the diffusion of renewable-energy-based power sources. To accommodate such appealing power supplies, distributed generation systems (DGSs) have experienced a steady increase in the number of sources [1,2], such as wind farms, photovoltaic power systems, energy storage systems, and microgrids. Many researchers and industry developers have turned their attention to such promising markets. The DGSs consist of primary energy resources (i.e., renewable energy, or energy storage), DC-DC converters (i.e., voltage level regulation), DC-AC inverters (i.e., grid-interactive device), and public utilities (i.e., grid).

However, the gradual change from a unidirectional power flow to a bidirectional power flow network endowed by small prosumers faces several new challenges, such as uncontrolled power flow, under- and overvoltage, quality of energy, higher harmonic circulation, which may trigger resonances, and improper tripping of protections. Moreover, it is extremely beneficial to overcome some of these problems, thereby taking full advantage of the electronic power capacity of distributed generators (DGs). In this way, DGs are becoming more proactive in power system, and the grid codes have been updated to accommodate the possible ancillary functions, such as the reduction of reactive circulation, harmonic compensation, fault isolation, voltage regulation, and low-voltage ride-through. As a practical example, the technical reference rules for connection of active users to the low-voltage network in Italy require reactive power injection under low-voltage ride-through for inverters with total output power higher than 11.08 kW [3].

In this context, DGSs may be understood as a sort of instrumentation similar to that defined in Chapter 1, but this time, with large-scale variables. The function of this instrument is to correct deviations in a higher level of power, totally distinct from the ordinary low-power instrumentation. As an example, Figure 7.1 shows a distributed generator-based photovoltaic (PV) power system. Similar to what was explained in Chapter 3, the voltage and current signals are measured through sensors and translated and conditioned by the transducer to the data acquisition system. The signal conditioner consists of a group of operational amplifiers arranged to process the sensed signals, as described in Chapter 2. The data acquisition system possesses a

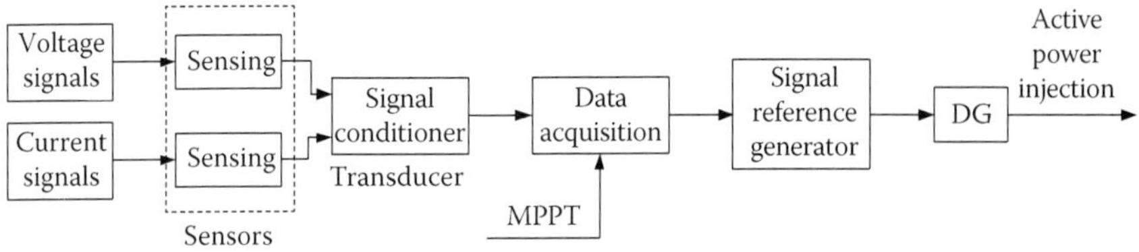

FIGURE 7.1 Instrumentation of distributed generator-based PV.

programmable software where the digital numeric values are manipulated according to a particular purpose, as the PV-based distribute generation (DG) aims at providing the maximum power extracted from PV modules via maximum power point tracking (MPPT) techniques. The signal reference generator is the algorithm responsible for interfacing the data acquisition output information to the DG's control scheme. Both blocks are detailed in Sections 7.3 and 7.2, respectively.

In 2015, the Pacific Gas and Electric Company [4] included in its new rules the definition of "smart inverter" that must be able to perform additional functions, such as those following commands from the distribution provider, i.e., central controller. All the new inverters to be connected to the distribution network must comply with such requirements. Here, the *smart inverters* may also be understood as instrumentation commanded via remote signals, as shown in Figure 7.2. However, the set point of active and reactive power generation comes from the distribution provider, which is the central controller.

Monitoring power quality takes an important role in this prospective distribution power system, mainly due to the flexible and scalable characteristics of bidirectional power networks, where loads and generators can plug in or disconnect at any time, and to their intermittency power capacity that often changes because of weather conditions, aging, and operational boundaries. A power quality monitoring instrumentation is shown in Figure 7.3, where some possible electrical disturbances are highlighted. Analysis of disturbances is processed using digital signal processing, and their influence on the power factor and harmonic analysis are discussed in Chapter 6.

Finally, a schematic diagram of an automonitoring distribution generator is shown in Figure 7.4, in which the instrumentation goal (i.e., active, reactive, and harmonic power references) is autosetting using the power quality monitoring of Figure 7.3.

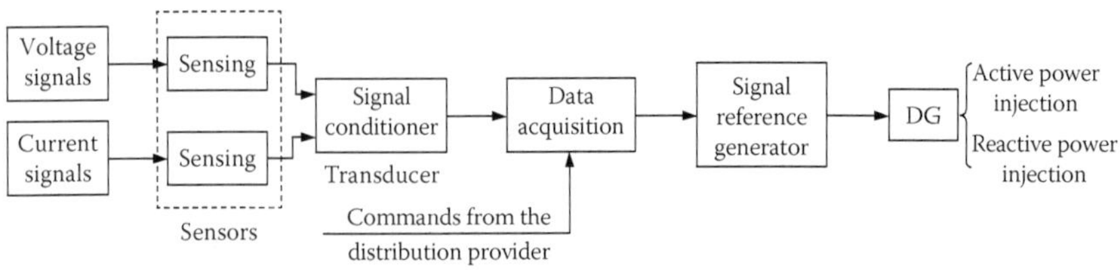

FIGURE 7.2 Instrumentation of distributed generator-based smart inverter.

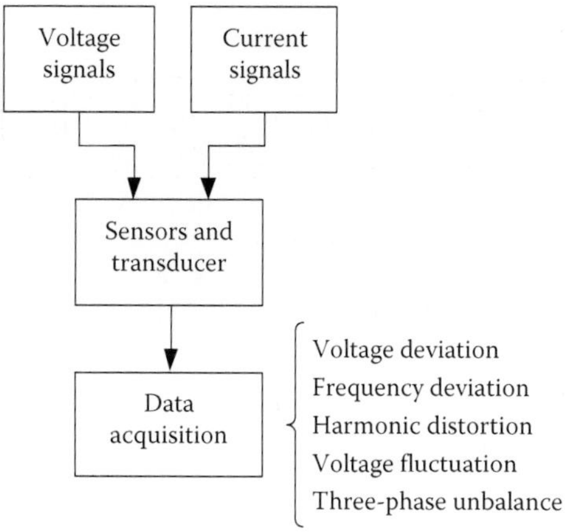

FIGURE 7.3 Power quality monitoring.

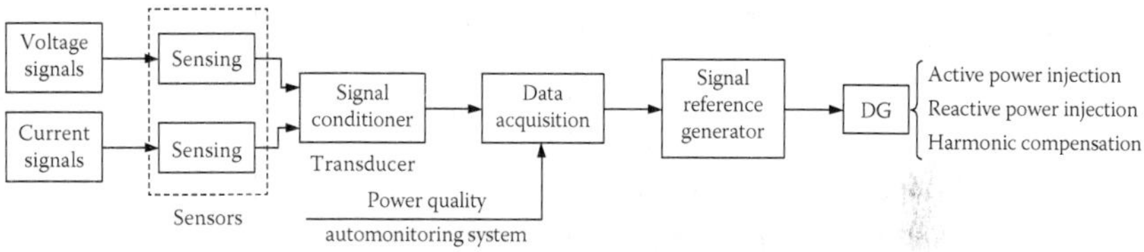

FIGURE 7.4 Instrumentation of distributed-generation-based automonitoring inverter.

7.2 GENERAL CONTROL SCHEME FOR DISTRIBUTED GENERATION

As discussed in the previous section, a high-power-level instrumentation involving power systems, as presented in this chapter, has a structure similar to a DG architecture and almost the same as used for the ordinary low-power-level instrumentation. It does not have a practical difference with the practical implementation, varying only the digital processing going through the data acquisition card. This section describes the control scheme of DG devised in the DG block of Figures 7.1, 7.2, and 7.4, which is equal to all DG applications. Section 7.3 depicts the signal reference generator that is also equal for all DGSs.

The conventional rotating generators are inevitably controlled as a voltage source, while DGs may act as a controlled voltage source or as a controlled current source, depending on the type of operation (grid connected or islanded), type of grid interface (i.e., control scheme), and applicable standards [5].

An overall power electronic structure of DGSs is shown in Figure 7.5, where the DC side consists of a primary energy resource, e.g., the wind, solar, or energy storage, and DC-DC converter, and the AC side is the public utility with variable loads. This general structure may act as a voltage source or as a current source, depending

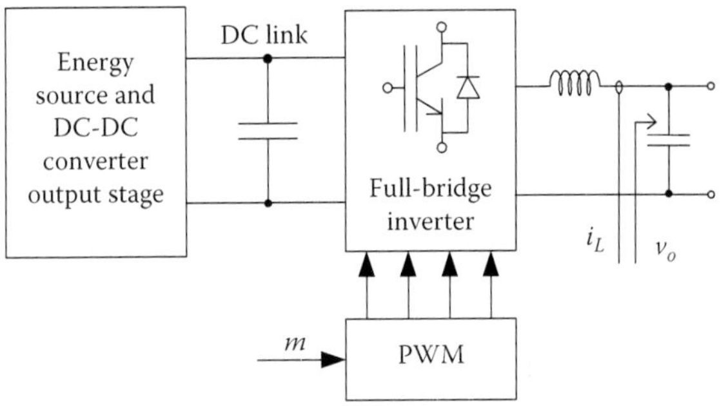

FIGURE 7.5 Overall structure of distributed generation systems.

on how the modulating quantity, *m*, is regulated to drive the full-bridge inverter through pulse-width modulation (PWM).

7.2.1 Current-Based DG

Most grid codes require DGs controlled as a current source because they have high output impedance, viewed by the grid, and hence less disturbing grid frequency and better system stability. However, such DGs do not have a grid-forming ability; therefore, they cannot be operated in stand-alone mode. Figure 7.6 shows a typical control scheme for current source-based DG, where i^* is the current reference provided from the signal reference generator block, i_L is the feedback inductor current of the inverter's LC output filter, and C_i may be any controller, such as proportional-integral (PI), hysteresis, and proportional plus resonance [6].

7.2.2 Voltage-Based DG

The voltage source-based DG may operate either in stand-alone or in grid-connected mode, applying the well-known droop control. This control method regulates the active and reactive power imitating the parallel operation characteristic of the zero-inertia synchronous generator. This type of DG was first proposed in Reference 7 for uninterruptible power supply (UPS), and therefore modified when used for the parallel operation of DGs, especially in microgrids [8]. However, another method exists to control DGs as a voltage source, as the virtual synchronous generator (or synchronverter) [9], virtual oscillator control [10], or indirect current control [11]. Figure 7.7

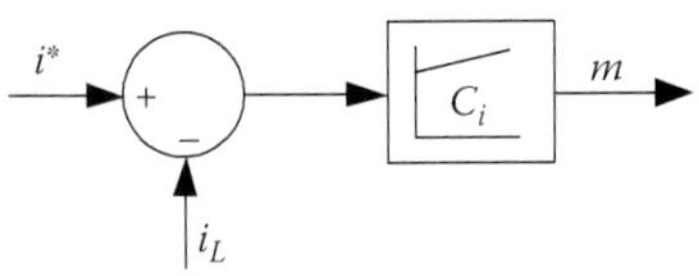

FIGURE 7.6 Controller of current loop, generation modulation index, typically used in current source-based DG.

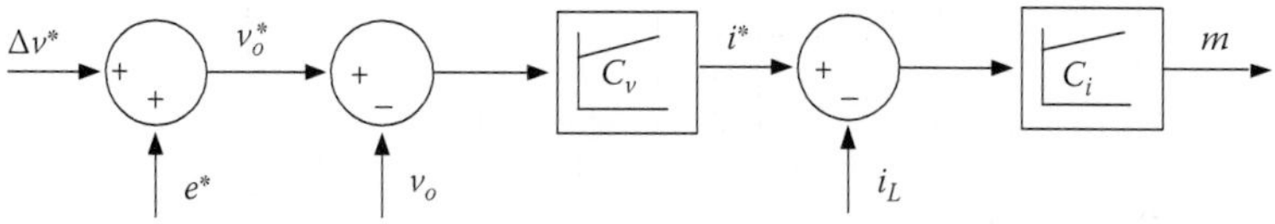

FIGURE 7.7 Loop controls for indirect current command for DGs.

shows the block diagram of an indirect current control applied to DGSs. Δv^* is the voltage reference that comes from the signal reference generator block responsible for controlling the active and reactive power, e^* is the nominal voltage reference, while C_i and C_v are, respectively, the voltage and current controllers.

7.3 SIGNAL REFERENCE GENERATOR

In Section 7.2, we devised the control scheme of current-based and voltage-based DGs, in which both schemes have a signal reference (i^* or Δv^*) as input. Such input reference aims to regulate, in steady state, the active and reactive power flow, and, possibly, the harmonics with a compensation purpose. So, this section shows a general generator of signal references to be applied to any DG type. Either i^* or Δv^* is created by the combination of in-phase and quadrature signal terms, and possibly nonfundamental frequency terms.

The input reference for the current-based DGs, i^*, in Figure 7.6, may be formed as follows:

$$i^* = i_a^* + i_r^* + i_h^* \tag{7.1}$$

such that i_a^* is the in-phase term, i.e., active current; i_r^* is the quadrature term, i.e., reactive current; and i_h^* is the nonfundamental frequency terms, i.e., harmonic current.

Similarly, the input reference for the voltage-based DGs, v_o^*, in Figure 7.7, may be created as follows:

$$v_o^* = e^* + \Delta v^* = e^* + \Delta v_{\parallel}^* + \Delta v_{\perp}^* + \Delta v_h^* \tag{7.2}$$

where

e^* is the nominal system voltage
$\Delta v_{\parallel}^*$ is the in-phase term
$\Delta v_{\perp}^*$ is the quadrature term
Δv_h^* is the nonfundamental frequency terms

7.3.1 Fundamental Frequency Terms

Equation 7.2 may be further extended to represent the nonlinear expression in fundamental frequency terms,

$$v_o^* = e^* + \Delta v_{\parallel}^* + \Delta v_{\perp}^* = V_n \sin\left(\omega t + \theta_g\right) + V_{\parallel} \sin\left(\omega t + \theta_{\parallel}\right) + V_{\perp} \sin\left(\omega t + \theta_{\perp}\right) \tag{7.3}$$

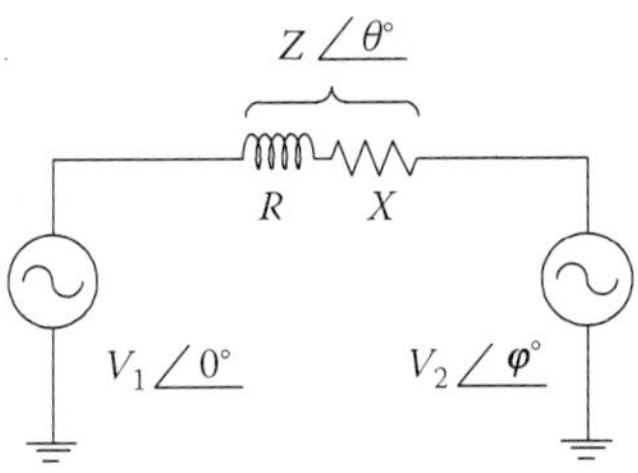

FIGURE 7.8 Equivalent circuit of voltage-based DG connected to the grid.

As can be inferred, to generate the signal reference, it is enough to track or estimate the frequency and phase of the main grid, and thereupon to regulate the magnitude value of each associating term. The frequency and phase of the grid may be tracked using a frequency synthesizer, such as frequency detector or phase-lock-loop algorithm, as described in Section 5.3.

Unlike current source inverters that have a well-defined control of the active and reactive power flow through the active and reactive currents, the voltage source inverters depend on the *R*/*X* ratio of the power line impedances. Let us consider the voltage sources V_1 and V_2 of Figure 7.8, apart from each other through an impedance, $Z\underline{|\theta}$. $V_1\underline{|0^\circ}$ is the utility and $V_2\underline{|\varphi}$ is the voltage-based DG.

The active and reactive power exchanged between both voltage sources are expressed as follows:

$$P = \frac{1}{Z}\cdot\left[\left(V_1V_2\cos(\varphi) - V_1^2\right)\cos(\theta) + V_1V_2\sin(\varphi)\sin(\theta)\right] \tag{7.4}$$

$$Q = \frac{1}{Z}\cdot\left[\left(V_1V_2\cos(\varphi) - V_1^2\right)\sin(\theta) - V_1V_2\sin(\varphi)\cos(\theta)\right] \tag{7.5}$$

Assuming that $X \gg R$, then $Z \approx X$ and $\theta \approx 90^\circ$, which simplifies the expressions to

$$P = \frac{V_1V_2\sin(\varphi)}{X} \quad \text{and} \quad Q = \frac{\left(V_1V_2\cos(\varphi) - V_1^2\right)}{X} \tag{7.6}$$

If the lag between the utility and DG is considered minimal, φ very small, then $\sin(\varphi) = \varphi$ and $\cos(\varphi) = 1$, which results in:

$$\varphi = \frac{PX}{V_1V_2} \quad \text{and} \quad \left(V_2 - V_1\right) = \frac{QX}{V_1} \tag{7.7}$$

If it is assumed that $X \ll R$, then $Z \approx R$ and $\theta \approx 0^\circ$, Equation 7.6 reduces Equations 7.4 and 7.5 to:

$$P = \frac{\left(V_1V_2\cos(\varphi) - V_1^2\right)}{R} \quad \text{and} \quad Q = -\frac{V_1V_2\sin(\varphi)}{R} \tag{7.8}$$

Considering the lag between the utility and DG minimal, φ very small, then $\sin(\varphi) = \varphi$ and $\cos(\varphi) = 1$, results in:

$$(V_2 - V_1) = \frac{PR}{V_1} \text{ and } \varphi = -\frac{QR}{V_1 V_2} \tag{7.9}$$

The simplified Equations 7.7 and 7.9 describe the effect of magnitude deviation and phase shift in the active and reactive power flow between any two parallel voltage sources. Note that the correspondence between these quantities is opposite, depending on the *R*/*X* ratio of power line impedances.

7.3.1.1 Generator of the In-Phase Reference Term

To generate the in-phase reference term to both DGs, i.e., i_a^* for current-based DG and $\Delta v_{\|}^*$ for voltage-based DG, Figure 7.9 shows the power closed-loop control scheme. Observe that the closed feedback is performed by means of the active power calculated through Equation 7.10. The DG's output measured voltage and current are those shown in Figure 7.5.

$$\overline{p} = \frac{1}{T} \cdot \int_{t-T}^{t} v_o \cdot i_L d\tau \tag{7.10}$$

The PI's output represents the conductance, *G*, or the magnitude of the in-phase term, $V_{\|}$, as in Equation 7.3, respectively, for the current-based and voltage-based DG. v_o signals to shape an in-phase signal reference proportional to the voltage waveform. For DGS application, P^* may come from any MPPT usually applied to DGs [12], or from the energy storage charging controller for energy storage applications. Positive and negative signals of P^* correspond to the charging and discharging power flow direction.

7.3.1.2 Quadrature Reference Term Generator

The quadrature terms, i.e., i_r^* for current-based DG and $\Delta v_{\perp}^*$ for voltage-based DG, may be generated as illustrated in Figure 7.10.

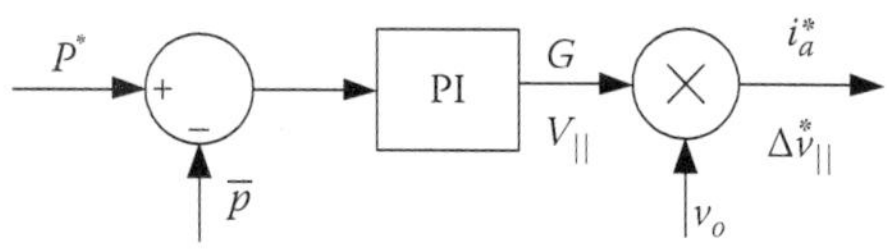

FIGURE 7.9 Generator of in-phase signal reference.

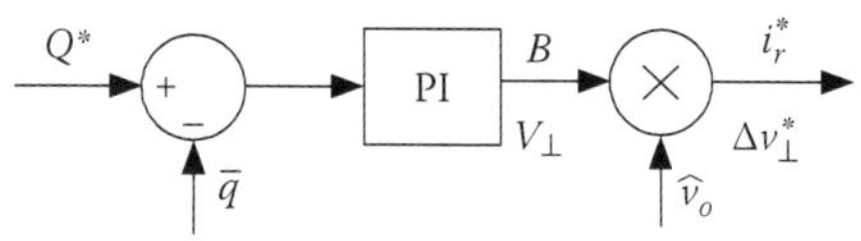

FIGURE 7.10 Generator of quadrature signal reference.

The reactive power at the DG's output can be calculated as follows:

$$\bar{q} = \frac{1}{T} \cdot \int_{t-T}^{t} \hat{v}_o \cdot i_L d\tau \quad (7.11)$$

where $\hat{v}_o$ is the output voltage shifted by 90°.

The PI's output represents the susceptance, B, or the magnitude of quadrature term, $V_\perp$, as in Equation 7.3, respectively, for current-based and voltage-based DG.

7.3.2 Nonfundamental Frequency Terms

The nonfundamental frequency terms can be generated similarly in fundamental frequency terms by only applying the concept of conductance (G_n) and susceptance (B_n) for the n-harmonic frequency, as introduced by [13]. Figure 7.11 shows how the nonfundamental frequency reference terms, i.e., i_h^*—for a current-based DG and Δv_h^*—the incremental harmonic voltage (as observed in Figure 7.11 for generating non-fundamental signal reference), can be generated for a voltage-based DG. The subscript "n" represents the specific harmonic order, e.g., third, fifth, and seventh.

Signals v_{on} and $\hat{v}_{on}$ can be generated by means of a frequency synthesizer, such as PLL, described in Section 5.3. The quantities $D_{\|n}$ and $D_{\perp n}$ can be calculated as follows:

$$D_{\|n} = V \cdot I_{\|n} \quad (7.12)$$

$$D_{\perp n} = V \cdot I_{\perp n} \quad (7.13)$$

The variable V is an rms value of voltage, while $I_{\|n}$ and $I_{\perp n}$ are, respectively, the rms values of the in-phase and quadrature current term to the n-harmonic, which can be typically based on a fast Fourier transform (FFT), addressed in Section 6.4.

The signal reference generator presented in this section aims to control the power flow in slow-dynamic closed loops by power and quantities measurement, as well as to collaborate to grid power quality improvements possible requested for future grid codes.

Figure 7.12 shows a diagram for a DG-based instrumentation scheme with three echelons: (1) power electronic converter corresponding to the hardware and drive, (2) current and voltage control scheme, and (3) signal reference generator. Current-based

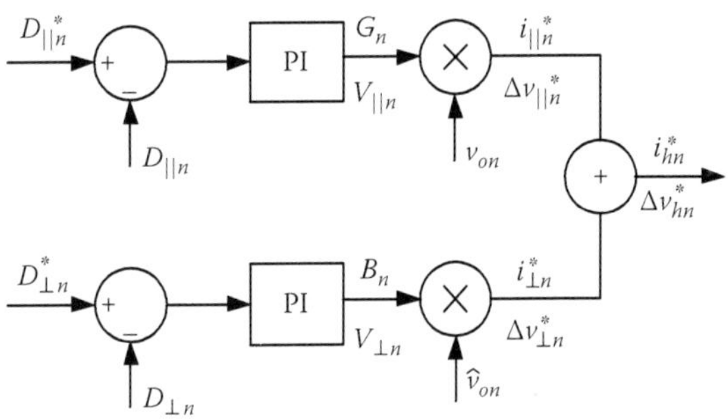

FIGURE 7.11 Generator of nonfundamental signal reference.

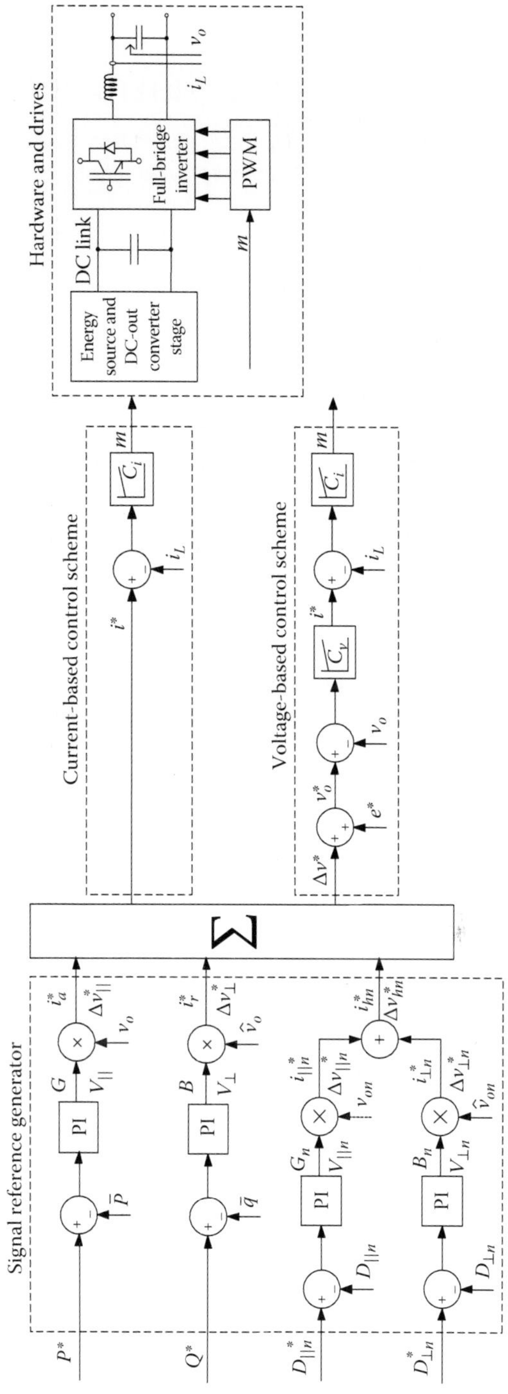

FIGURE 7.12 Complete control scheme of a DG-based instrumentation.

control or voltage-based control scheme must be chosen, depending on the desired characteristic for DG.

7.4 POWER QUALITY STANDARDS APPLIED TO DG

When there is an arbitrary connection of small-size generators across distribution network, there is a potential impact on the grid power quality. Some of the major power quality issues in distribution generation are as follows: (1) the voltage fluctuation due to intermittency of the renewable energies, e.g., run-of-a-river, solar, and wind, (2) impacts on the protection system because of possible bidirectional power flow, (3) overvoltages caused by a high feed-in active power, and (4) higher harmonic orders and resonances possibly triggered by high-order filters of grid-connected inverters.

Thus, to enlarge the installation of DGS in this new paradigm, all the concerns related to safety, reliability, and power quality must be circumvented. It has been current understanding that using surplus capability of DG is advanced with power electronic systems, making possible smart-grid solutions when approaching these challenges. Therefore, the DGs may be understood as instrumentation responsible for injecting active power from the primary energy source into the grid and, simultaneously, whenever possible, to perform some ancillary service, such as reactive compensation, voltage support, and harmonic mitigation to improve the system power quality and efficiency.

Figure 7.2 shows the smart inverter instrument in which the previous control scheme and the signal reference generator can be impressed. For this instrumentation, the distribution provider should send the active and reactive power references, as shown in Figures 7.9 and 7.10. Handling both commands, the smart inverter can inject the active power into the grid, and simultaneously regulate the reactive power flow, either by reactive compensation or voltage support purposes, enhancing the power factor at the point of common coupling and the overall voltage profile of the distribution network.

7.4.1 Harmonic Circulation

To regulate and reduce the circulation of high-order harmonics through the grid, which may influence other devices or even trigger resonances, most of the distribution system operators are adopting international standards or defining their own. Conventionally, harmonics have been compensated using tuned passive filters and active filters installed along the utility structure itself. Recently, there was an appeal to perform harmonic compensation by the DG consumers themselves.

One of the first developed standards about harmonics was the IEEE 929-2000, mainly based on PV systems. This standard was replaced by the IEEE 1547, which deals with all types of generation. IEEE 1547 has been currently revised to attend novel developments in the distributed generation field. The IEC 61727-2004 is another standard for PV systems network. The EN 50438 is the European standard for micro generation connected in parallel with the network. The VDE-AR-N 4105:2011 (VDE) is the German standards for low-voltage network, while the BDEW-2008 is for medium-/high-voltage network. In summary, similar standards have been developed around the world, some of them shown in Table 7.1.

TABLE 7.1
General Standards for Distributed Generation

	System Type	Voltage	Power
IEEE 1547	Sync., and async. Machines, and inverter	Primary/secondary distribution voltage	≤10 MVA
VDE-AR-N 4105	All	LV (<kV)	<100 kVA
BDEW 2008	All	MV (1–66 kV)	No limit
IEC 61727	Photovoltaic	LV	≤10 kVA

TABLE 7.2
Maximum Harmonic Current Distortion in Percent of Current

Individual Odd Harmonics	$h < 11$	$11 \leq h < 17$	$17 \leq h < 23$	$23 \leq h < 35$	$35 \leq h$	THD
max. (%)	4.0	2.0	1.5	0.6	0.3	5.0

In order to control harmonic propagation in networks, all the grid-connected inverters must comply with the current standard, for example, the one suggested in Table 7.2, for the maximum limits of odd harmonic current allowed to a grid-connected inverter on IEEE 1547. The even harmonic currents are limited to 25% of the odd harmonic limits of Table 7.2.

Thus, on the basis of IEEE 1547 Std., a harmonic monitoring or a power quality monitoring, like that shown in Figure 7.3, may be used with the automonitoring DG instrumentation of Figure 7.4 to perform selective power quality improvements on the system. Considering the control scheme and signal reference generator of Sections 7.2 and 7.3, the automonitoring instrumentation should provide the in-phase and quadrature signal references, as shown in Figures 7.9 and 7.10, and the nonfundamental frequency term, as shown in Figure 7.11. Handling these reference terms, the automonitoring inverter instrumentation can inject active power into the grid, regulate the reactive power flow either by reactive compensation or voltage support purpose, and mitigate harmonic levels. Such instrumentation could be managed in an optimal fashion as in Reference 14.

7.4.2 Power Factor and THD Concerns in DG

So far, we have seen that the power quality analysis is fundamental to the recent development in distribution of power systems based on DGs, for which power factor and harmonic analysis are important quantifiers. Both have been discussed in Chapter 6. However, they have to be carefully applied to DGs, since both power quality analyzers are only fraction quantities and they are usually applied as a merit quantifier for power quality standards. Ideally, the power factor should be unitary, while the total harmonic distortion (THD) should be null.

The power factor value depends on the active power and the nonactive power (as in Equation 6.50), while the THD value depends on the fundamental signal magnitude and the nonfundamental signal magnitude (as in Equation 6.51). When DGs operate at unity power factor, that is, only injecting active power in the grid, they contribute to a performance increase on the grid's operation, since the utility does not need to supply reactive power to compensate any reactive needs of the DG with improved THD. Using a layer of communication with the DG and the utility is possible to also implement programmable reactive power from the DG, but there are no accepted standards for such functions yet accepted by the industry. Such reduction in the power factor value or an increase in the THD value occurs because part of the active power drained by the loads supplied by the utility is now provided locally by the DG.

To exemplify that effect, Figure 7.13 shows the PV-based inverter instrumentation of Figure 7.1, simulated by a current source in parallel with a nonlinear load. The DG injects an in-phase sinusoidal current of 10 A_{pk} after 0.28 s (see Figure 7.14). Observe that the power factor and THD at the grid side before the instant 0.28 s represent the

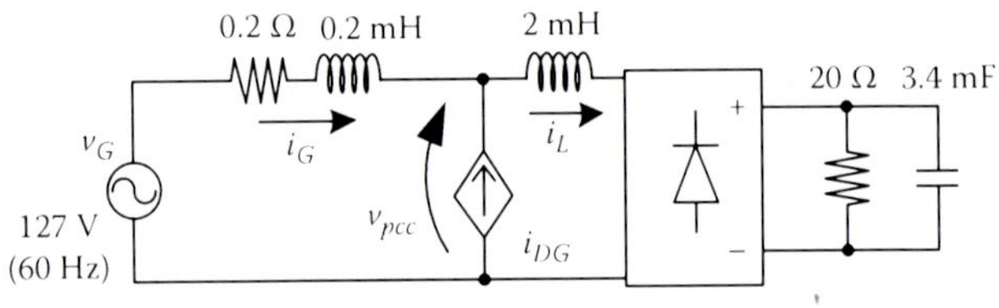

FIGURE 7.13 Electrical circuit with DG generating unity power factor.

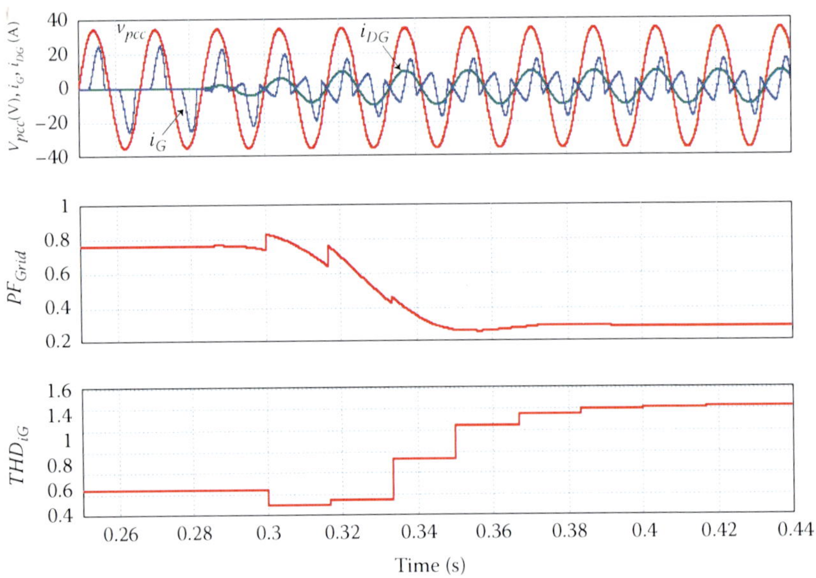

FIGURE 7.14 Power factor reduction and THD increase due to DG active power generation.

load power quality analyzers, respectively, 0.77% and 64.4%. After 0.28 s, during the DG active power injection, the power factor is reduced to 0.29, and the THD increased to 140%, even with a DG sinusoidal current and in-phase with the voltage.

7.5 DISTRIBUTED GENERATOR BASED ON INSTRUMENTATION: CASE STUDIES

This section shows the operation of the three instrumentations based on DG through MATLAB®/Simulink® simulations [15]: PV-based inverter, smart inverter, and auto-monitoring DG. The following sections enlighten the reader about the use of distribution generation systems as an instrumentation aiming to generate power from renewables and, simultaneously, to assist the power quality of the system performing by either voltage support or current compensation.

7.5.1 Instrumentation of DG-Based PV

The PV-based inverter instrumentation of Figure 7.1 aims to extract the maximum power from PV modules and inject it into the grid. Figure 7.5 shows the power electronic part of this instrumentation, in which the DC converter stage, i.e., a boost converter, by MPPT technique controls quickly the output current of PV modules injecting power into the DC link. For the sake of the DC link voltage regulation, the power balance at the DC link's capacitor must be guaranteed. Therefore, the full-bridge inverter must inject into the grid the same amount of power extracted from the PV modules.

Figure 7.1 illustrates the MPPT technique combined with the DC converter to regulate the maximum power tracking and provide at the same time the maximum power value reference (P^*_{mppt}) to the DG's data acquisition. Thus, applying such power reference to the signal generator of Figure 7.9, and therefore to the control scheme of Figure 7.6, the PV-based inverter instrumentation may be implemented similarly as in Figure 7.13. Figure 7.15 shows the simulation results where the rise time and the active power/current overshoot are set by the generator's PI controller of the in-phase signal reference of Figure 7.9. Observe that the DC link's capacitor is usually designed to support a significant range of voltage deviation, allowing a slower dynamic response of the inverter than the DC converter. Of course, the time response of the active power closed loop must be sped up during application of the feed-forward technique.

7.5.2 Instrumentation of DG-Based Smart Inverter: Voltage-Based DG

The smart inverter instrumentation of Figure 7.2 must inject active power and at the same time regulate the reactive power flow to the grid following commands from the distribution provider, i.e., the central controller (CC). Such commands are the active (P^*_{cc}) and reactive (Q^*_{cc}) power references applied to the signal reference generator shown in Figures 7.9 and 7.10, respectively.

The electronic power structure of this instrumentation is shown in Figure 7.5, and its control scheme is shown in Figure 7.7. As discussed in Section 7.3.1, the active and reactive power flow control of this voltage-based DG depends on the *R*/*X* ratio of

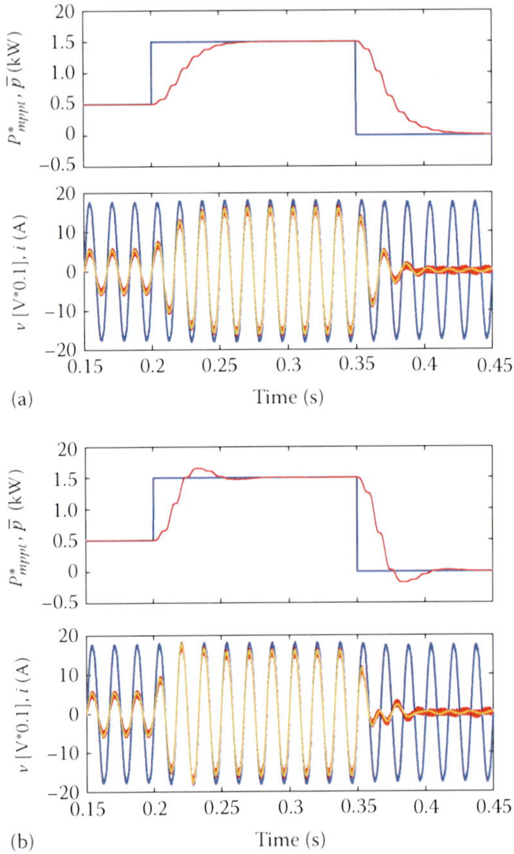

FIGURE 7.15 Simulation results of a PV-based inverter instrumentation with different settings of the PI controller from the in-phase signal generator: (a) $K_P = 0$, $K_I = 0.014$ and (b) $K_P = 0$, $K_I = 0.007$.

the power line impedances. The simulation results shown in Figure 7.16 were generated considering distinct line impedance values, such as high R/X ratio and low R/X ratio. Note that the power/current response changes with the line impedance value, indicating that this system requires an adjustable or autosetting controller for power closed loops. Furthermore, observe that in this case, the DG will control only active and reactive powers, while each harmonic current order generated by load will flow through the least impedance path.

7.5.3 Instrumentation of DG-Based Automonitoring Inverter: Current-Based DG

Finally, the automonitoring-based DG in Figure 7.4 must inject the available active power from the primary energy side into the grid, and particularly operate as an active power filter compensating current disturbances, such as reactive and harmonic circulation. Again, the electronic power side is shown in Figure 7.5, the control scheme is shown in Figure 7.6, and the signal reference is generated as described in Section 7.3.

The automonitoring instrumentation, as the one shown in Figure 7.3, should measure voltage and current at the point of common coupling and by the power quality standards or user requirements define the compensation set points (P^*_{am}, Q^*_{am}, $D^*_{\|n\_am}$, $D^*_{\perp n\_am}$). To demonstrate the harmonic level control, or harmonic compensation, the automonitoring instrumentation is connected in parallel with a nonlinear load, as in Figure 7.13, and controlled through the previous power set point in order to ensure unity power factor at the grid side.

Figure 7.17 shows the simulation results of an automonitoring instrumentation. Figure 7.16a shows the $\bar{p}$ and $D_{\|3}$ dynamic responses and steady-state performance by tracking their corresponding reference. Variables $\bar{q}$ and $D_{\perp 3}$ present similar responses, but they are not shown here. Figure 7.17b shows the current of a nonlinear

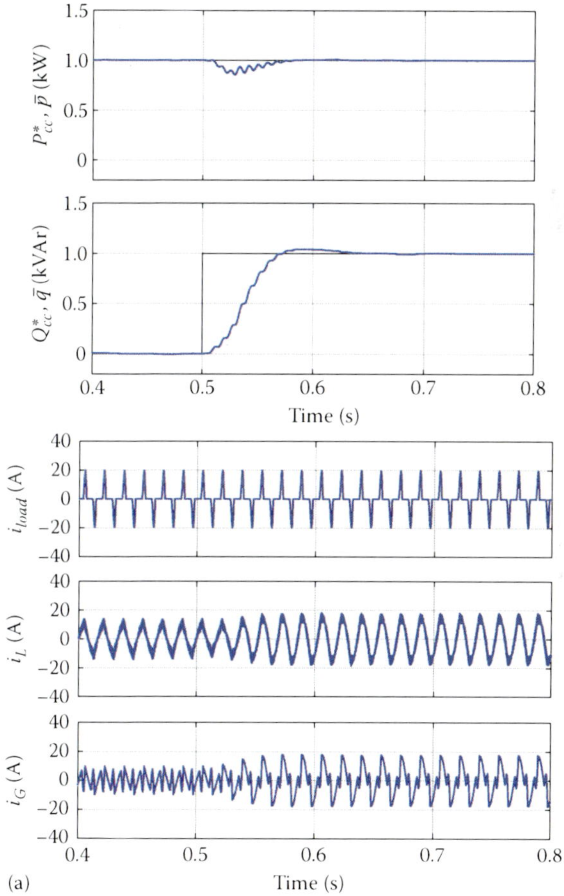

FIGURE 7.16 Simulation results of *smart inverter* instrument with different *R*/*X* ratios of the power line impedances: (a) high *R*/*X* ratio—$R/X \approx 5$. *(Continued)*

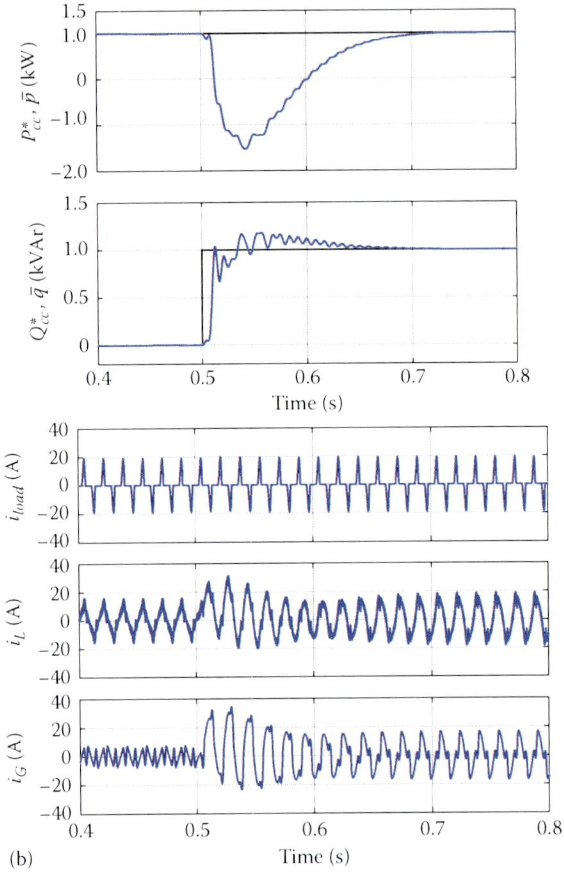

(b)

FIGURE 7.16 (*Continued*) Simulation results of *smart inverter* instrument with different *R*/*X* ratios of the power line impedances: (b) low *R*/*X* ratio—*R*/*X* ≈ 0.15.

load, the DG's output current, and the grid side current and voltage. After 0.2 s, the DG generates 1 kW, in which the load absorbs part of it and the remaining part flows into the grid. Note that after 0.2 s the current is shifted by 180° from voltage, indicating power flow from the DG into grid. The harmonics are controlled by an almost sinusoidal current provided by DG. After 0.35 s, the harmonic compensation is released, cleaning the grid current. The selective generator of the nonfundamental signal reference provides i^*_{h3}, i^*_{h5}, i^*_{h7}. Figure 7.17c shows $I_{\parallel 3}$ and $I_{\perp 3}$ of grid side being reduced to zero and consequentially the THD of the grid current decreasing to 14.8%. Such compensation could better reach a performance index whenever wider harmonic orders are taken into account. Moreover, there are many other options to control DGs connected to the grid, and here control schemes based on natural frame (*abc*) focused on power converters in AC microgrids are described.

$$C_i(s) = \frac{K_P s + K_I}{s}$$

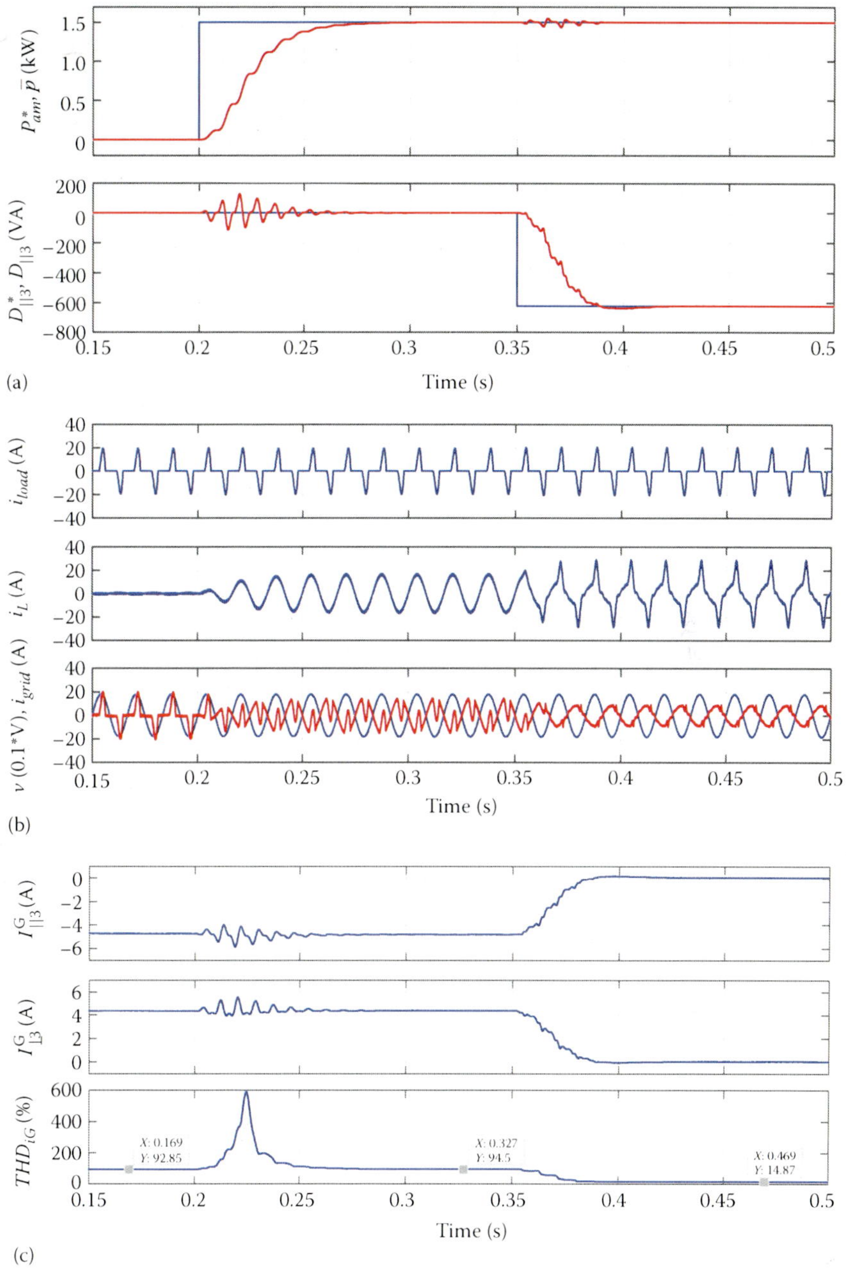

FIGURE 7.17 Simulation results of an automonitoring DG instrumentation: (a) $\bar{p}$ and $D_{\|3}$ tracking their corresponding references; (b) load, DG, and grid currents and voltage; and (c) $I_{\|3}$, $I_{\perp 3}$, and THD of grid side.

EXERCISES

7.1 Implement using a power electronics simulation software, e.g., MATLAB/Simulink or PSIM, a voltage source converter (VSC) operating through three-level PWM and controlled as a current source. Such circuit with its corresponding parameters is shown in Figure 7.18.

a. Design the PI's gains based on its open-loop transfer function. Figure 7.19 shows the closed-loop control scheme. A crossover frequency of 1 kHz (using a switching frequency of 12 kHz) and a phase margin of 60° are suggested.

where

- $G_i(s)$ is the output inductor transfer function between inductor current and pulsed input voltage:

$$G_i(s) = \frac{i(s)}{v_p(s)} = \frac{1}{Ls + R} \tag{7.14}$$

- $G_{conv}(s)$ is the converter transfer function between pulsed voltage and duty cycle:

$$G_{conv}(s) = \frac{v_p(s)}{d(s)} = 2 \cdot V_{CC} \tag{7.15}$$

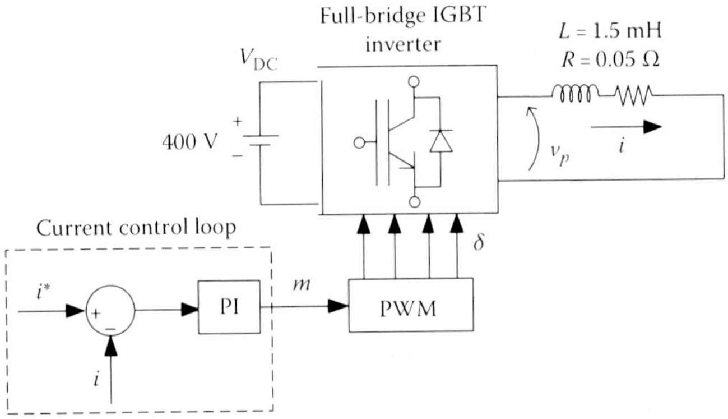

FIGURE 7.18 VSC operating through PWM controlled as current source.

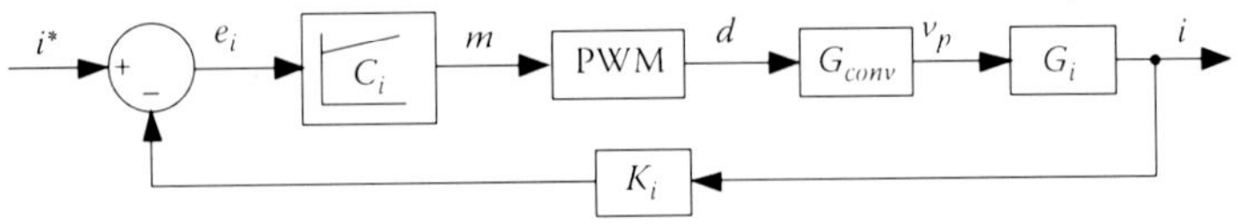

FIGURE 7.19 Closed-loop control scheme of controlled current source.

- $PWM(s)$ is the PWM transfer function between duty cycle and PWM's modulator:

$$PWM(s) = \frac{d(s)}{m(s)} = \frac{1}{c_{pk}} \tag{7.16}$$

- $C_i(s)$ is the PI transfer function between PWM's modulator and current error:

$$C_i(s) = \frac{K_P s + K_I}{s} \tag{7.17}$$

- K_i is the current transducer gain.

b. Evaluate the designed controller applying a sinusoidal signal as a current reference, i^*, with the following frequencies: 180 Hz, 300 Hz, 420 Hz, and 660 Hz. Compare the inductor output current with its reference, measuring the deviation of magnitude and phase between them.

7.2 Implement using a power electronics simulation software, e.g., MATLAB/Simulink or PSIM, the previous current source PWM-VSC connected to the grid, as shown in Figure 7.20. The parameters are shown in the figure itself.

a. Implement a power generation strategy for MPPT-PV based DG in the signal reference generator sub-system as an open-loop active strategy. Use the same PI's gains designed previously, and (7.18) to generate the signal reference, such that P^* is the active power reference provided by MPPT technique, V_{pcc} is the rms value of PCC voltage, while v_{pcc} is its instantaneous measured value.

Evaluate such open-loop active power strategy under sinusoidal and nonsinusoidal grid voltage.

$$i^* = \frac{P^*}{V_{pcc}^2} \cdot v_{pcc} \tag{7.18}$$

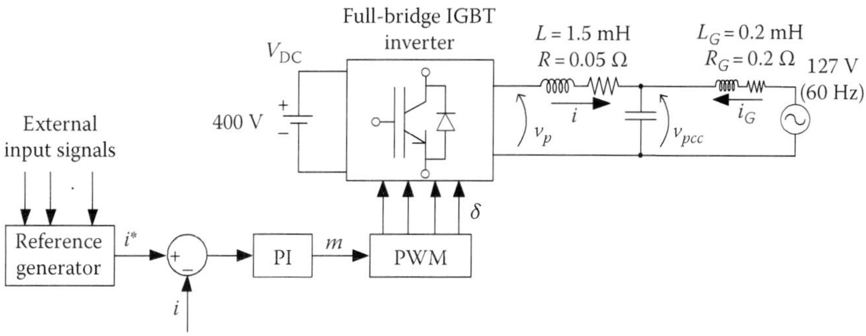

FIGURE 7.20 VSC operating through PWM controlled as current source connected to the grid.

b. Based on Figure 7.9, implement a closed control loop to regulate the active power injection. Again, P^* is provided by MPPT technique.
   Evaluate such closed-loop active power strategy under sinusoidal and non-sinusoidal grid voltage.
c. Based on Figure 7.10, implement a closed control loop to regulate the reactive power injection. Q^* is provided by the central controller for a DG-based smart inverter. Such control loop may have the same PI's gains of previous problem regardless of whether quantities have the same order of magnitude or not.

7.3 Implement using a power electronics simulation software, e.g., MATLAB/Simulink or PSIM, a voltage source converter (VSC) operating through three-level PWM and controlled as a voltage source. Such circuit with its corresponding parameters is shown in Figure 7.21.

a. Design the PI's gains based on its open-loop transfer function. Figure 7.22 shows the closed-loop control scheme. A crossover frequency of 600 Hz (using a switching frequency of 12 kHz) and phase margin of 60° are suggested.
   where
   - G_v is the output capacitor transfer function between capacitor current and capacitor voltage (i.e., capacitor impedance):

$$G_v(s) = \frac{v_{pcc}(s)}{i_c(s)} = \frac{1}{Cs} \tag{7.19}$$

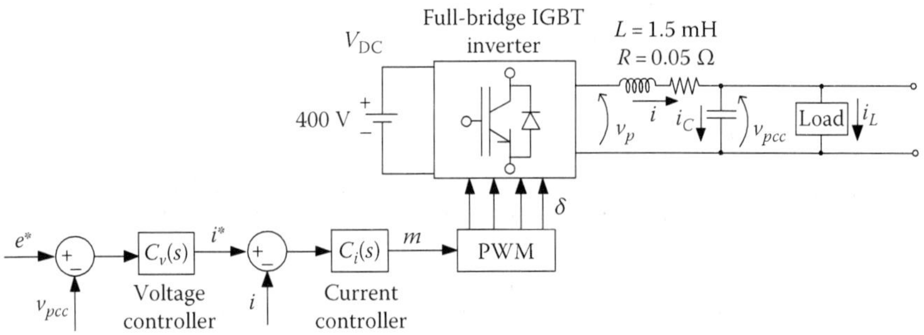

FIGURE 7.21 VSC operating through PWM controlled as voltage source.

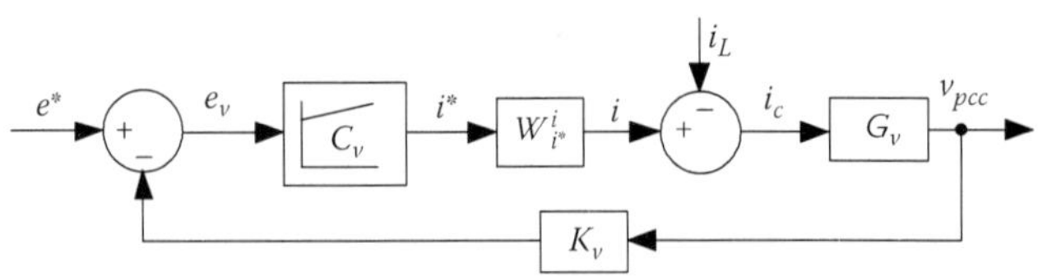

FIGURE 7.22 Closed-loop control scheme of controlled voltage source.

- $W_{i^*}^i$ is the inner current control loop which can be modeled as low-pass filter tuned at the crossover frequency (ω_c) of current control, and the static gain given by current sensor gain:

$$W_{i^*}^i(s) = \frac{i(s)}{i^*(s)} = \frac{1/K_i}{\dfrac{1}{s \cdot \omega c} + 1} \tag{7.20}$$

- C_v is the voltage controller transfer function:

$$C_v(s) = \frac{K_P s + K_I}{s} \tag{7.21}$$

- K_v is the voltage transducer gain.

b. Evaluate the designed controller applying it to supply a heavy linear load, no load, and a heavy nonlinear load.

REFERENCES

1. U.S. Energy Information Administration, International energy outlook 2016, DOE/EIA-0484(2016), Washington, DC, 2016.
2. Farret, F.A. and Simoes, M.G., *Integration of Alternative Sources of Energy*, 2nd edn., IEEE-John Wiley & Sons, New York, 2017.
3. Comitato Elettrotecnico Italiano, Reference technical rules for the connection of active and passive users to the LV electrical utilities, *Norma Italiana CEI*, Milano, Italy, 0-21, 2016.
4. Pacific Gas and Electric Company, Electric Rule No. 21, Generating facility interconnections, Wilseyville, CA, Section H.2.i, 2015.
5. Rocabert, J., Luna, A., Blaabjerg, F., and Rodriguez, P., Control of power converters in AC microgrids, *IEEE Transaction on Power Electronics*, 27(11), 4734–4749, November 2012.
6. Buso, S. and Mattavelli, P., *Digital Control in Power Electronics*, 2nd edn., Morgan & Claypool Publisher, San Rafael, CA, 2016.
7. Chandorkar, M.C., Divan, D.M., and Adapa, R., Control of parallel connected inverters in standalone ac supply systems, *IEEE Transaction on Industry Applications*, 29(1), 136–143, January/February 1993.
8. Debrabandere, K., Bolsens, B., Van den Keybus, J., Woyte, A., Driesen, J., and Belmans, R., A voltage and frequency droop control method for parallel inverters, *IEEE Transaction on Power Electronics*, 22(4), 1107–1115, July 2007.
9. Zhong, Q.C. and Weiss, G., Synchronverters: Inverters that mimic synchronous generators, *IEEE Transactions on Industrial Electronics*, 58, 1259–1267, April 2011.
10. Johnson, B.B., Dhople, S.V., Hamadeh, A.O., and Krein, P.T., Synchronization of parallel single-phase inverters with virtual oscillator control, *IEEE Transactions on Power Electronics*, 29(11), 6124–6138, November 2014.
11. Kwon, J., Yoon, S., and Choi, S., Indirect current control for seamless transfer of three-phase utility interactive inverters, *IEEE Transactions on Power Electronics*, 27(2), 773–781, February 2012.
12. De Brito, M.A.G., Galotto, L., Sampaio, L.P., De Azevedo Melo, G., and Canesin, C.A., Evaluation of the main MPPT techniques for photovoltaic applications, *IEEE Transactions on Industrial Electronics*, 60(3), 1156–1167, March 2013.

13. Czarnecki, L.S. and Pearce, S.E., Currents' Physical Components (CPC)—Based comparison of compensation goals in systems with nonsinusoidal voltags and currents (Invited Lecture), International School on Nonsinusoidal Currents and Compensarion (ISNCC 2010), Łagow, Poland, June 20–23, 2010.
14. Brandao, D.I., Guillardi, H., Morales-Paredes, H.K., Marafão, F.P., and Pomilio, J.A., Optimized compensation of unwanted current terms by AC power converters under generic voltage conditions, *IEEE Transactions on Industrial Electronics*, 63(12), 7743–7753, July 2016.
15. Simoes, M.G. and Farret, F.A., *Modeling Power Electronics and Interfacing Energy Conversion Systems*, IEEE/Wiley, Hoboken, NJ, 2016, ISBN: 978-1-119-05826-7, 2016.

8 Fuzzy Logic and Neural Networks for Distributed Generation Instrumentation

8.1 INTRODUCTION

In the twentieth century, particularly after the advent of computers and advances in mathematical control theory, attempts were made for augmenting the intelligence of computer software. There were several studies and developments for furthering capabilities of logic, using models of uncertainty with adaptive learning algorithms. The initial developments in neural networks were made during the 1950s. Also, a very radical and fruitful idea was initiated by Lotfi Zadeh in 1965 with his publication of the paper "Fuzzy Sets" [1]. When an engineering system becomes very complex, the lack of ability to measure or to evaluate its features is usually associated with a lack of definition of very precise modeling. Also, many other uncertainties and incorporation of human expertise make it almost impossible to explore a very precise model for a complex real-life system. Either by using neural networks for dealing with complex and huge data or with a mathematical approach that bundles thinking, vagueness, and imprecision, the field of artificial intelligence became very solid with fuzzy logic and neural networks, and a strong foundation for what currently is considered in the twenty-first century smart control, smart modeling, intelligent behavior, and artificial intelligence. This chapter discusses some basics of fuzzy logic and neural networks, with a few successful applications in the area of energy systems, power electronics, power systems, and power quality. These systems have the following characteristics:

- Parameter variation that can be compensated with designer judgment
- Processes that can be modeled linguistically but not mathematically
- Setting with the aim to improve efficiency as a matter of operator judgment
- When the system depends on operator skills and attention
- Whenever one process parameter affects another process parameter
- Effects that cannot be attained by separate Proportional-integral-derivative (PID) control
- Whenever a fuzzy controller can be used as an advisor to the human operator
- Data-intensive modeling (use of parametric rules)
- Parameter variation: temperature, density, impedance
- Nonlinearities, dead-band, and time delay
- Cross-dependence of input and output variables

There are typically three frameworks, i.e., three paradigms, that can be used for energy conversion systems, with artificial-intelligence-based computation: (1) a function approximation or input/output mapping, (2) a negative feedback control, and (3) a system optimization. The first one is the construction of a model using either heuristics or numerical data; the second one is the comparison of a set point with an output that can be either measured or estimated with a function that minimizes the error of the set point with the output; and the third one is a search for parameters and system conditions that will maximize or minimize a given function. Fuzzy logic and neural network techniques make the implementation of those three paradigms robust and very reliable for practical applications.

8.1.1 Fuzzy Logic Systems

Every design starts with the process of thinking, i.e., a mental creation, where people will use their linguistic formulation, with their analysis and logical statements about their ideas. Then, vagueness and imprecision are considered as empirical phenomena. Scientists and engineers try to remove most of the vagueness and imprecision of the world by making accurate mathematical formulation using laws of physics, chemistry, and nature in general. Sometimes, it is possible to have precise mathematical models, with strong constraints on nonidealities, parameter variation, and nonlinear behavior. But it is very common that precise models are very difficult to build using mathematical formulation.

Fuzzy control has a lot of advantages when used for optimization of alternative and renewable energy systems. The parametric fuzzy algorithm is inherently adaptive because the coefficients can be altered for system tuning. Thus, a real-time adaptive implementation of the parametric approach is feasible by dynamically changing the linear coefficients using a recursive least-square algorithm repeatedly on a recurrent basis. Adaptive versions of the rule-based approach could be implemented by changing the rule weights (Degree of Support), or the membership functions are recurrently possible. The disadvantage of the parametric fuzzy approach is the loss of the linguistic formulation of output consequents, sometimes important for industrial plant/process control environment.

Processes that require learning, or probabilistic reasoning because of uncertainty and systems that are ill-modeled because those are associated with imprecision, or are described by ad-hoc models and descriptions can be approached by Fuzzy logic. Zadeh defined fuzzy logic as "computing with words." Such a methodology has the following features:

- Are applicable to nonlinear systems
- Have the ability to deal with nonlinearity
- Follow more human-like reasoning paths than classical methods
- Utilize self-learning
- Utilize yet-to-be-proven theorems
- Robust in the presence of noise, errors, and imperfect data

Fuzzy and neuro-fuzzy techniques became efficient tools in modeling and control applications. There are several benefits in optimizing cost-effectiveness because fuzzy logic is a methodology for handling inexact, imprecise, qualitative, fuzzy, verbal information in a systematic and rigorous way, such as temperature, wind speed, humidity, and pressure. A neuro-fuzzy controller generates, or tunes, the rules or membership functions of a fuzzy controller with an artificial neural network approach.

8.1.2 Neural Network Systems

Artificial neural networks (ANNs) consist of a computing paradigm usually inspired by biological neural networks, such as the ones modeling the central nervous systems of animals, or the brain neurons, spine connections, or cerebellar/cortex models. They are powerful algorithms capable of estimating or approximating functions that can depend on a large number of inputs, where their interaction is unknown. ANNs are formulated as systems of interconnected "neurons," mathematically represented by connections with numeric weights, which can be tuned based on data, consequently making neural networks adaptive to inputs with learning capabilities.

Applications of neural networks for alternative and renewable energy systems show a lot of advantages. The reason is that installation costs are high, and optimization on the sizing and design is dependent on the availability of the renewable energy, which by its nature is intermittent. There are efficiency constraints, and it becomes important to optimize the efficiency of electrical power transfer, even for the sake of relatively small incremental gains, to amortize installation costs within the shortest possible time.

8.2 APPLICATIONS OF ARTIFICIAL NEURAL NETWORK IN INDUSTRIAL SYSTEMS, ENERGY CONVERSION, AND POWER SYSTEMS

ANNs have a wide range of applications that span across science, art, engineering, data analysis, and many diverse fields. It has many advantages over conventional modeling approaches, since its special structure makes it a suitable alternative to classical statistical modeling techniques, provided the data are available for training the network. Several optimization problems in industrial plants, system identification, and complex, uncertain nonlinear systems can be modeled and controlled by ANN methodologies. ANN can perform nonlinear modeling and filtering of the system data and coupled nonlinear relations between independent and dependent variables without their dynamic equations. It is also a cost-effective and reliable approach for condition monitoring, where data related to the condition of the system can be classified and trained for data analysis. ANN can be applied in examining condition-based maintenance, detecting anomalies, and identifying faults.

A typical problem in industrial systems is the calibration of sensors and improving the downtime of the process because of sensor failures. ANN has been considered as an outstanding solution for sensor fault monitoring plus related complexities in modeling and control of nonlinear systems. The real data obtained from an industrial system can be used to develop a simple ANN model with very high prediction accuracy. In control design, a neural network may directly implement the controller; in such a case, the NN will be trained as a controller based on some specified criteria.

It is also possible to design a conventional controller for an available ANN model where the data of factories and plants may include noise or some corruption, or they are inaccurate or incomplete because of faulty sensors. Particularly in aged and aging systems where maintenance is not frequent, such as rural areas with older power distribution, the ANN has the capability to work considerably well for such data that might be noisy or incomplete. ANN can also learn from incomplete and noisy data. The development of ANN requires less formal scientific personnel, and it is not necessary to have a professional with advanced math or statistical knowledge. If the data sets and appropriate software are available, newcomers in the field can handle the neural network design and implementation process. It also has the capability of dealing with stochastic variations of the scheduled operating point with online processing and classification.

In addition to applications of ANN to industrial systems, it has many general advantages such as simple processing elements, fast processing time, easy training process, and high computational speed. Other characteristics of ANN include capturing any relation and association, exploring regularities within a set of patterns, and having the capability to be used for very large number and diversity of data and variables. It provides a high degree of adaptive interconnections between elements and can be used where the relations between different parameters of the system are difficult to uncover with conventional approaches. ANN is not restricted by a variety of assumptions such as linearity, normality, and variable independence, as many conventional techniques are. It can even generalize the situations for which it has not been previously trained. It is believed that the ability of ANN to model different kinds of industrial systems in a variety of applications can decrease the required time on model development, thus leading to a better performance compared with conventional techniques.

8.2.1 Applications of Fuzzy Logic and Neural Networks in Distributed Generation Systems

Energy conversion systems have two features that define requirements for advanced control systems: (1) unconstrained energy systems and (2) constrained energy systems. In fact, any energy source is constrained because there are only finite energy resources in nature. However, several constrained systems are simplified to be unconstrained; this prevents a very complex modeling and decision-making in the balance of the system. For example, a large power system will have several large power plants, supplying electrical power to a distribution system. The distribution company will sell such power and will care about its reliability and quality. The users will just pay the fees and tariffs, believing that such electrical power is always available and the electrical energy supply is not bounded. This is a simplification, but it works well for the old paradigm of centralized power plants.

Constrained energy systems have a finite energy and very often a finite maximum power (which means finite maximum derivative of energy). There are two types of constrained systems, the ones based on fossil fuel (gas, coal, oil, hydrogen), where a certain amount of the input fuel will convert energy through a thermodynamic cycle (usually Rankine or Brayton, or a fuel cell), with internal losses and maximum conversion efficiency; and the ones based on renewable energy (wind, solar, tidal, geothermal). Such renewable energy systems can be sustainable as long as the amount of energy conversion is less than the recovery of such energy by the environment. Renewable energy sources are constrained. Therefore, their power, or their derivative of specific energy should be optimized. This characteristic means that a convex function defines the amount of power conversion dependent on the usage. For example, a wind turbine will have a peak power that depends on the tip-speed ratio and the output load; a photovoltaic system will have a peak power that depends on the solar irradiation, temperature, and the equivalent impedance across its terminals; a hydropower system will have a turbine control dependent on the mechanical shaft energy converted by an electrical generator. Therefore, the system performance depends on finding the peak power operating point for such constrained renewable energy system.

The optimal system performance depends on the coherent operation of components; for example, an engineer will understand that a compressor with heat exchangers and a throttle will make up a heat pump. But the operation of a thermodynamic system such as a heat pump requires information, measurement, and control of the compressor, which depends on refrigerant pressure; a temperature measurement, which is taken by a controller to evaluate how much heat is required; and a very intricate understanding of physics, chemistry, electrical engineering, and mechanical engineering to make such a heat pump operate at its maximum efficiency. Therefore, several factors must be taken into consideration, and efficient energy conversion for electrical power systems can be advanced by using fuzzy logic or neural networks.

8.3 FUZZY LOGIC AND NEURAL NETWORK CONTROLLER DESIGN

Classic and digital controls have several methods for designing controllers used in dynamic systems. All of them require a mathematical formulation (transfer function or a state-space formulation) for the system controller to be designed. Some of those methods are as follows:

- *Proportional-integral-derivative (PID) control*: Over 90% of the controllers in operation at present are PID controllers (or at least some form of PID controllers like a P or PI or an I+P controller). This approach is often viewed as simple, reliable, and easy to understand. Sometimes, fuzzy controllers are used to replace PID, but it is not clear yet if there are real advantages.
- *Classical control*: Lead-lag compensation, Bode method, Nyquist method, root-locus design.
- *State-space methods*: State feedback, observers.
- *Optimal control*: Linear quadratic regulator, use of Pontryagin's minimum principle or dynamic programming.
- *Robust control*: H2 or H∞ methods, quantitative feedback theory, loop shaping.
- *Nonlinear methods*: Feedback linearization, Lyapunov redesign, sliding mode control, backstepping.

- *Adaptive control*: Model reference adaptive control, self-tuning regulators, nonlinear adaptive control.
- *Stochastic control*: Minimum variance control, linear quadratic Gaussian (LQG) control, stochastic adaptive control.
- *Discrete event systems*: Petri nets, supervisory control, infinitesimal perturbation analysis.

These control approaches have a variety of ways to utilize information from mathematical models. Sometimes, they do not consider certain heuristic information early in the design process, but use heuristics when the controller is implemented to and tuned (tuning is invariably needed since the model used for the controller development is not perfectly accurate). Unfortunately, when using some conventional control approaches, engineers may become somewhat removed from the control problem itself and become more involved in their mathematics. Such a heavily oriented mathematical approach may achieve the development of unfeasible control laws or conditions that are not possible to meet in real life. Sometimes, conventional control designers ignore useful heuristics because such ones not fit into the proper mathematical framework. Fuzzy logic and neural network approaches go a long way in real-life understanding instead of heavily math-oriented control methodologies. Fuzzy logic and neural networks allow heuristics and learning from past case studies or numerical data, retrofitting an excellent performance controller, which most of the time will be superb when compared with heavily mathematical control design approaches.

For example, an induction motor has a very complicated instantaneous model based on a decoupled *d-q* equations modeling, with real-time trigonometric Park and Clarke transformations, where an instantaneous inverse model is resolved mathematically to control torque and flux with virtual *d-q* currents. Such controller response is reverse-calculated in real time to generate the pulse-width modulation of all transistors in a power electronics inverter that commands the induction machine. It seems that fuzzy logic and neural networks are natural solutions for induction motor speed control, optimization of flux, and signal processing of nonlinear functions, i.e., the three areas described earlier. A fuzzy logic speed control can be designed for an induction motor or a DC motor drive (speed control), as depicted in Figure 8.1, which shows the input signals for the fuzzy logic control: *e* (error) and *ce* (change in error) and the output is *du* (a derivative of output control). Figure 8.2 shows the fuzzy sets and their corresponding membership functions; fuzzy sets are linguistically defined

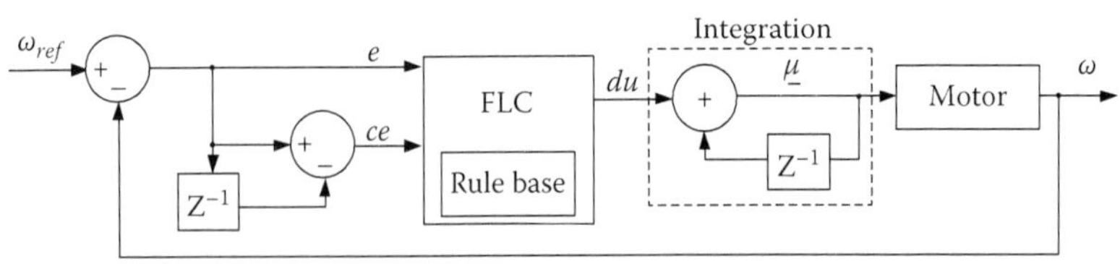

FIGURE 8.1 Fuzzy logic speed control system showing the input of error and change in error with output of the controller through accumulative summation to feedback the command for the electric motor.

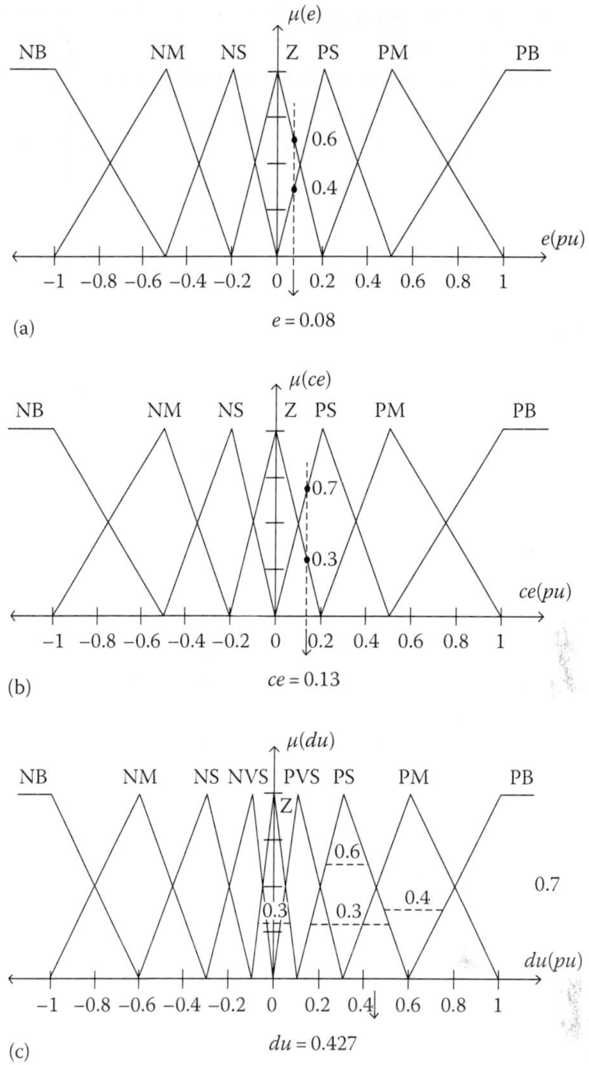

FIGURE 8.2 Fuzzy logic control membership functions with their associated linguistic variables: (a) error, (b) change in error, and (c) change in output.

in Tables 8.1 and 8.2. Such fuzzy controller can also be used in similar PI-like control loops.

The range of fuzzy variables is called "universe of discourse." In Figures 8.1 and 8.2, the universe of discourse is expressed in per unit. Such normalization allows the controller to be fine-tuned with scaling gains for *e*, *ce*, and *du*. Assuming seven membership functions for each E_{pu} and CE_{pu}, there are a total of 49 possible rules. The output DU_{pu} is considered to have nine membership functions. Table 8.2 depicts

TABLE 8.1
Fuzzy Linguistic Variables for Fuzzy Control

NB	= Negative Big
NM	= Negative Medium
NS	= Negative Small
NVS	= Negative Very Small
ZE	= Zero
PVS	= Positive Very Small
PS	= Positive Small
PM	= Positive Medium
PB	= Positive Big

TABLE 8.2
Fuzzy Controller for a Motor Drive Speed Control Loop

		Error (E_{pu})						
		NB	**NM**	**NS**	**ZE**	**PS**	**PM**	**PB**
Change in error (CE_{pu})	**NB**	NVB	NVB	NVB	NB	NM	NS	ZE
	NM	NVB	NVB	NB	NM	NS	ZE	PS
	NS	NVB	NB	NM	NVS	ZE	PS	PM
	ZE	NB	NM	NVS	ZE	PVS	PM	PB
	PS	NM	NS	ZE	PVS	PM	PB	PVB
	PM	NS	ZE	PS	PM	PB	PVB	PVB
	PB	ZE	PS	PM	PB	PVB	PVB	PVB

a fuzzy logic controller with 49 possible combinations, which can be used for speed control of a motor drive, where the derivative of the output must be accumulated (or integrated) to build up the torque control command. The top row and the left column indicate the fuzzy sets for the variables E_{pu} and CE_{pu}, respectively, and each cell in that table gives the output variable DU_{pu} for an AND operation of those two inputs.

In several fuzzy logic control systems, some rules may not appear, either because they have been determined in situations that do not exist, or because the output should not change, and if there is no rule, the output does not change. There are some other ways to implement a fuzzy PI or a fuzzy PD, but this rule table mostly works well as long as good scaling factors are retrofit for a particular application; there are some previous simulation studies and some trial-and-error tweaking on the controller. For example, a typical rule in the matrix is like:

$$\textbf{\textit{IF}}\ E_{pu} = PS\ \textbf{\textit{AND}}\ CE_{pu} = PM\ \textbf{\textit{THEN}}\ DU_{pu} = PB$$

General considerations in designing such a fuzzy controller are expressed by the following meta-rules:

1. If both E_{pu} and CE_{pu} are zero, then maintain the present control setting, i.e., $DU_{pu}=0$.
2. If E_{pu} is not really zero but it is approaching zero at a good rate, i.e., CE_{pu} has a good polarity and value, then maintain the present control setting, i.e., $DU_{pu}=0$.
3. If E_{pu} increases, then change the control signal DU_{pu} to bring the plant output back to the conditions where error should be zero, i.e., reverse the trend and make signal DU_{pu} dependent on the magnitude and sign of E_{pu} and CE_{pu} in order to force E_{pu} toward zero.

The rule matrix and membership functions of the variables are associated with the heuristics of general control rule operation; such heuristics would be the same way, as an expert would try to control the system if he or she was in the feedback control loop himself or herself. The rules are all valid in a normalized universe of discourse, i.e., the variables are in per unit. For a simulation-based system design, the controller tuning can be done with the MatLab Fuzzy Logic Toolbox, or LabVIEW is another nice environment for such implementation. It is also possible to develop the whole structure of the controller using C language compiled code. For advanced design, it is possible to use neural network or genetic algorithm techniques for fine-tuning the membership functions, implementing an adaptive neuro-fuzzy inference system (ANFIS). Such details are outside the scope of this chapter. This fuzzy speed control algorithm can be numerically explained and clarified with the following step-by-step procedure:

1. Sample set point speed ω_r^* and actual shaft speed ω_r.
2. Computer error, change in error, and then per-unit values as follows:

$$E(k)=\omega_r^*-\omega_r$$

$$CE(k)=E(k)-E(k-1)$$

$$E_{pu}=\frac{E(k)}{GE}$$

$$CE_{pu}=\frac{CE(k)}{GCE}$$

3. Suppose the fuzzy logic control is written in compiled language. Therefore, the commands are sequentially executed. This step must identify in the universe of discourse where such variables, error, and change in error are located. Two indexes I and J for E_{pu} and CE_{pu} can be used to define which fuzzy sets are triggered by the sampling of the error and the change in error. If the membership functions are simple triangular or trapezoidal shapes,

each index I and J will help to decide a linear equation to be used to calculate the membership function evaluation in the next step.

4. Calculate the degree of membership for each fuzzy set for E_{pu} and CE_{pu} by applying the correct equations identified in step 3, defining exactly which rules in the rule table will be used (usually four are triggered, assuming full overlapping of membership functions).
5. Identify the four valid rules; all the rules are stored in a chain of IF-THEN-ELSE statements, or maybe in a look-up table, or a similar functional implementation. It is necessary to calculate the individual degree of truth for those four valid rules. Normally, it is the AND operator used, which can be implemented either as MIN of two membership functions or by the multiplication of the values of membership functions of the input variables. In this step, four rules will each have their degree of truth, to be used next for making up the fuzzy output and then calculate the defuzzification.
6. Compose the output fuzzy set and get it prepared for defuzzification by the center of gravity, or use a simplified height defuzzification method, which is a simple weighted average of the output fuzzy singletons weighed by the degree of truth of each rule.
7. Since the output is a change or a derivative, integration or accumulative sum must be performed, as indicated by Figure 8.1.

A neural network can also be used in the control of a nonlinear system, as portrayed in Figure 8.3. This topology is based on model reference adaptive controller (MRAC). A reference model is assumed for the nonlinear plant, which can be a linearized model around an operating point of the set point. Two neural networks are used in

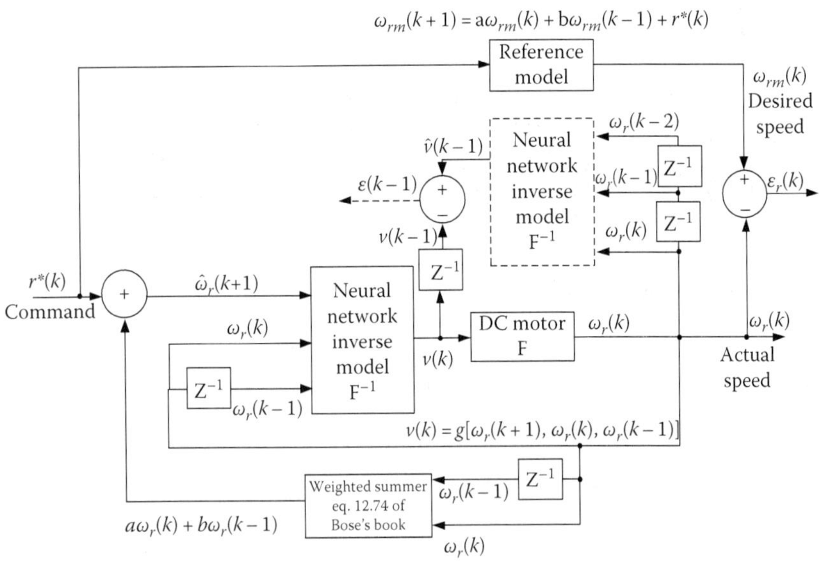

FIGURE 8.3 Neural-network-based inverse dynamics adaptive speed control.

this scheme; one is online training and learning the inverse model of the output/input function F^{-1}. After the training converges, the neural network weights of this inverse model are transferred to the neural network that controls the system; this swapping can be made in timing that is compatible with the bandwidth of the overall plant.

In this neural-network-based inverse dynamics adaptive control approach, the applied voltage can be considered a function of one-step ahead angular speed reference $\omega_r(k+1)$, with current angular speed $\omega_r(k)$ in addition to the previous one $\omega_r(k-1)$. Therefore, Equations 8.1 and 8.2 are applied to support the neural network inverse model:

$$v(k) = g\left[\omega_r(k+1), \omega_r(k), \omega_r(k-1)\right] \tag{8.1}$$

$$v(k-1) = g\left[\omega_r(k), \omega_r(k-1), \omega_r(k-2)\right] \tag{8.2}$$

Figure 8.3 shows in the dashed box the neural-network-based inverse model (F^{-1}) of the machine, which is first trained off-line by the input/output data generated by the machine. The network has no dynamics as it is not recurrent type, and it can be implemented with a three-layer feed-forward network with five or maybe more hidden layer neurons where the time-delayed speed signals are generated from the actual machine speed by a delay line, or just a shift command and allocated memory. The network output signal $\hat{v}(k-1)$ is compared with $v(k-1)$, which is generated with one-step time delay from the machine input. The training is conducted until the error signal $\varepsilon(k-1)$ becomes acceptably small. Therefore, backpropagation can be used for training, but it must be coded properly in order to use $\varepsilon(k-1)$ for convergence. Once the inverse model (F^{-1}) is trained successfully, it is placed in the forward path, by downloading the trained weights of the dashed box network into the solid line box network, which will become the inverse-model control of the plant, capable of cancelling the dynamics of (F). This inverse-modeling control brings an acceptable stable closed-loop control, but it might be sluggish in response. However, very complex plants could be controlled with such approach.

The system of Figure 8.3 has a comparison of the actual output response of a reference model. The reference model should be a linearized version of the plant to allow a tracking signal $\varepsilon_r(k)$, which should be theoretically zero, or at least bounded. If the tracking signal indicates a steady amplitude increase, it means the plant may have parameter variation (such as thermal or saturation effects on a machine), making it necessary to update the inverse model with the online training, plus adding a feedback loop for stability, which is a weighted summer that corrects the set point. In extreme cases, the reference model can be substituted by another average model valid in another operating point. Therefore, the MRAC neural-network-based system should have a supervisory control to make sure that the system is always functioning properly.

8.4 FUZZY LOGIC AND NEURAL NETWORK FUNCTION OPTIMIZATION

Optimization is always an important activity for real-world applications. When a designer wants to consider a system that attains the best characteristic or responds to what is an "optimal response," it is necessarily a methodology that allows optimization. Mathematically, optimization is defined as the search for a combination of parameters commonly referred to as decision variables, which minimize or maximize a certain quantity, typically a scalar, which is called a score or cost, assigned by an objective function or cost function. Also, there is a set of constraints, which provide boundaries on decision variables or may define regions of nonfeasibility in the decision variable space. An optimization can be done off-line, such as studying all the possible configurations of renewable energy with several sizes of wind turbines, PV arrays, batteries, hydrogen, and diesel, to establish the best configuration of a predesigned system, which will be implemented. It may be a study of the best configuration for a rocket to be lifted up and reach a certain goal in the outer space. The optimization can also be required online, and, typically, mathematical programming has many implementation issues with online optimization. However, as told by Don Knuth, a great computer scientist, "...We should forget about small efficiencies, say about 97% of the time: premature optimization is the root of all evil." Therefore, using heuristics, fuzzy logic, and neural networks, an approximate optimal point can be established, and as long as this optimal condition is online and close to the real optimal solution, that is the ideal practical implementation. For example, pumping fluid will have a flow that increases, but the pressure decrease and the power for pumping initially go up, reaching a maximum and decreasing as a throttle continues opening. There is an optimum pump speed for the corresponding maximum of power flow. Although thermodynamic analysis, mass flow rate, and energetic modeling will be very complex, the idea of hill climbing can be used, i.e., a heuristic way of searching the maximum could be based on the following metarule: "If the last change in the input variable (x) has caused the output variable (y) to increase, keep moving the input variable in the same direction; if it has caused the output variable to drop, move it in the opposite direction." Such a rule can be easily implemented in any online optimization, or it can be further developed in a rule table of fuzzy statements. It can also be used to program a neural network, which will learn the parameters that minimize a certain cost function using training and adaptation algorithms.

One important application of online optimization is for flux programming of induction generator-based energy systems. Figure 8.4 depicts a fuzzy logic optimization control system where a search for the best flux operating point and induction generator will be made based on the fuzzy inference of the DC-link power generation (with inverter included) and the last command of the flux quadrature current. Usually, machines operate with rated flux to have maximum developed torque per ampere and optimum transient response. However, in renewable energy applications, such as wind or hydro power, when the mechanical shaft torque is light, excessive flux will increase the quadrature current component of the machine stator current, increasing the copper loss. By programming the quadrature current for a lower value, the total stator current will decrease, thereby decreasing the copper loss, and the flux will also decrease, decreasing the core loss. However, there is a minimum value for the flux that will keep the system stable, and so a search can be done based on heuristics: measuring the generated power, for example, at the DC-link, the quadrature current is decreased as long as the generated power increases, but when the generated power starts to decrease, the flux search is reversed. Of course, a certain oscillation around

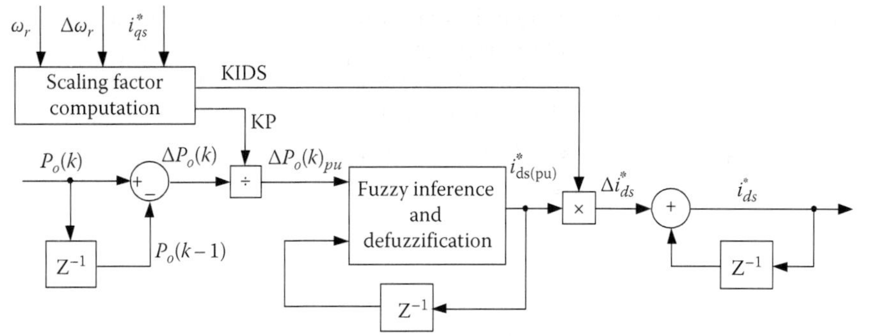

FIGURE 8.4 Fuzzy logic optimization control system where a search for the best flux operating point and induction generator will be made based on the fuzzy inference of the DC-link power generation (with inverter included) and the last command of the flux quadrature current.

the optimal point is expected, but a fuzzy logic control can be made to have large adaptive steps for the beginning of the search, and small progressive steps as the best operating point is reached. Flux optimization requires two variables to be controlled, they are the change variation of power at the DC-link $\Delta P_{d\,(pu)}(k)$ and change of variation of d-axis induction generator flux component $\Delta i^*_{ds(pu)}(k)$. Figure 8.5 shows the seven asymmetric triangular membership functions, comparing the variation of power $\Delta P_{d\,(pu)}(k)$ with the last variation of quadrature current, i.e., the previous one, $\Delta i^*_{ds(pu)}(k-1)$. Table 8.3 shows the corresponding rule table for this fuzzy controller, in which a typical rule is:

$$\boldsymbol{IF}\,\Delta P_{d(pu)}(k) = Positive\ Small(NS)\,\boldsymbol{AND}\,\Delta i^*_{ds(pu)}(k-1) = Negative(N)$$

$$\boldsymbol{THEN}\ \Delta i^*_{ds(pu)}(k) = Negative\ Small(NS)$$

The basic assumption is that if the last control action indicates an increase of DC-link power, the search follows to the same direction as before, in accordance to the last control action, and the control magnitude should be somewhat proportional to the measured DC-link power change. When the control action results in a decrease of P_d, i.e., $\Delta P_d < 0$, the search direction must be reversed. At steady state, the operation oscillates around the optimal point, with a very small step size.

The use of artificial intelligence for function optimization has been successfully used for the wind and solar applications. The principles of peak power tracking control for wind energy will be discussed herein, but similar principles can also be applied to photovoltaic arrays. The large energy capture of variable-speed wind turbines makes the life-cycle cost lower, but it is required that a control system programs the wind turbine to operate at its maximum power energy conversion operating conditions. Figure 8.6 shows the torque-speed curves of a wind turbine at different wind velocities. At a particular wind velocity (V_{W1}—point A), if the turbine speed ω_r decreases from ω_{r1}, the developed torque increases, reaches the maximum torque

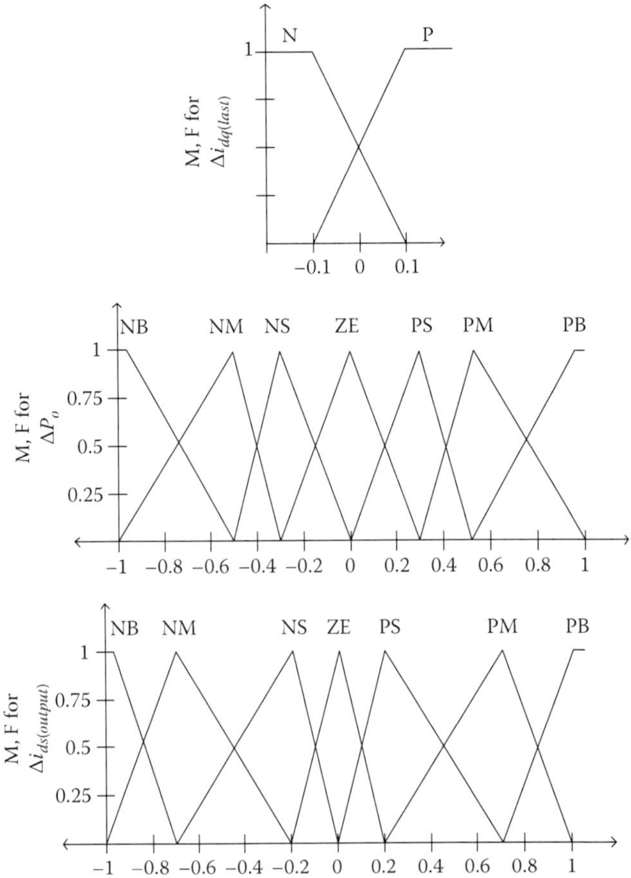

FIGURE 8.5 Fuzzy logic membership functions with their associated linguistic variables for change in power and change in flux quadrature current, where the asymmetrical functions help the convergence of the searching and online optimization.

TABLE 8.3
Fuzzy Optimization Search of Best Induction Generator Rotor Flux

		Last Flux Change $\Delta i^*_{ds(pu)}(k-1)$	
		N	**P**
Power change $\Delta Pd_{(pu)}(k)$	**PB**	NM	PM
	PM	NS	PS
	PS	NS	PS
	NS	PS	NS
	NM	PM	NM
	NB	PB	NB

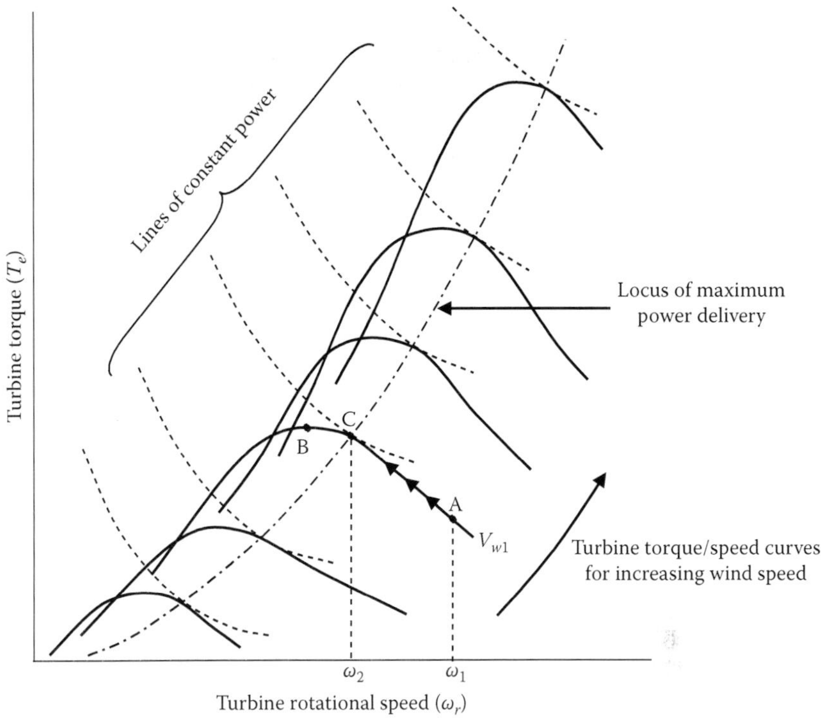

FIGURE 8.6 Torque-speed curves of fixed-pitch wind turbine for different wind velocities, where the maximum power locus delivery intercepts the curves at the peak power point tracking set point.

at point B, and then decreases at lower ω_r. Superimposed on the family of curves there are curves of constant power lines, indicating the points of maximum power output for each wind velocity. Therefore, as the wind velocity varies, the turbine speed should change to get the maximum power delivery, optimizing the aerodynamic efficiency. For example, in Figure 8.6 for point A, or for point B, an intelligent based controller should seek point C, which is the one that gives the maximum power conversion. Of course, a family of power curves could be plotted against the turbine rotational speed, and for that particular set of curves, the algorithm would search the apex of the curve. Figure 8.6 shows that the peak torque for a particular wind turbine will not necessarily be the one that maximizes the power conversion.

The fuzzy logic control for optimizing a wind energy system will have an implementation block diagram like the one depicted in Figure 8.7, with fuzzy membership functions given in Figure 8.8 and fuzzy inference rule table in Table 8.4. It is an extension of the method employed for searching the flux, with the difference that power will be maximized, instead of minimizing copper and core losses. A certain oscillation around the optimal point is expected, but a fuzzy logic control can be made to have large adaptive steps for the beginning of the search, and small progressive steps as the best operating point is reached. Two variables should be the inputs

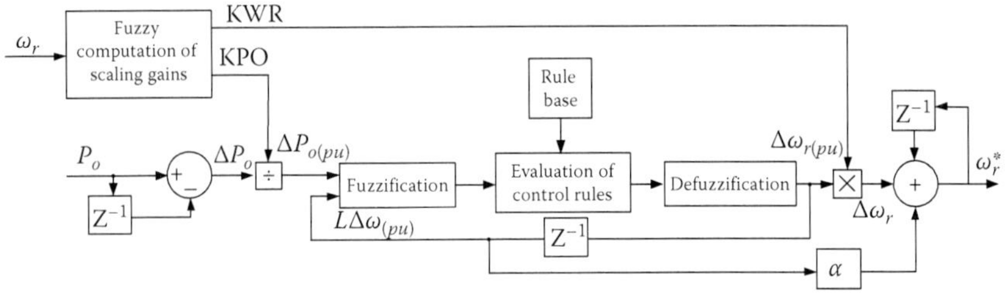

FIGURE 8.7 Search of the best shaft angular speed for a wind turbine based on the fuzzy inference of the grid converter power generation compared with the last command of the angular velocity.

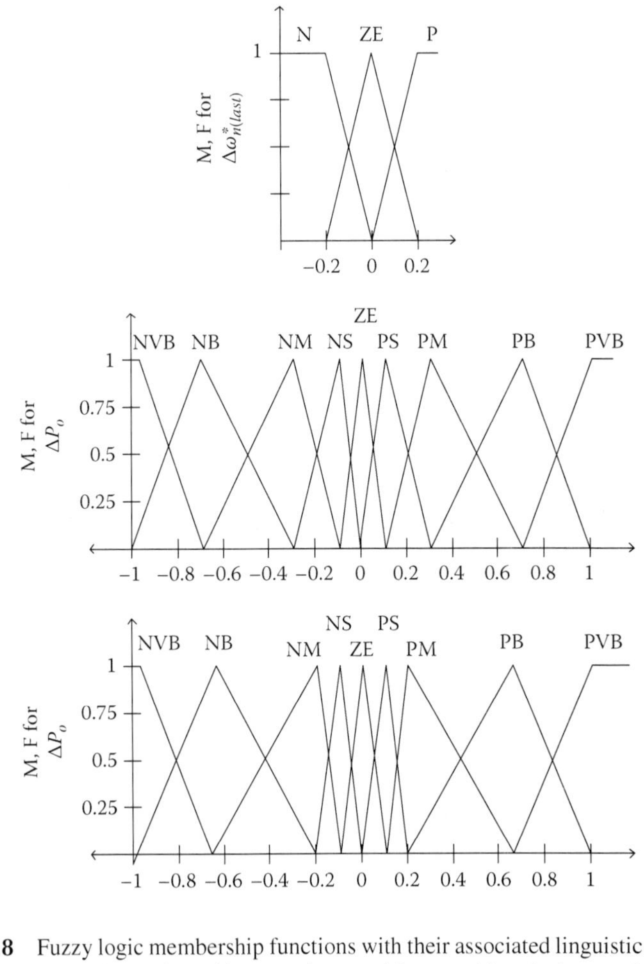

FIGURE 8.8 Fuzzy logic membership functions with their associated linguistic variables for change in power and change in turbine angular speed for searching the best peak power point of the wind turbine with online optimization.

TABLE 8.4
Fuzzy Rules for Optimization of Wind Turbine Power for Variable Wind Velocity

		Last Speed Change $\Delta\omega^*_{r(pu)}(k-1)$		
		P	**ZE**	**N**
Power change $\Delta Po_{(pu)}(k)$	**PVB**	PVB	PVB	NVB
	PBIG	PBIG	PVB	NBIG
	PMED	PMED	PBIG	NMED
	PSM	PSM	PMED	NSM
	ZE	ZE	ZE	ZE
	NSM	NSM	NMED	PSM
	NMED	NMED	NBIG	PMED
	NBIG	NBIG	NVB	PBIG
	NVB	NVB	NVB	PVB

of such a fuzzy controller the change, the output power at the grid (including the whole inverter system), P_0, i.e., for ΔP_0 positive, with the last $\Delta\omega^*_r$, we can define this variable as $L\Delta\omega^*_r$, the search is continued in the same direction. If, on the other hand, $+\Delta\omega^*_r$ causes $-\Delta P_0$, the direction of search must be reversed. The speed oscillates by a small increment when it reaches the optimum condition. The normalized variables $\Delta P_{0\,(pu)}(k)$, $\Delta\omega^*_{r(pu)}$, and $L\Delta\omega^*_{r(pu)}$ are described by membership functions, as shown in Figure 8.8. In a search of peak power of wind turbine, there is possibly some wind vortex and torque ripple that may trap the search in a nonlocal minimum, so some amount of $L\Delta\omega^*_{r(pu)}$ is added to the current set point, similar to a momentum factor used in a neural network. The scale factors KPO and KWR, in Figure 8.7, are a function of the generator speed and the turbine, and some fine-tuning might be necessary with the scaling, in order to make the system sensitive to the power variation with the turbine angular speed variation.

Figure 8.9 shows how a power search will operate, indicating the search as wind velocity changes. If initially the wind is at velocity V_{W4}, the power output will be at point A if the generator speed is ω_{r1}, the fuzzy logic control will alter the speed in steps on the basis of an online search until reaching speed ω_{r2} where the output power is maximum at point B. If this system freezes the operating point at ω_{r2} for steady-state conditions, a next search of the best induction generator flux takes over, and the system is brought to the operating point at C. Now, if the wind velocity changes to V_{W2}, the output power will jump to point D; one fuzzy controller will bring the operating point to E by searching the speed until arriving to ω_{r4}; the system is locked on this angular speed conditions for steady state; and another fuzzy controller will search the best flux of the induction generator, bringing the operating point to F. A similar discussion can be made for a decrease of wind velocity to V_{W3}. Figure 8.10 shows a complete fuzzy logic-based control and optimization of an induction generator-based wind turbine grid-connected energy system. A double PWM back-to-back voltage

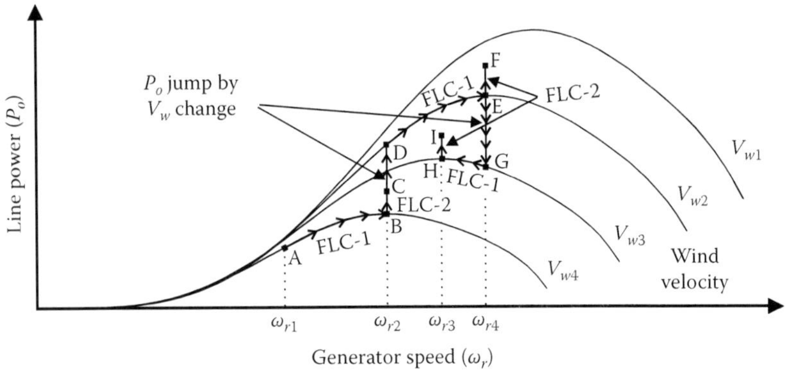

FIGURE 8.9 Wind turbine peak power point search with online optimization.

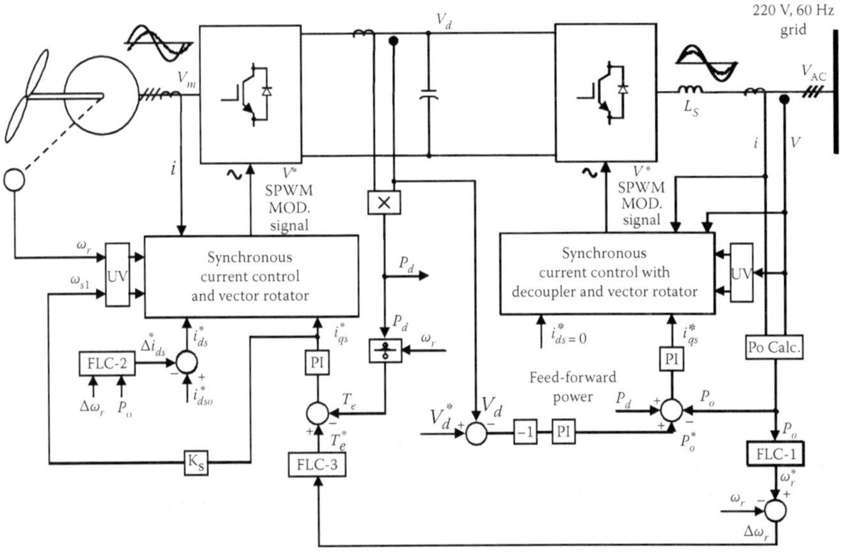

FIGURE 8.10 Fuzzy logic-based control and optimization of an induction generator-based wind turbine grid-connected energy system.

source converter connects the induction generator to the utility grid. The machine side converter is controlled with field-oriented, or *d-q* vector, control. The torque command is impressed with a fuzzy logic controller, defined as FLC-3, which is the same as discussed in Section 8.3.

There are two online fuzzy optimization controllers in Figure 8.10. FLC-1 searches the generator speed to maximize the output power, whereas FLC-2 searches the machine excitation current to optimize generator efficiency at light power generation, and the online function optimization fuzzy controllers are same as described previously. The advantages of such fuzzy controllers are obvious. They will accept noisy and inaccurate signals, they will provide adaptively decreasing step size in the search that leads to fast convergence, they provide robust control on the machine shaft against wind turbine vibration and mechanical resonance, and, additionally, wind velocity information is not needed and the system is insensitive to parameter variation. The principles of FLC-1, FLC-2, and FLC-3 have been tested with analysis, simulation, and implementation, as discussed in the further reading section at the end of this chapter.

8.5 FUZZY LOGIC AND NEURAL NETWORK FUNCTION APPROXIMATION

Real-world function approximation problems are modeling system solutions, which can be algebraic solutions, as a mapping of input to output, or state-space solutions, as memory-based equations, where the output will depend on the internal states plus past inputs. Mathematical discussions are usually relevant when function approximation involves incomplete information, high dimensionality, nonlinearities, and noise. Function approximation is the problem of finding a function that approximates a target of another function, based on a sample of observations taken from such unknown target function. In machine learning, the function approximation formalism is used to describe general problem types in pattern recognition, classifications, clustering, and curve fitting. A statistical approach will use a probability density function (PDF) supporting statistical machine learning and density estimation, or even Bayesian algorithms. A neural network is a very simple way for easily learning a function that relates a huge data set of input variables versus output variables. Neural networks can be used to support energy forecasting, load-flow modeling of large power systems, learning of nonlinear functions in power electronics and power systems, estimation of ill-modeled systems, for example, temperature variation effect of induction motor rotor resistance, nonlinear response of capacitors, loss modeling of transformer core, lifetime expectation of protection circuits, and so many other applications in which it is usually very difficult to find a function approximation using pure mathematical theory. Function approximation can be useful in several problems related to signal processing in power electronics, power systems, and power quality. One example is the estimation of the distorted wave. Power converters are characterized for generating nonsinusoidal voltage and current waves, and it is often necessary to determine their parameters such as total RMS current, fundament rms, active power, reactive power, distortion factor, and displacement factor. These parameters can be measured by electronic instrumentation (hardware and software) or estimated by a mathematical model, FFT analysis, and so on. Fuzzy logic principles can be applied for fast and reasonably accurate estimation of these parameters due to their enhanced nonlinear mapping (or pattern recognition) property. A fuzzy logic-based pattern recognition can be applied for the first time to power electronics, where the estimation of a diode rectifier line current wave is discussed and studied with simulation analysis of two methodologies, the Mamdani method and the Takagi-Sugeno approach [2]. The main idea is to observe the pulsed nonlinear current waveforms and use the width (W) and height (H) for each pulse. For a single-phase rectifier, there is one pulse per semicycle of voltage line, while for the three-phase rectifier, for each semicycle of the phase voltage, the line rectifier current has two pulses. When using the Mamdani method, or Type I, several rules can be designed such as:

$$\textbf{\textit{IF}}\ H = PMS\ \textbf{\textit{AND}}\ W = PSB\ \textbf{\textit{THEN}}\ I_s = PMM, I_f = PSB\ and\ DPF = PMS$$

where the power factor will be numerically calculated as $PF = DPF\frac{I_f}{I_s}$, i.e., each rule gives multiple outputs. Further reading references at the end of this chapter will make it possible to understand and compare the development and accuracy of a fuzzy TS (also called Type II, estimation), where a rule would read as:

$$\textbf{\textit{IF}}\ H = PMS\ \textbf{\textit{AND}}\ W = PSB$$

$$\textbf{\textit{THEN}}\ I_s = a_0 + a_1H + a_2W, I_f = b_0 + b_1H + b_2W\ \ and\ \ DPF = c_0 + c_1H + c_2W$$

where the linear coefficients can be found with numerical examples based on experimental data, fitted with the least-square method. It is obvious that Type II approach is more precise and has a more compact rule table than Type I.

Neural networks are very powerful algorithms for real-world approximation problems, which can be algebraic solutions or state-space ones. Power electronics, power systems, and power quality have several signal processing phenomena, involving incomplete information, high dimensionality, nonlinearities, and noise. Neural networks have been applied for various control, identification, and estimation in power electronics and drives; some of these applications are as follows:

- Single or multidimensional look-up table functions
- Converter PWM
- Neural adaptive PI drive control
- Delay-less filtering
- Vector rotation and inverse rotation in vector control
- Drive MRAC
- Drive feedback signal estimation
- Online diagnostics
- Estimation of distorted waves
- FFT signature analysis of waves

Figure 8.11 shows the block diagram of a direct vector-controlled induction motor drive, with a feed-forward neural-network-based estimator.

A backpropagation neural network has been trained to estimate rotor flux (ψ_r), unit vector ($\cos\theta_e$, $\sin\theta_e$), and torque (T_e) by solving Equations 8.3 through 8.10. A DSP-based estimator was used for comparison with the neural network. Such network is feed-forward trained with backpropagation, with instantaneous mapping, so the machine terminal voltages were initially integrated by a hardware low-pass filter in order to generate the stator flux signals. These variable frequencies and variable magnitude sinusoidal signals have been used to calculate the output parameters, with a topology with three layers, where the hidden layer has 20 neurons, as indicated

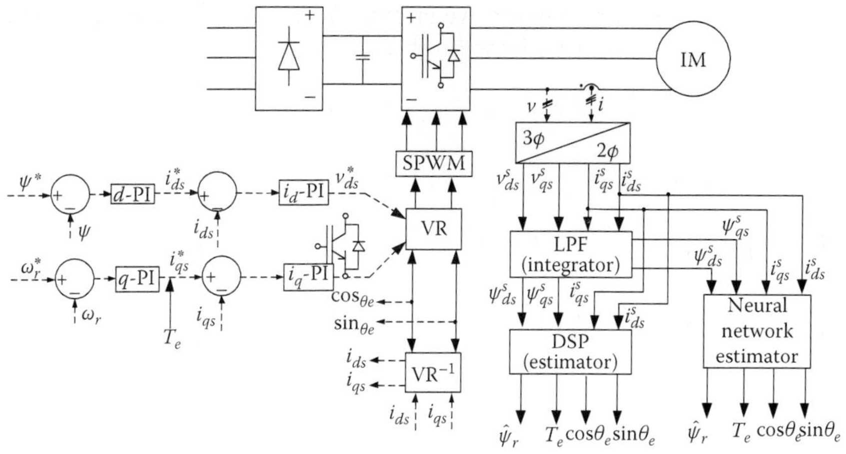

FIGURE 8.11 Modern adjustable speed vector control with neural network estimation.

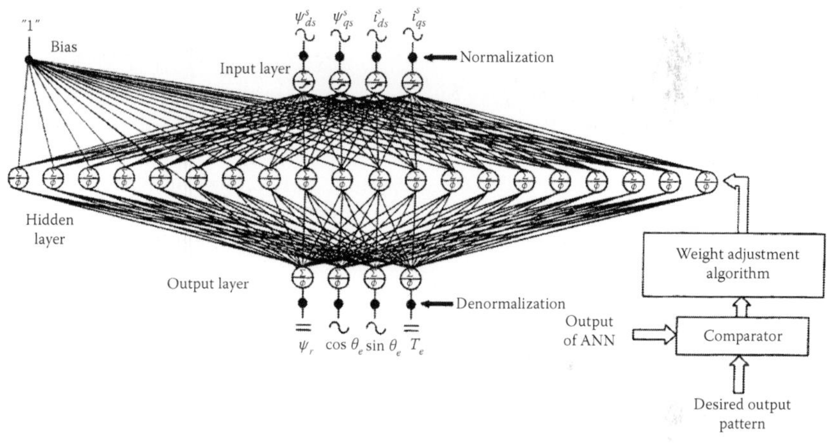

FIGURE 8.12 Neural network for estimation of vector control motor drive signals.

in Figure 8.12. The input layer neurons have linear activation features, but the hidden and output layers have a hyperbolic tangent-type activation function, in order to allow bipolar outputs. This network is capable of correctly and accurately tracking torque, flux, and unit vector signals; they were tested with high and low inverter frequency, working satisfactorily for closing the loop of an adjustable speed modern induction motor drive.

$$\psi^s_{dm} = \psi^s_{ds} - i^s_{ds} L_{ls} \tag{8.3}$$

$$\psi^s_{qm} = \psi^s_{qs} - i^s_{qs} L_{ls} \tag{8.4}$$

$$\psi_{dr}^{s} = \frac{L_r}{L_m}\psi_{dm}^{s} - L_{lr} i_{ds}^{s} \tag{8.5}$$

$$\psi_{qr}^{s} = \frac{L_r}{L_m}\psi_{qm}^{s} - L_{lr} i_{qs}^{s} \tag{8.6}$$

$$\hat{\psi}_r = \sqrt{\left(\psi_{dr}^{s}\right)^2 + \left(\psi_{qr}^{s}\right)^2} \tag{8.7}$$

$$\cos\theta_e = \frac{\psi_{dr}^{s}}{\hat{\psi}_r} \tag{8.8}$$

$$\sin\theta_e = \frac{\psi_{qr}^{s}}{\hat{\psi}_r} \tag{8.9}$$

$$T_e = \frac{3}{2}\left(\frac{P}{2}\right)\left(\psi_{dr}^{s} i_{qs}^{s} - \psi_{qr}^{s} i_{ds}^{s}\right) \tag{8.10}$$

REFERENCES

1. Zadeh, L.A., Fuzzy sets, *Infection Control*, 8, 338–353, 1965.
2. Simoes, M.G. and Bose, B.K., Applications of fuzzy logic in the estimation of power electronic waveforms, *Conference Record of the 1993 IEEE Industry Applications Conference Twenty-Eighth IAS Annual Meeting*, vol. 2, Toronto, Ontario, Canada, pp. 853–861, 1993, DOI: 10.1109/IAS.1993.298999, URL: http://ieeexplore.ieee.org/stamp/stamp.jsp?tp=&arnumber=298999&isnumber=7405.

BIBLIOGRAPHY

Bose, B.K., Neural network applications in power electronics and motor drives: An introduction and perspective, *IEEE Transactions on Industry Electronics*, 54(1), 14–33, February 2007, DOI: 10.1109/TIE.2006.888683, URL: http://ieeexplore.ieee.org/stamp/stamp.jsp?tp=&arnumber=4084644&isnumber=4084635.

Godoy Simões, M. and Bose, B.K., Fuzzy neural network based estimation of power electronics waveforms, *Revista da Sociedade Brasileira de Eletrônica de Potência*, 1(1), 64–70, June 1996.

Godoy Simões, M., Blunier, B., and Miraoui, A., Fuzzy-based energy management control: Design of a battery auxiliary power unit for remote applications, *IEEE Industry Applications Magazine*, 20(4), 41–49, July–August 2014.

Horikawa, S., Furuhashi, T., Okuma, S., and Uchikawa, Y., Composition methods of fuzzy neural networks, *IECON'90. 16th Annual Conference of IEEE* vol. 2, Industrial Electronics Society, Pacific Grove, CA, pp. 1253–1258, 1990, DOI: 10.1109/IECON.1990.149317, URL: http://ieeexplore.ieee.org/stamp/stamp.jsp?tp=&arnumber=149317&isnumber=3941.

Kim, M.H., Godoy Simões, M., and Bose, B.K., Neural network based estimation of power electronic waveforms, *IEEE Transactions on Power Electronics*, 11(2), 383–389, March 1996.

McCulloch, W.S. and Pitts, W., A logical calculus of ideas immanent in nervous activity, *Bulletin of Mathematical Biophysics*, 5, 115–133, 1943.

Meireles, M.R.G., Almeida, P.E.M., and Godoy Simões, M., A comprehensive review for industrial applicability of artificial neural networks, *IEEE Transactions on Industrial Electronics*, 50(3), 585–601, June 2003.

Simões, M.G., Fuzzy logic and neural network based advanced control and estimation techniques in power electronics and ac drives, *Doctoral dissertation*, Adviser: B.K. Bose, The University of Tennessee, Knoxville, TN, 1995.

Simões, M.G. and Bose, B.K., Neural network based estimation of feedback signals for a vector- controlled induction motor drive, *IEEE Transactions on Industry Applications*, 31(3), 620–629, May/June 1995.

Simões, M.G., Bose, B.K., and Spiegel, R.J., Design and performance evaluation of a fuzzy logic based variable speed wind generation system, *IEEE Transactons on Industry Applications*, 33(4), 956–965, July/August, 1997.

Sousa, G.C.D. and Bose, B.K., A fuzzy set theory based control of a phase controlled converter dc drive, *Transactions on Industry Applications*, 30(1), 34–44, January/February. 1994.

Sousa, G.C.D., Bose, B.K., and Cleland, J.G., Fuzzy logic based efficiency optimization of an indirect vector controlled induction motor drive, *IEEE Transactions on Industry Applications*, 42(2), 192–198, April 1995.

Weerasooriya, S. and El-Sharkawi, M.A., Identification and control of a DC motor using back-propagation neural networks, *IEEE Transactions on Energy Conversion*, 6(4), 663–669, December 1991.

Zadeh, L.A., The calculus of fuzzy if/then rules, *Proceedings of the Theorie und Praxis, Fuzzy Logik*, Springer-Verlag, Berlin, Germany, 1992.

9 Instruments for Data Acquisition

9.1 DATA ACQUISITION BY COMPUTERS

In the last few decades, there has been an enormous change in the use of digital computers in all branches of knowledge, particularly engineering, that have evolved into automation and Internet-based data acquisition for instrumentation purposes. Electronics, and particularly analog circuits, was well established for monitored process and data acquisition procedures, and the electronic instrumentation field was already very strong during the 1970s. However, the advent of modern computers allowed a comprehensive integration of highly complex and versatile mathematical algorithms to be used in data analysis and real-time decision-making. Therefore, the previous analog-based instrumentation required a reformulation on how to integrate with digital computers, then microprocessor, digital signal processor, microcontrollers, networking, and the current state of the art in the IoT (Internet of Things). In the beginning of the 1980s, with incorporation of digital signal processing techniques, with ever-growing number-crunching power of computers, better implementation of software-based user interfacing and data acquisition by computers became ubiquitous. In the twenty-first century, it is not possible to imagine any electrical, mechanical, civil, or chemical engineer or a scientist or technician without the foundations and understanding of such an important and revolutionary field that emerged in the last few decades: computer-based signal processing enhanced automation and systems monitoring and control.

9.1.1 Fundamentals of Digital Signal Processing

We should understand the use of processors in the data transfer from a physical process to the computer by learning the architecture basis of a data processor, as represented in Figure 9.1. The crossed vertical and horizontal line segments shown in Figure 9.1a represent a bus connecting each bit of the central processing unit (CPU) to the corresponding bit of a random-access memory (RAM) and read-only memory (ROM). The RAM memories are those where data can be written and read from, or be replaced or turned off at any time. The ROM memories are those where the program data are registered to control the processor functions, and, therefore, they are not to be erased along with the program execution or when disconnecting the DC source [1–3].

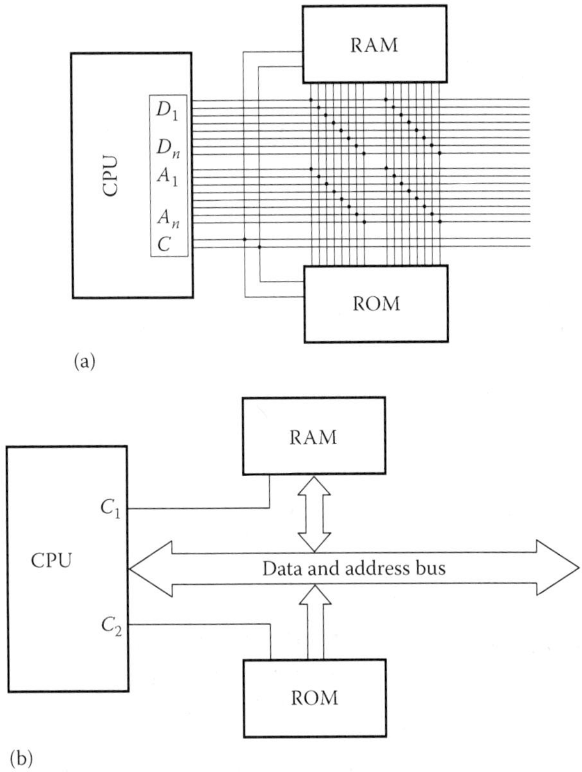

FIGURE 9.1 Elementary computer: (a) basic computer hardware for an 8-bit wire data and address bus and (b) oriented arrows to represent data and addresses.

The representation by arrows in Figure 9.1b indicates data flow along the buses, usually a multiple of 2 (4, 8, 16, 32, etc.), that depends on the processor hardware design. The bits of a data bus are assigned by $D_1, D_2, \ldots, D_n$, and the bits of the address bus are represented by $A_1, A_2, \ldots, A_n$. There also exist control bus bars, assigned by C $(C_1, C_2, \ldots, C_n)$. These buses, generically called data bus, constitute the CPU data access buses to the memories and other peripherals, and vice versa.

The memories store computer inputs and output data, usually in a RAM, but there are other options such as the program data (ROM) and CPU final and intermediary results (registers). The address bus lets the CPU locate these data in the memory. The control bar is to control the CPU data flow and processing. This, in turn, is in charge of executing the basic processing functions. Such functions include control of the data bus, control of the program, state verification of program event indicative "flags," movement of the stack pointer guiding the program execution clock through registers, the accumulator, and, then, the inner heart of the computer, which is the arithmetic and logical unit (ALU).

The control of the data bus has more or less the same function as a traffic guard controlling the automobile flow (data) at a crossing of several roads. It is used to guide the data and to establish the instant these data can be transferred from CPU to memory or any other peripherals, as we see below. The program control serves to destine the data and to establish the instant these data should be transferred between the registers and accumulator, and it helps in the execution of operations realized by ALU. The flags serve to indicate programming events such as if the sign resulting from an operation is positive or negative; if a value is larger, smaller, or equal to other in a logical comparison; and if it should have an interruption in the normal processing to assist an external call from CPU determined by some external event. The stack pointer indicates the program memory address where the instruction is contained to say what should be accomplished according to the instruction contained in the program memory in that indicated position. The registers are the immediate memory used by CPU to execute internal operations and to establish the results of logical comparisons.

The most important unit of a CPU is the ALU because it is responsible for the execution of all operations and programmed decisions. In a simplistic way, it can be said that everything it does is to implement a "hardware" summation controlled by the program contained in the memory. This sum operation can be then implemented by a summation of a positive number with a negative one. The product is the summation of the multiplied as many times as the times specified by the multiplicand. Conversely, the division is the sum of the dividend by the negative of the divider by so many times as necessary until turning the result is as small as possible. A logical decision is defined by the signal resulting from the sum of a positive number with a negative number. The result of this comparison alters the state of a "flag," which can be either zero or one. All these operations are executed with the help of an accumulator and the registers. Other forms of product and division execution are also available in hardware to speed these operations up. The results of ALU are then kept in the memory through data buses associated to address buses, as discussed previously.

A computer design with CPU and memory is functional when it is possible to intervene in the program or the data input and output. Therefore, several other units around CPU and memory are necessary to allow the data observation (monitor screen), to alter the program or data (Bluetooth, keyboard, or mouse), to communicate with computer networks or peripheries (modem), or to enlarge the capacity of the RAM and ROM memories (hard disk, zip drivers, CD, DVD, USB drivers, etc.) for mass storage data.

Figure 9.2 shows the basic connection outline of data for the most ordinary peripheral of a computer adapted to instrumentation, which does not differ in almost anything from the conventional. It is observed in this figure that some connections are bidirectional while others are not, meaning that in the former, the data flows in one direction or in another, while in the latter, the data flows just in the direction indicated by the oriented arrows. So, the role of the CPU control bus is fundamental, because it decides which unit will send data and which one will receive them, disabling all the others during each transfer.

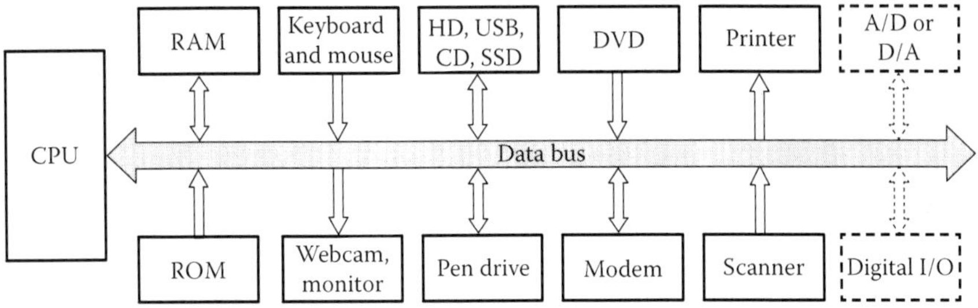

FIGURE 9.2 Peripherals of a computer architecture.

9.1.2 The Data Acquisition Board

Data acquisition boards, as they are known, reunite circuits to transfer from the inside of computers all types of analogical signals from the physical processes to be studied for storage, performance analysis, and processing. Once the data are acquired, they are inside of the computer, and so any processing and manipulation of the data will be allowed. Processing may involve the selection/classification of data, registration in tables, printing/plotting, process driving, and an endless list of possibilities that can be created by human mind.

For specific applications, an interface can be designed to connect the computer data bus to a process through external cables. Almost all computer motherboards nowadays have "slots" (female connectors) readily available to receive circuits or interface boards. As all signals in a computer are digital, and therefore subject to deformations due to spurious capacitances, the external cables should not be very long since they degrade communication signals and cause bad interpretation to a terminal equipment. An alternative for that is the construction of an interface that is partially in the computer and partially out of the computer. That is shown on the right of Figure 9.2, where there are two very special units regarding instrumentation. These are the units of the coming input and output data from processes or experiments to be measured, worked out or monitored. These signals can be alphanumeric (letters, numbers, states, and symbols) or real-time inputs. For variable control of experiments and processes, such as voltage, current, and temperature, it is necessary to use analog-to-digital converters, A/D, so that the computer can understand them, and digital-to-analog converters, D/A, so that the process to be controlled can understand the computer, as discussed in Chapter 8.

Figure 9.3 illustrates how data transfer operates in between the processor and the process using a command OUT (to output data) or another similar for data output. This command makes the processor arrest a word during a data output setting the data flag in the Data Available condition and, soon after that, receiving the "flag" Data Accepted coming from outside of the interface. The command IN (data input) works in a similar way, with the "flag" indicating that the "buffer" is empty and that it is ready for data reception. The input line "strobe" is used to allow fixation ("latch") of data when these are valid. These functions are available in most of the commercial D/A and A/D converters according to each application. The most known data

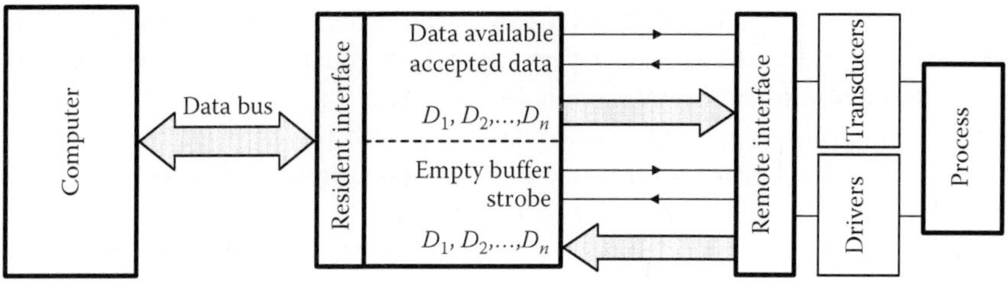

FIGURE 9.3 Interface between computer and process.

acquisition boards are manufactured by the Hewlett Packard Co., Texas Instruments Inc., Gould, Hitachi, and National Instruments, offering wide programming support (hardware and software).

Many are the computer applications and functions using signal converters and digital inputs and outputs as well as timers and other special gates. These relationships between process and computer have been advancing so much that new interface circuits were created by gathering all these special functions and peripherals in a single board called data acquisition board. These boards, usually, can directly accept data "slots" (connectors) to the motherboard's bus and therefore accessed and controlled. With this, the computer receives the process data and may carry out all its processing capacity, besides being able to intervene in the external process connected to it with this purpose. Figure 9.3 also illustrates the relationship between the computer and the process controlled through the interface board.

Data acquisition boards allow a large number of measurement instruments and adapt them for many purposes in a single computer. It is usually necessary to have sensors and transducers to transform the variable being measured (temperature, pressure, position, speed, acceleration, voltage, current, power, radiation, energy, state, etc.) in an electric signal. Formulas, algorithms, and process mathematical relationships, once converted appropriately, can establish output variables to be interpreted by human beings, instruments, or machines. As an indication of the data output, a computer monitor can be used in such a way that, once programmed, it shows results from a desired process or situation in the most familiar way the user may want it. Look-up tables and graphs, scales, numbers, and report forms can be easily displayed on a computer screen with proper data interpretation, allowing great precision. Printed reports, transmission through Internet, WhatsApp, mobile telephones, and other usual computer modern communication forms are commercially available nowadays.

Finally, it should be said that data can be communicated between computer and process in series or in parallel. The reduced wiring of the former is compensated by the high-speed data transfer of the latter.

9.1.3 Microcontrollers

A microcontroller is a type of dedicated controller containing a microprocessor and several peripherals in a single chip. The data, addresses, and control bars are all embedded in the same chip having very special advantages: weight and volume reduction, immunity to dust and/or humidity, minimization of bad connections, programming flexibility, high precision, noise minimization by the extremely reduced distances between the internal connections, and high performance and reliability, among others. Such characteristics are inherent to the instruments based on microcontrollers besides having extremely reduced cost.

There exist a wide range of commercial microcontrollers from the most varied origins and characteristics. The most well-known microcontrollers are made by Intel, Motorola, Texas Instruments, and Microchip, but in the last few years there has been a widespread design of other dedicated microcontrollers, particularly with the advent of the IoT.

9.1.4 Example of Data Acquisition

An example of data acquisition to obtain the no-load characteristics of transformers and electrical machines, in general, is given in Figure 9.4. This figure represents an automatic process to obtain the magnetization characteristic of iron cores through a computer using interface IEEE488. The block diagram shows the electric connections and computer drivers as implemented by one of the authors of this book in the Center of Excellence in Energy and Power Systems (CEESP) of the Federal University of Santa Maria (UFSM) [4–7].

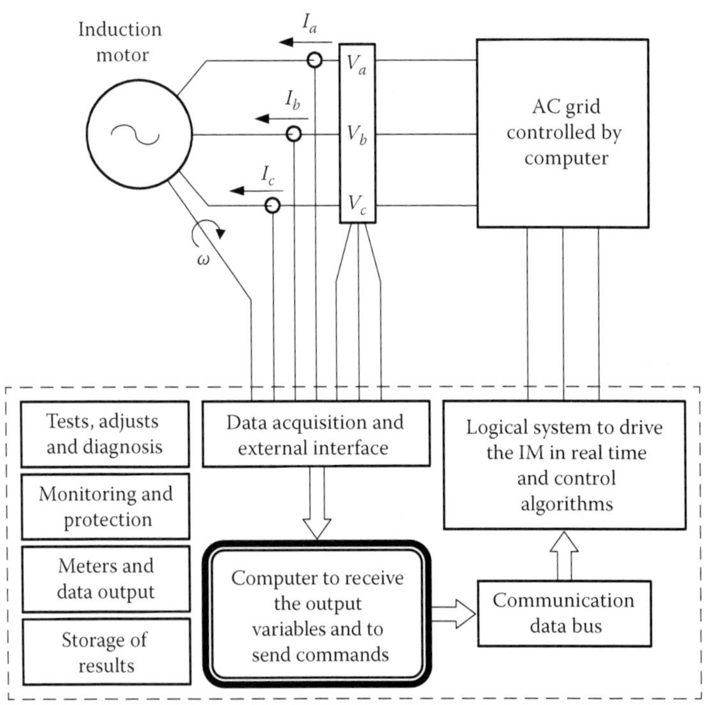

FIGURE 9.4 UFSM model to obtain computer data (IEEE488).

Table 9.1 presents the data of the induction motor (IM) tested by the CEESP acquisition system and an automatic computer driver as suggested in Figure 9.4. The algorithms of data acquisition and control are based on the formulas of Table 9.2.

In this example, the amplifiers with optical isolation measure phase voltages V_a, V_b and V_c and, in turn, phase currents I_a, I_b, and I_c are measured by Hall effect devices. As the read values are instantaneous ones, there is no need to read the phase angles, because classic circuit formulations can obtain these data.

The method suggested in this example to obtain the magnetization curve uses a minimum number of data as presented in Table 9.3 used to determine parameters

TABLE 9.1
Plata Data of the Tested Motor

Power	10 hp	Connection Δ	220 V; 26 A
Isolation	B. IP54	Connection Y	380 V; 15 A
Regime	S1	Frequency	60 Hz
I_{ph}/I_{rated}	8.6	Rotation	1765 rpm
Category	N	SF	1.0

TABLE 9.2
Data from the Motor Test

Test	Current	Voltage	Measured Values	Formulas for K_i
1.	$I_{m1}=0.08$	65.00	$a=V_{g1}/I_{m1}=812.5$	$K_1=(c-K_3)\left(\frac{a-b}{b-c}\right)^{49/24}=425.05$
2.	$I_{m2}=0.40$	248.33	$b=V_{g2}/I_{m2}=620.8$	$K_2=\frac{49}{24}\frac{\ell n\left(\frac{b-c}{a-b}\right)}{I_{m3}^2}=-4.0455$
3.	$I_{m3}=0.56$	290.00	$c=V_{g3}/I_{m3}=517.9$	$K_3=\frac{ac-b^2}{(a+c)-2b}=398.33$

TABLE 9.3
Measurements from the No-Load Tests of the MI Described in Table 9.1

Test	V_g (V)	I_m (A)
1.	65.00	0.08
2.	248.33	0.40
3.	290.00	0.56

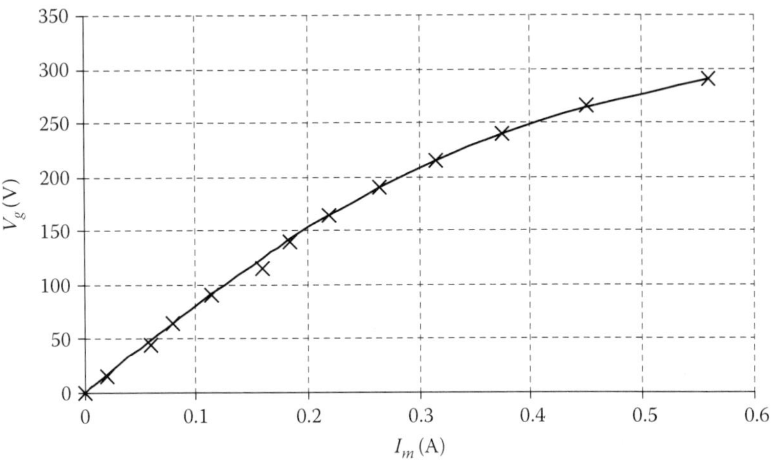

FIGURE 9.5 Magnetization curves obtained by digital instrumentation.

K_1, K_2, and K_3, and for adjustment of the magnetization curve with the aid of the following equation:

$$V_g = I_m\left(K_1 e^{K_2 I_m^2} + K_3\right) \tag{9.1}$$

where

V_g is the gap voltage of the magnetic core
I_m is the magnetization current

The maximum current displayed in Table 9.3 was taken as the test reference, since it is the highest possible current, without putting in danger the motor integrity. The other values were obtained by numerical interpolation to satisfy the relationship 1:5:7 for the values foreseen in the method described above to obtain K_1, K_2, and K_3, as shown in Table 9.2, together with the necessary testing formulas. The no-load curves foreseen by Equation 9.1 and the values obtained in laboratory tests (marked with X) are plotted in Figure 9.5. These are the effective values of the motor phase currents and voltages.

9.2 SIGNAL PROCESSORS FOR INSTRUMENTATION

9.2.1 Digital Signal Processors

Digital signal processors (DSPs) were initially designed in the 1970s but became commercially available in 1979. Since then, DSPs became diversified for several advanced applications with ever-increasing enhanced performance. They are specifically designed for real-time computing in applications of high-speed digital processing for large volumes of data. Many of them are used in modems (modulation-demodulation), voice synthesizers/encoders, voice recognition, and image processing systems.

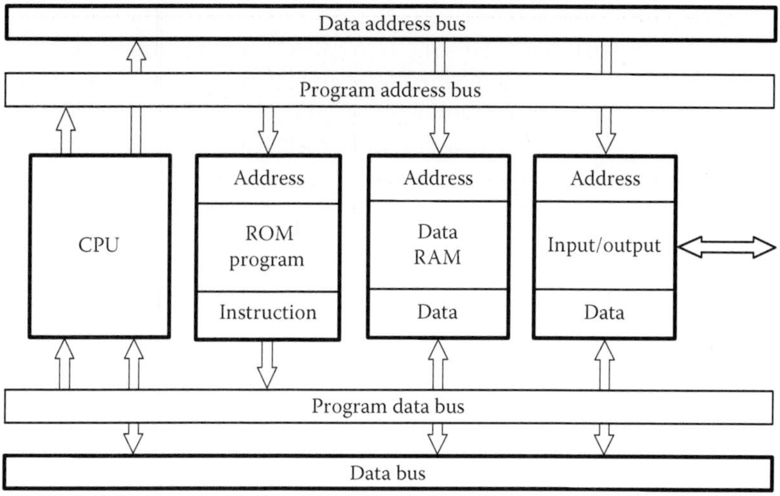

FIGURE 9.6 Architecture Harvard.

In most DSPs built with an architecture Harvard, data and instructions occupy memory separately and move in separate buses, as displayed in Figure 9.6. This is commonly used in 32-bit and 64-bit microprocessors. With such double structure, the controller can bring simultaneously one datum and one instruction.

The "pipeline" transfer operation of instructions and data is possible in the architecture Harvard DSP, resulting in a high rate of transfer instructions in the process. The "pipeline" can be of 2–4 depth levels, depending on the architecture. To optimize the processing speed, some operations usually are more computer time-consuming than others, such as multiplication and displacement, and because of that, they may be accomplished in hardware. In most recent DSPs, the execution speed is still higher since they use several independent units, multiple sets of buses, and additional units of memory, such as cache of instructions, registration file, and memory of double access.

The DSP operation is optimized in such a way that most of the instructions are executed in a single cycle. In some of them, it can be accomplished by an ALU parallel multiplication and operations with integer data or of floatation point, everything in a single cycle. This multiply/accumulate operation is fundamental in more optimized DSPs, and it is used in most algorithms of signal and control processing that can be expressed as a summation of polynomial products. Among these are digital filtering, PID controllers, Fourier analysis, wavelets, and classification of large volumes of data. The majority of DSP has the capacity of repetition blocks of the program to reduce the number of instruction cycles. Other typical functions depend on each manufacturer: Texas Instruments, Analog Devices, AT&T, Motorola, NEC, and others.

9.2.2 Parallel Processing

Many instruments for measurement and driving systems need to be quite fast and, to work appropriately, they need compatible signal processing. Therefore, the use of processors in parallel and multiprocessed systems is increasingly becoming common. The effect of processors operating in parallel through partitioning its control functions increases the processing speed through higher sampling rates of the measures and easier execution control in a multitask environment (multiple simultaneous programs). These processing systems are still expensive, and the development tools in real-time paralleling are still not completely matured, thus demanding larger time for the development of measurement algorithms and control.

9.2.3 Transputers

Transputers are devices based on microprocessors with internal memories and communication links, specially designed for parallel processing. They were developed mostly for applications in data processing (language OCCAM). There was an evolution in its use from ASICs (application-specific integrated circuits) since they were used in data acquisition for high-speed instrumentation. They are an evolution of VLSI (very large-scale integration) to allow adaptation of instrument designs during the CI development to adapt it to the designer's needs. The ASIC's complexity may vary thoroughly, from a simple logical interface for data acquisition to a complete DSP or RISC processor (reduced instruction set computing), or to a neural or fuzzy logic controller [8].

The ASIC technology uses CMOS and BiCMOS with sizes smaller than 0.5 μm. With ASIC using 0.8 μm CMOS, devices with more than 25.000 gates, generally defined as NAND, can be manufactured. With ASIC using 0.5 μm CMOS, it is possible to reunite 600.000 gates. ASIC CMOS are available in standard cells and gate arrangements.

BiCMOS, or bipolar CMOS, offers higher operational speeds for the cost of a more complex process and lower densities of gates. They combine CMOS and bipolar technology using "sea of gates" and could operate at a frequency up to 100 MHz and densities lower than 150.000 gates. With the 0.5 μm BiCMOS technology, ICs can concentrate up to 300.000 gates.

ASICs can use analogical and digital signals all mixed in the same chip. Analogical cells include operational amplifiers, comparators, D/A and A/D converters, sample-and-hold circuits, voltage references, and RC active filters. Logical cells include gates, counters, registers, microsequencers, PLACs (programmable logic array), RAM, and ROM memories. Interface cells include input and output (I/O) parallel gates of 8–16 bits, as well as serial synchronous gates and UART (Universal Asynchronous Reception Transmission). There are, also, commercially available RISC and DSP cores in mega cells for the design of advanced and customized processors.

Although the technology of transputers did not achieve mass production and acceptability during the 1990s, there have been some similar developments, particularly by Cypress Semiconductors. One very promising real-time parallel controller is the Parallax P8X32A Propeller chip, introduced in 2006. It is a multicore architecture parallel microcontroller with eight 32-bit RISC CPU cores, with Propeller Assembly language, and Spin programming language and "Propeller Tool and Spin interpreter" designed by Parallax's cofounder and president Chip Gracey. Parallax Inc. released Propeller P8X32A Verilog and top-level HDL files under the GNU General Public

License 3.0. Chip Gracey promised to release Propeller II, which might be the best implementation of a transputer type of microcontroller when it becomes a product ready for use.

9.2.4 Design of Instruments with ASIC

To manufacture an ASIC technology, instrument passes through the following stages:

1. Simulation and logical design
2. Checking of the placement of components, routes, and connectivity
3. Mass production of the prototype

The final user can enter the design process following semistandard, semicustomized, or totally customized ways, depending on the instrument application [8–10].

9.2.5 FPGA (Field-Programmable Gate Arrays) in Instrumentation

FPGA integrates a class of ASIC that differs from the mask-programmable gate array in that its programming is made by the final user in its place without the IC masking steps. They consist of arrangements of logical blocks that can be program-connected to implement different instruments. The logical blocks are based on the following:

- Equal transistors
- Small basic gates (NAND and OR-exclusive with two inputs)
- Multiplexers
- Look-up tables
- Wide fan-in AND-OR structures

The FPGA programming is electrically made by programmable keys implemented by one of the three main technologies: static RAM, antispindle technology, and floating gate technology.

The static RAM technology uses passage transistors with a state controlled by one bit of RAM. In case it uses SRAM, the data will be written into a static RAM.

The antispindle technology puts in short circuit certain previously selected connections by the user. An antispindle is a two-terminal device that irreversibly changes a discharge connection to a low resistance when electrically programmed by high voltage.

In the floating gate technology, the key is a floating transistor gate that can be turned off by injecting charge into the floating gate. The charge can be removed, later on, by exposing the gate to ultraviolet light (UV), which is the same as in the EPROM technology or by using an electric voltage (technology EEPROM).

The design of an FPGA also consists of three steps:

1. Simulation and logical design
2. Checking of the placement of components, routes, and connectivity
3. Programming

Current FPGAs offer an equivalent complexity of 20.000 conventional gates (arrangements), and the clock frequency can go from 40 to 60 MHz.

The main advantage of FPGA on programmable ASIC with the mask is the possibility of cheap and fast reprogramming.

9.2.6 PLDs (Programmable Logic Devices) in Instrumentation

PLDs are uncommitted arrangements of AND or OR gates that can be organized to accomplish dedicated functions through the interconnection of these gates. Most recent PLDs have additional elements such as the following:

- The macro cell of logical outputs
- Own clock
- Safety fuse
- Buffers with a tri-state output
- Feedback of the programmable output

Most common PLDs are PAL (programmable array logic) and GAL (generic array logic).

The PLD programming can be made by fuses (in PAL) or by EEPROM or SRAM reprogrammable technologies. Current PLD may have 8.000 gates and frequencies up to 100 MHz.

The main advantages of PLD compared with FPGA are the speed and ease of use without incurring in engineering costs. In contrast, the size of PLD is smaller than that of FPGA.

9.3 COMPUTER-BASED SYSTEMS INSTRUMENTATION

9.3.1 Essential Components

To allow the use of a computer to control a group of measuring instruments, it is necessary to match the equipment under test and the computer. The primordial differences are as follows:

- Variable signals to the equipment under test, too high frequencies, mechanical systems, and high voltages
- The computer of which it is an integral part of the software, the data output, and input
- Interface to execute the communication [11–15]

9.3.2 Tests Operated by Computer

Complexity, duration, and repetitively demanded details in the execution of certain tests, for instance, a test of electronic components, may demand an automatic computer operation. For this, dedicated interfaces were developed for general instrumentation and tests. The most important of them is the parallel DIPI IEEE488 (digital interface for programmable instrumentation) for parallel buses. Initially, it was designed for 8-bit computers; however, it may be adapted for another number of bits in larger computers. Another more recent interface is the RS485, which has been much used for communication between process and computer with 32-bit serial data or more.

9.3.3 The Parallel Interface IEEE488

Figure 9.7 illustrates the analysis of a device under test (DUT) with the interface IEEE488.

The IEEE488 is an 8-bit parallel word transmission bus. Several bits of STATUS are also transmitted, but on separate lines from the main bus. This interface is for short distance data transmissions (inside of a room), not being suited for telephone, or other low noise communication means at a distance. It can be used, typically, for less than 15 pieces of equipment in one or two parallel linked racks, at up to 3 or 4 m of distance using whip connections.

The response time of individual systems of the test varies, depending a lot on whether they are of mechanical, electric, or electronic. Equipment, as the frequency synthesizer, may also have very slow signals. The logical levels used in interface IEEE488 are shown in Table 9.4.

In logic levels, the presence of high frequencies and other noises, where the reactance of cables influences more than their resistances, is common, and, for this, cables of low inductance should be used. To immunize the data acquisition against

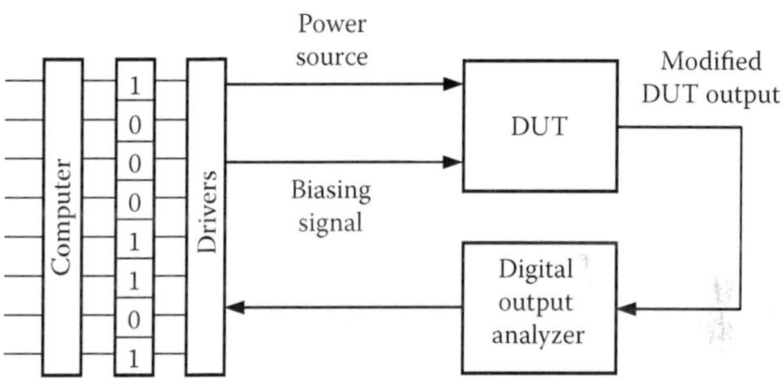

FIGURE 9.7 Example with the interface IEEE488.

TABLE 9.4
Logical Levels Used by IEEE488 (TTL)

Logical Level	Electrical Variable	Immunization Level
0	<0.8 V	<0.5 V
1	>2.0 V	>2.4 V
Open	<48 mA	<0.5 V
		Suitable for logical drivers

noises, logical drivers should be used with good reference levels on ground. These levels should not vary more than some hundreds of mV. The immunization levels refer to the values recommended for drivers and typically for the common collector transistor base, which needs a current above 48 mA or a base-emitter voltage of 0.5 V in case the "drivers" are in the high impedance state.

Note that an AND gate should be used to establish synchronization between the bit signals DAV, NRFD, and NDAC so as to avoid the data bus transfer to be faster than the slowest unit. Figure 9.6 displays the connection details of this gate.

The pins of IEEE488 are shown in Table 9.5 with the numeric position of each bit. The terminology of pins is in Table 9.6.

The messages are transferred to bits of the resident interface, as represented in Figure 9.8, starting from the devices "talkers" (announcers that emit digital signals) to "listeners" (i.e., they receive digital signals) under the controller's command.

TABLE 9.5
Pin Identification for the IEEE488

Pin #	Function	Pin #	Function
1.	Data	13.	Data
2.	Data	14.	Data
3.	Data	15.	Data
4.	Data	16.	Data
5.	EOI	17.	REN
6.	DAV	18.	gnd
7.	NRFD	19.	gnd
8.	NDAC	20.	gnd
9.	IFC	21.	gnd
10.	SRQ	22.	gnd
11.	ATN	23.	gnd
12.	Shield	24.	Logical content

TABLE 9.6
Pin Description of the Interface IEEE488

Bit Pin	Meaning	Description
DAV	Data valid	Waits for the device's settling time
NRFD	Not ready for data	Receiving unit is not ready to receive data
NDAC	Data is not accepted	When high, the receiver does not accept data, in spite of being ready
ATN	Attention	It puts into action the bus devices that should answer
IFC	Interface clear	Bus in the quiescent state
SRQ	Service requested	To request or interrupt service; example: to interrupt test if $I > I_{max}$
REN	Remote enable	It allows the choice of two alternative sources of data programming
EOI	End or identify	As a talker, conversation ends; as a listener, communication begins

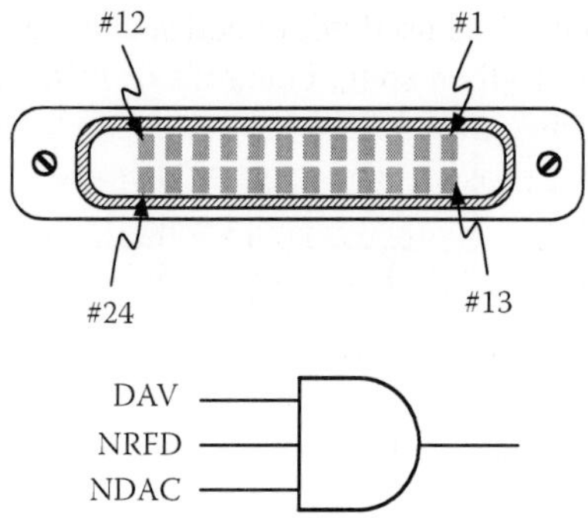

FIGURE 9.8 Interface pins of the IEEE488.

Some equipment are "talkers," other "listeners," and still other both. For example, a signal generator is usually a listener if it is adjusted by a computer. It receives signals to behave in this or that way. Voltmeters are "talker," because they can indicate values to be emitted to the computer; in other words, they inform what was measured. A frequency counter behaves like both, i.e., it indicates values, and they can be adjusted by a computer in time or frequency.

In Figure 9.8, the external interface connected to the resident interface of IEEE488 is also represented. Several standard circuits can be used in this external interface as we will see below. In all of them, it is always good to remind the basic rule for connecting digital circuits: the device input can never stay open, and the output can never be physically grounded or short-circuited.

9.3.4 Example

9.3.4.1 Frequency Counter (Hardware)

The frequency counter can be a "talker" or "listener" accordingly to the transceiver demand. The messages received by the computer are decoded and then generated a command to the counter so it can work as a "talker" or a "listener." Figure 9.9 represents the block diagram of this assembly using standard IEEE488.

Assuming a digital frequency counter then, the data control and values will be just put on the bus, one at a time. According to every case, it is used as a shift register or a multiplexer associated with interface circuits that match the electric requirements of the data bus.

9.3.5 Input and Output Drivers as Open Collectors (Transceiver)

The input and output drivers use a "pull-up" or "pull-down" resistor to supply well-defined voltage levels in case the "drivers" are in the high impedance state (maximum current of 48 mA and voltage lower than 0.5 V on ground). Figure 9.10 illustrates this driver.

Gates with Schmitt-trigger can be used to prevent the excessive noises at the input. The receiver has yet to contain a diode to clamp negative cycles of the voltage

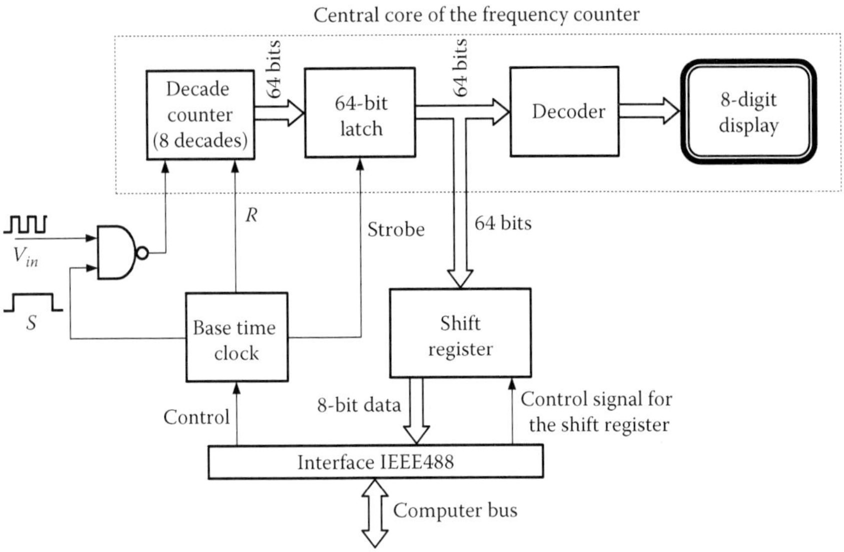

FIGURE 9.9 Frequency counter.

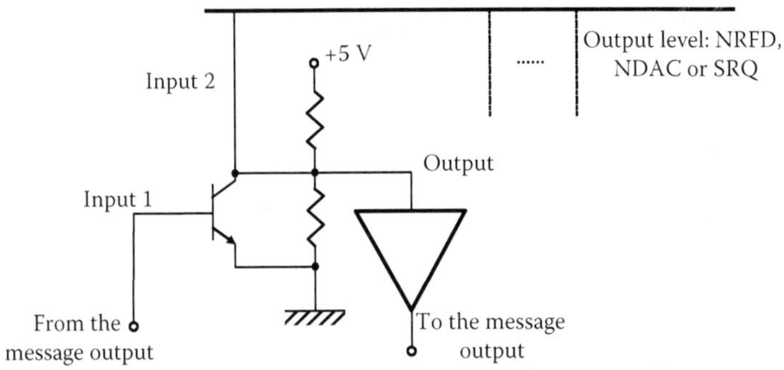

FIGURE 9.10 Bus transceiver (driver with open collector).

generated by the short-response times of signals, besides to prevent damages and attenuate line signals. Almost all TTL gates and Schmitt-triggers contain this diode.

9.3.6 Output Drivers

The output drivers use a 3-state bus transceiver, as displayed in Figure 9.11. The logical states of this bus are shown in Table 9.7.

In the interface IEEE488, the cables should have 24 shielded drivers with more than 85% efficiency, i.e., $V_{out}/V_{in}>0.85$. Of these, 16 drivers are used for data and states, while the remaining ones are for the return of signals, grounding, and screening.

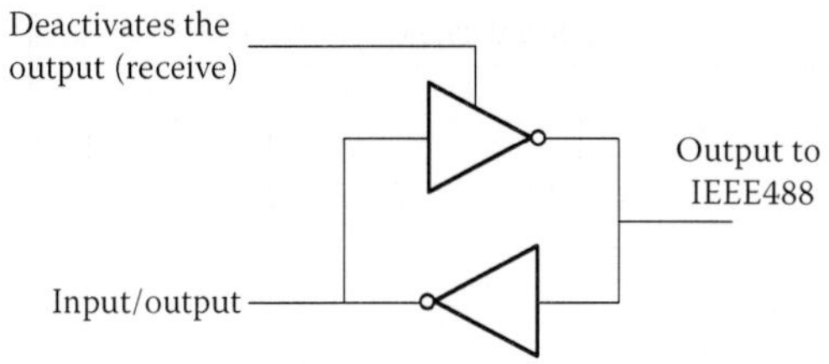

FIGURE 9.11 The 3-state bus transceiver.

TABLE 9.7
The Logical States of the 3-State Bus Transceiver in Figure 9.11

Logical State	Comments
0	It allows current <48 mA and voltage <0.5 V.
1	In cases where the current is not enough to charge the line capacitance for when resistors have to supply the current of the cable capacitances, the 3-state driver can supply itself even with high currents.
∝	It does not load the line except through the terminal resistors.

The lines DAV, NRFD, NDAC, IFC, ATN, EOI, and SRQ are coiled with the ground wire to avoid interferences between them and ground and to make the stray capacitances to be minimum. Typically, the coiled pairs are put in the center of the cable with the data drivers arranged around.

9.3.7 Serial Interface IEEE485

The standard EIA/TIA-485 or, simply, RS485 allows a transmission line to be shared in split-line or multidrop modes. They can share up to 32 drivers/receptors in the multidrop network whose characteristics are similar to another well-known standard, the standard RS422, just that in this case the range of common mode voltage is expanded from −7 V up to +12 V. As a bus driver can be disconnected or put in the high impedance state, it should support this range of common mode voltages. Figure 9.12 displays a typical multidrop network of two wires using the standard 485. Note that an end line resistor is used at the beginning and the end of the line but not in intermediary line drops. These endings should just be used in the cases of high data transfer rates and very long lines. A line of the ground signal should also be used for the standard 485 to maintain the common mode voltage between −7 V and +12 V that the receiver will receive. In the following sections, this interface is used to discuss some topics related to data communication between the process and the computer.

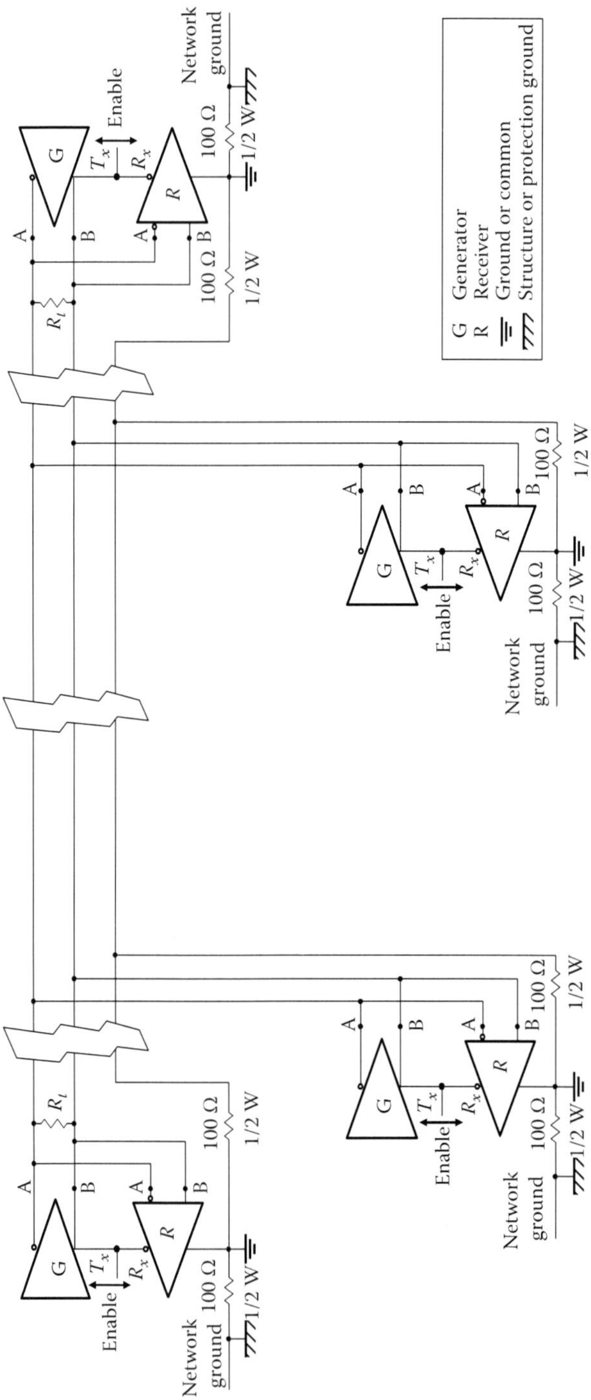

FIGURE 9.12 Typical multidrop 2-wire network with standard 485.

9.3.8 USB Connectors

The Universal Serial Bus (USB) is the fastest connector for data transfer (see Figure 9.13), in which any computer has one or more USB gates (connectors). These USB ports allow connecting several external devices, ranging from mice to printers. The operating system also supports USB interface, so the device driver installation is quick and easy. Compared with other ways of device connection (including parallel ports, serial ports, and special cards installed inside the machine cabinet), USB devices are straightforward. USB 3.0 hit the market in 2009, commercially known as USB SuperSpeed (SS). It is characterized mainly by an increase in data transfer speed because it has more connections than in model 2.0, which were 4, while in model 3.0, they reach 9, being able to receive and send data at the same time. Generally, motherboards with USB 2.0 connections can use the benefits of USB 3.0. Table 9.8 is a table comparing the most popular USB cables.

In the past, printers were interconnected by specific and bulky parallel ports. External devices such as ZIP drives would require a high-speed connection to the computer using the parallel port, but personal computers would usually have low speed. Serial ports provided another typical connection in old computers. Modems used the serial port, just as many printers and a variety of devices such as laptops, Palm Pilots, and digital cameras. Most computers have at most two serial ports and are often very slow. Devices that earlier needed faster connections came with their cards, which fit into expansion slots inside the computer case. Unfortunately, the number of expansion slots is limited, and, in some cases, a skilled technician is

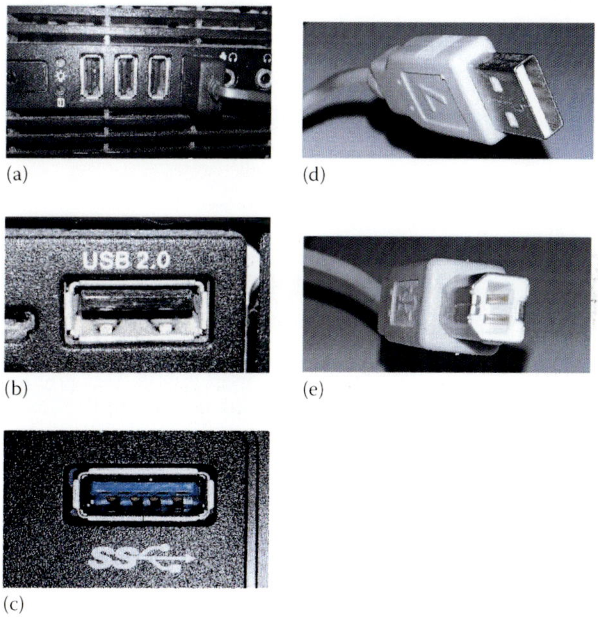

FIGURE 9.13 USB typical connectors: (a) female connector 1.0, (b) female connector 2.0, (c) female connector 3.0, (d) male connector type "A," and (e) male connector type "B."

TABLE 9.8
Comparison Chart of the Most Common USB Connections

USB Type	Connector	Speed
1.1	Standard	12 Mbps
2.0	A	480 Mbps
2.0	B	480 Mbps
3.0		4.8 Gbps
3.1 Gen 1	SuperSpeed (SS)	4.8 Gbps
3.1 Gen 2	SuperSpeed (SS)	9.7 Gbps

required to install the software. The goal of USB is to end these difficulties. The USB provides a single, standardized, and easy way to connect up to 127 devices on a computer. Currently, almost all peripherals are available in a version for USB. Examples of USB devices that can be purchased nowadays include the following:

- Printers
- Scanners
- Mouse
- Pen drive
- Joysticks
- Consoles for flight simulators
- Digital cameras
- Webcam
- Devices for the acquisition of scientific data
- Modems
- Speakers
- Phones
- Video phones
- Data storage devices
- Network connections

To connect a new device, the operating system autodetects and requests the disk drive. If the device has already been installed, as in most of the cases, the computer activates and starts the communication process. USB devices can be connected and disconnected at any time.

9.3.8.1 USB Hub

Latest computers have typically two or more USB ports, but with so many USB devices on the market, generally it is not enough. Suppose on a computer there is a printer, a scanner, a webcam, and a network connection, all with standard USB. If the computer has only two connections, the easiest solution is an inexpensive USB hub. The USB standard supports up to 127 devices, and the USB hub also is a part of this standard (see Figure 9.14).

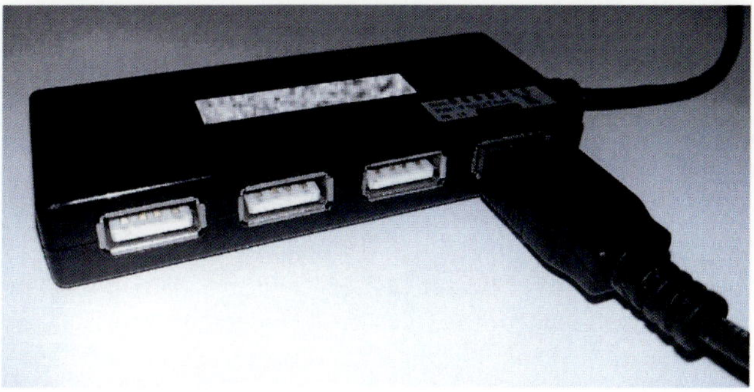

FIGURE 9.14 A typical USB four-port hub supports up to four connections of type "A."

A hub typically has four new ports, but it is not limited to this capacity. The hub is to be plugged into a computer, and then the devices (or other hubs) can be turned on. Chaining up several hubs can have many USB ports available on a single computer.

There are hubs that work with or without power. The USB standard allows devices to get their power from the USB connection. Obviously, a high-power device like a printer or scanner must have its power supply. However, low-power devices, such as mice and digital cameras, get their power from the bus to simplify them. The computer supplies the energy (up to 500 mA at 5 V). With many self-powered devices (like printers and scanners), the hub does not need to be fed if none of the devices connected to the hub needs additional power. If there are multiple devices that do not have their power, like mice and cameras, then an energy-powered hub is required. In this case, the hub has its converter providing power to the bus, so that the devices do not overload the computer's power supply.

9.3.8.2 USB Process

When the host (computer) starts, it checks all the devices connected to the bus and assigns an address to each. This process is called enumeration. The devices are also enumerated when they connect to the bus. The host also finds each device, connected to the hubs within 5 m.

The types of USB data transfer modes are as follows:

- Interruption used in devices such as a mouse or keyboard, which do not send much data.
- Bulk used with devices like a printer receiving data in large packets. A data block is sent to the printer (in 64-byte fragments) and checked to ensure it is correct.
- Isochronous streaming device (such as speakers and microphones) uses this mode. In this mode, data flow between the device and the host in real time, and there is no error correction.

The host can also send commands or verify parameters with control packets. As devices are enumerated, the host keeps a general registration of bandwidth demanded by all isochronous devices and may interrupt the device processes whenever required. They can consume up to 90% of data transfer rate of the available bandwidth. After 90% of use, the host denies access to any other isochronous or interrupt device. At least 10% of the transfer of large data packets and control use any other remaining bandwidth.

The USB divides the available bandwidth into frames, and the host controls these frames. Frames contain 1500 bytes, and a new frame starts every millisecond. During a frame, isochronous and interrupt devices get a slot, so the necessary bandwidths are guaranteed. The transfer of data packets and control use any space left.

The most common USB technology, so far, is the USB 2.0, USB 2.1, USB 3.0, and USB 3.1. The USB 2.0 has a maximum data transfer rate of 60 megabytes per second, works with current up to 500 mA at 5 V, and sends or receives data one at a time. In contrast, the most recent USB 3.0 has a maximum data rate of 600 megabytes per second, with current up to 900 mA at 5 V, and it can send and receive data simultaneously. Then, the USB 3.0 is faster and more efficient, can supply more devices, and covers all the previous features of USB 2.0.

9.3.8.3 USB Features

A USB cable has two wires for power (+5 V and ground) and a twisted pair to carry data (see Figure 9.15). On the power wires, the computer can provide up to 500 mA/900 mA of power at 5 V. Devices connected to the USB port rely on the USB shielded cable to carry power and data. Inside a USB cable, there are two wires for power, +5 V (red), at least one ground wire (brown), and twisted set (yellow and blue) wires to carry the data.

USB devices are hot-swappable (hot exchangeable), i.e., they can be connected and disconnected at any time. The various USB devices can be placed in sleep mode (hibernate) by the host computer when the computer enters the power-saving mode.

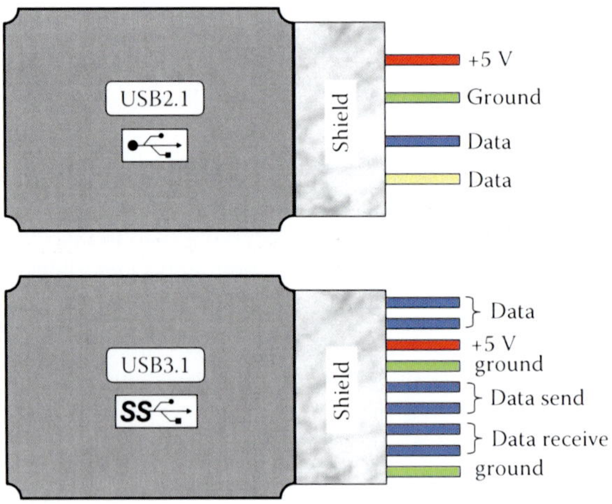

FIGURE 9.15 Cables inside the USB connector and symbols.

9.3.9 Signals of Data Transmission

The data transmission lines can be balanced or not. In drivers with no balanced line, each signal transmitted in a data transmission system appears at the interface connector as a voltage signal on the ground signal. For instance, a transmitted datum (TD) from a device DTE appears in the pin 2 on 7 (ground) in a connector DB-25. This voltage will be negative if the line is disabled or alternates between negative and positive levels, when the data are ordered in between +3 V to +12 V and −3 V to −12 V.

In the case of balanced and differential lines, the voltage produced by the driver appears in a pair of line signals that transmit just one signal. A driver of a balanced line produces a voltage from 2 to 6 V between the output terminals A and B and it has a ground connection to signal C. Despite the ground signal connection being important, it is not used by a balanced line receiver to determinate the logical state of a data line. The balanced line driver may also have an input signal called "enable" necessary to connect driver 485 to its output terminals, A and B. If the "enable" signal is low, the driver is considered disconnected from the transmission line. The disconnected or disabled condition of the line driver is usually known as "third state" to distinguish it from states "1" and "0."

The receivers of a balanced differential line sense the voltage state in terminals A and B of the transmission line beside a third ground signal, C, which is necessary to make the appropriate interface connection. If the differential input voltage is higher than +200 mV, the receiver will have a specific logical state across the output terminal. If the voltage is inverted to less than −200 mV, the receiver will create an opposite logical state at the output terminal. The input voltages of a balanced line receiver should be in the range from 200 mV to 6 V to take into account for the attenuation across the transmission line. Figure 9.16 displays an outline of connections to determinate signals "+" and "−" in correspondence with the standard EIA signals "A" and "B" of the RS485 bus. Note that, in the deactivated condition, it is possible to determine which terminal is A and which one is terminal B with one or two DVM multimeters, as shown in Figure 9.16.

The standard RS485 can also be connected in the 4-wire data mode plus one ground wire. In this 4-wire network, one is the master and the others are slaves, and the master may communicate with all slaves. In pieces of equipment using a mixed

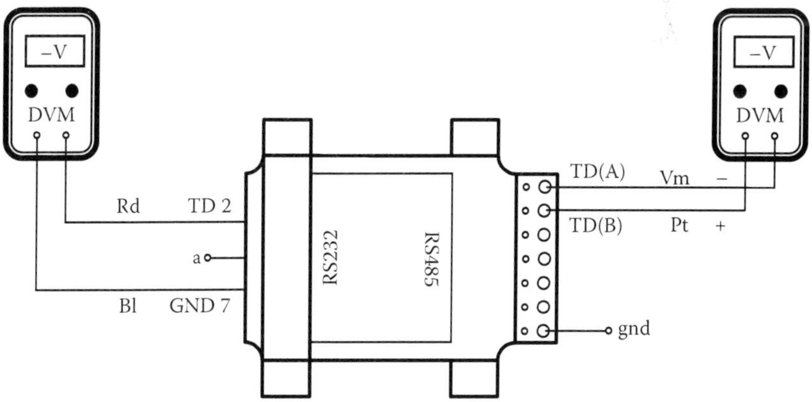

FIGURE 9.16 Convention of signal correspondences between standards RS232 and RS485.

communication protocol, the 4-wire mode has advantages, because no slave can hear the answer of the other slaves to the master. Therefore, a slave cannot give an incorrect answer for any other slave's knot.

In a standard RS485, any driver whose knot is not transmitting a signal has to be able to be disconnected. In particular, in the converter from RS232 to RS485 and a serial board RS485, this can be made by using a high logical control signal in the RTS line from an asynchronous serial gate that enables driver RS485. A low level of this signal puts it in the tri-state, allowing other knots to use the same data line. The timing of these states to control the gate is made by software and not by the standard converter.

When network RS485 is used in the 2-wire split-line multidrop mode, the receiver in each knot will be connected to the line. Many times, the receiver can be configured to receive the echo of its data transmission. This can be good in some systems but very bad in others; therefore, it is better to ensure how the function will be connected in the standard converter.

9.3.10 RS485 Network Biasing

In the disabled state, all knots of network RS485 are in the mode listener (signal reception). All drivers are in the tri-state. If voltage levels at the receiving inputs are below –200 mV, the logical level at the output will be at the value of the last received bit. To assume, then, that the levels are in state "1," it should use a biasing resistor to raise the state (pull-up, typically, to 5 V) in data B line or to lower the state (pull-down, to ground level) across data line A in such a way to force these lines to the disabled condition (see Figure 9.17). Observe that in the case of the 4-driver RS485, biasing resistors should be put in the receivers [8].

The value of the biasing resistors depends on the ending and number of knots of the system. The goal is to generate a DC biasing current through the net to maintain a minimum of 200 mV across data lines A and B.

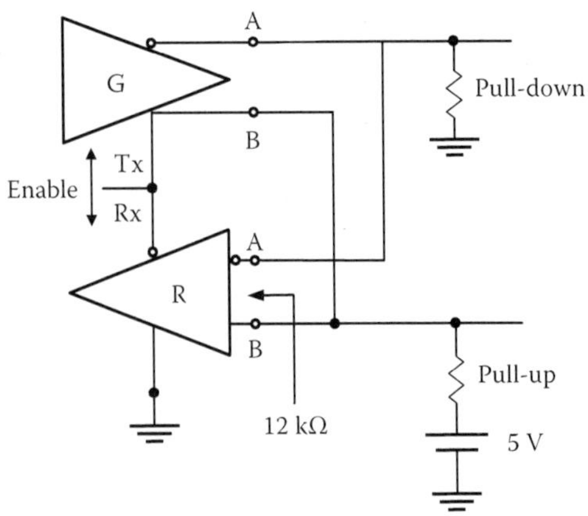

FIGURE 9.17 Biasing resistors of network RS485.

9.3.11 Calculation of Biasing Resistors

Let us suppose we wish to dimension the biasing resistor of a network RS485 for 10 μs with two ending resistors of 120 Ω each. Each RS485 knot has a load impedance of 12 kΩ. Therefore, 10 Ω resistors in parallel represent a load of 1200 Ω. Besides, the two ending resistors represent an additional load of 60 Ω. This value in parallel with the load impedance of each knot gives 57 Ω. To maintain a minimum of 200 mV across terminals A and B, a biasing load current of 35 mA is necessary. As a consequence, to obtain this current from a 5 V source, a total of 1428 Ω or less is necessary. Subtracting 57 Ω, which was already part of the load, results in 1371 Ω. By putting half of this value as a pulling up resistor for 5 V, and the other half as a pulling down resistor to ground, the maximum value of 685 Ω is obtained for each one of the biasing resistors.

9.4 DIGITAL CONTROL

Any acquisition of digital signal needs two functions: those of the device/process and those of the interface. The functions of the device/process are the variables that depend on the type of equipment being tested. For instance, a signal generator can have functions of signal width and modulation, which are not applied to frequency counters. The interface functions are defined for the system, and they are always the same ones for any test instrumentation connected to the data bus, including determination of data states, zeroing, and deletion.

When the source is energized, or when the microprocessor indicates that the deletion interface is put in logical state "1," all units are initialized. The microprocessor adjusts all devices with a deletion message (multiple lines). The processor sends the feeding source address and data to the device. The power is then supplied to the device, and the test begins. The source keeps waiting for new addressing, and the processor sends the address of the signal generator and the width and frequency data. Figure 9.18 represents the "status" of the output bit of time instants during a signal transfer. Table 9.9 illustrates the status of the clock signals from 0 to 5.

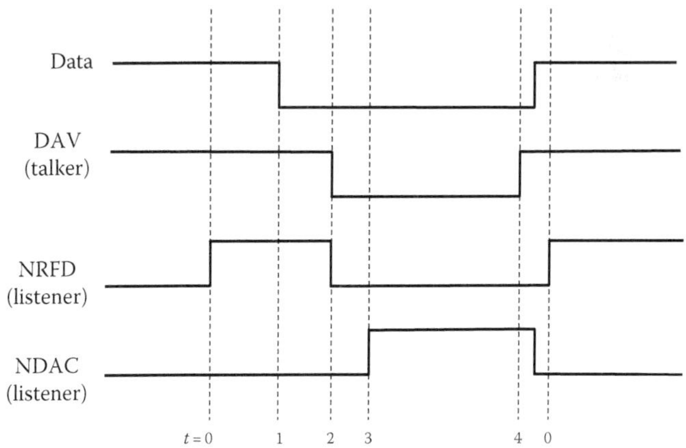

FIGURE 9.18 Output bit status for instants 0, 1, 2, 3, 4.

TABLE 9.9
Output Bit Status along Clock Signals from 0 to 5

Time Stamp	Status
0	Listener ready to receive data
1	Data in the line
2	Talker indicates data valid
3	Data accepted by the listener
4	Talker indicates that the data are not necessary anymore and may be replaced

The operation of the interface IEEE488 is completely asynchronous, requiring a finished task before beginning the other. The processing speed can vary from the lowest possible to as high as 1 Mb/s so that it is the maximum specified for this example.

9.4.1 Instrumentation Controlled by Computer

The compatibility between computer and process should be observed in all stages of the hardware and software so as not to have surprises during the control process. We can divide the compatibility analysis into three interfaces:

1. Signal interfaces according to their characteristics should be adapted to the device/process under test to assist compatibilities as type of signal (electric, mechanic, electronic, etc.), level of signal (much beyond or below this side of the maximum limits accepted by the data bus of the computer or electronic interface), and variation speed (position, frequency, etc.).
2. The equipment interfaces for reception of signals in the data bus from the external interface board.
3. Software and hardware interfaces, depending on each manufacturer, can be completely incompatible.

9.4.2 Programmable Logic Controller

The programmable logic controllers (PLCs) are digital computers used mostly for industrial and automation processes. They are designed for multiple input and output arrangements and aim to implement specific functions of control and monitoring. Figure 9.19 shows a general architecture of PLC, where its hardware structure

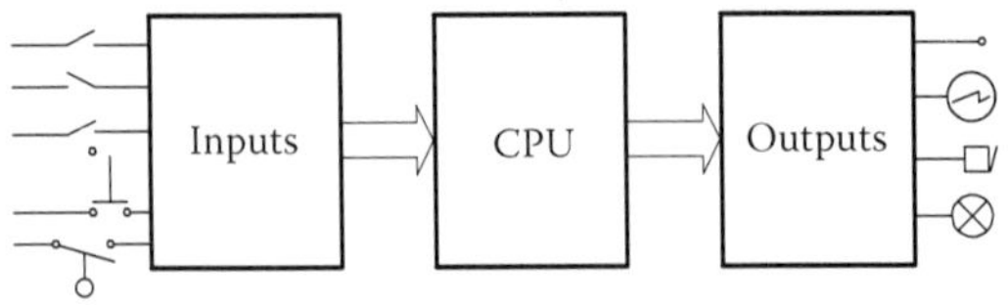

FIGURE 9.19 Programmable logic controller architecture.

consists of the power supply, CPU, and input/output (I/O) modules. The inputs are continuously monitored and processed by the CPU, through the control instructions stored in an internal memory, and then the outputs are updated; this procedure occurs cyclically. Thanks to the universal hardware structure, PLC can be applied to many sorts of applications and communicate with other devices [15,16].

More specifically, the application principles of PLC consist in gathering, continuously, input data from buttons, switches, and measuring devices and provide output commands to actuate machines and automation processes. In a PLC, the inputs are processed at high speed by a set of interconnected and sequential control instructions (*program*). From there, they are stored in non-volatile memory and programmed by the user to perform certain control and monitoring functions. At the end of its loop operation, the PLC performs the housekeeping stage, which includes communication with external devices and hardware diagnostics. Figure 9.20 shows the cyclic application and operation principles of PLC.

Based on the IEC 61131-3 programming industrial automation systems, five standard programming languages have emerged for PLC control instructions: *ladder diagram logic*, *function logic diagram*, *sequential function chart*, *structure text*, and *instruction list*. Among these languages, the ladder diagram programming is the most commonly used. The ladder programming language originated from the ladder logic, and it is based on electrical circuit diagrams of relay logic and is used from simple binary functions to complex mathematical expressions. It is usually applied to the continuous or parallel execution of multiple nonsequential operations, or processes requiring *interlock*.

The function block diagram (FBD) is a high-level graphical programming language based on standardized Boolean logic, where the blocks are "wired" together to establish a sequence. Its main advantage is the simplicity being more suitable for simpler programs consisting of digital inputs. The sequential function chart (SFC)

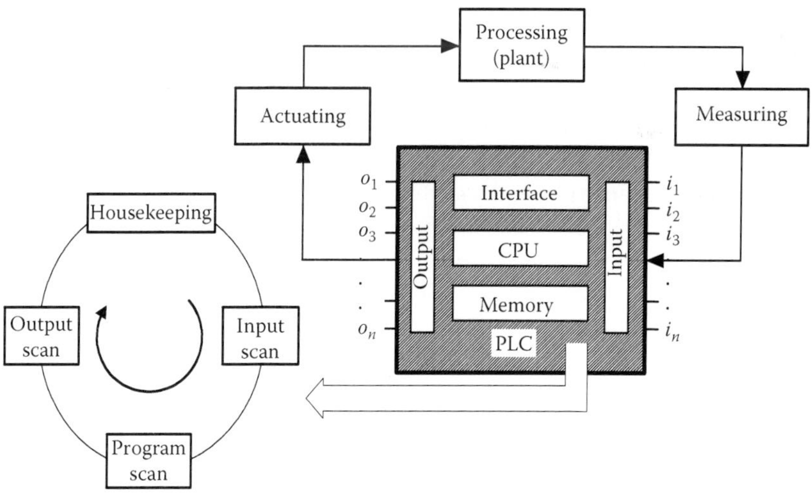

FIGURE 9.20 Application and operating principles of a PLC.

is also a high-level graphical language, which resembles the flowcharts. Its operating principle is based on the sequential action of programmed boxes. The latter are programmable blocks where the programmer can write codes through any sort of language, e.g., C/C++. SFC provides a quick and direct view of an implemented process.

The structured text is a high-level textual programming language, which remembers Pascal, BASIC, and C languages. This programming language is based on the Boolean logic, IF, WHILE, and case selector instructions. The instruction list is a sequential textual language based on patronized commands, like Assembler language. The main disadvantage of textual languages is that they are more complex than graphical languages, and verification of the code is harder, hindering in finding the possible code errors.

PLC is widely applied to industrial and automation processes, such as processing control, manufacturing automation, integration of automation systems, manufacturing and assembly lines, building automation, and power plant control, wherever controlling, automation, and monitoring functions are required. The tasks of PLC have been rapidly multiplied: timer and counter functions, memory setting and resetting, and mathematical computing operation that can be executed by any of today's PLCs. Some of the functions and their usual applications are as follows:

- *Sequencing*: used for machine tools, e.g., Computer Numerical Control (CNC), to control axis position, torque, forward speed, and acceleration, among others.
- *Proportional-Integral-Derivative (PID) controller*: commonly used in metal-mechanical industry, chemical industry, petrochemical industry, textile factory, and power plant generation to control quantities such as position, rotation, speed, temperature, pressure, flow, force, voltage, and power, among others.
- *Interlocking*: used for production and automated assembly lines, frequently substituting the original commands based on relays. It is usual in high-risk applications, where reliable systems are required, such as petrochemical industry, nuclear power plants, and chemical industry.

The digital processing controls based on PLC present many advantages over relay logics. They are modular and can be configured individually, and they are endowed with the ability to change and replicate the operation while collecting and communicating information; both characteristics facilitate the system's expansibility. Besides, they are more reliable, smaller, lighter, easy to communicate with other devices, robust, and suitable for the industrial environment.

Besides the advantages mentioned above, the PLCs possess high flexibility, thanks to their electrical-based characteristics that allow them to combine many technologies. Many sorts of input technologies can be used, e.g., switches, pushbuttons, and sensing devices (e.g., proximity, rotation, speed, temperature, pressure, flow, and photoelectric sensors), since their outputs provide a voltage signal within the allowable voltage range of PLC's input. Commercial signal converters, as analog-digital (A/D) converters, may be used to handle the voltage levels between sensor's output and PLC's input. In a similar way, many output technologies can be used, e.g., valves, motor starters, solenoids, and actuators (alarms, stack lights, counter, pumps, printers, fans, current regulators, etc.), since their input voltages match with the PLC's output voltages. Control relays can be used to handle voltage and power levels between PLC's output and actuator's input. Therefore, any sensing and actuator, undependable of their technologies, powers, and voltages, can be combined through a PLC, as shown in Figure 9.21.

The LOGO PLC from SIEMENS shown in Figure 9.22 is suitable for small-scale automation tasks. It is a flexible option to replace large numbers of conventional switching and control devices. The LOGO modules consist of a display, digital and analog modular expansion, and communication units, among others.

Some LOGO modules, e.g., OBA 7, provide Ethernet ports through devices that can be connected as master/slave or master/master architecture. All LOGO modules can be programmed by the *Soft Comfort* software, which allows all the five aforementioned standard programming languages. LOGO modules provide some basic

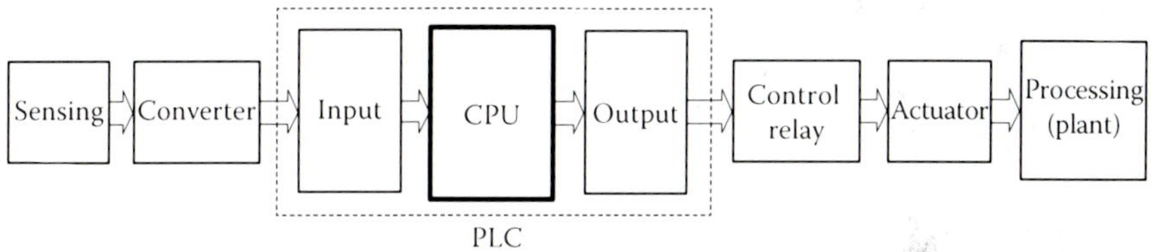

FIGURE 9.21 General PLC application to different types of input/output technology.

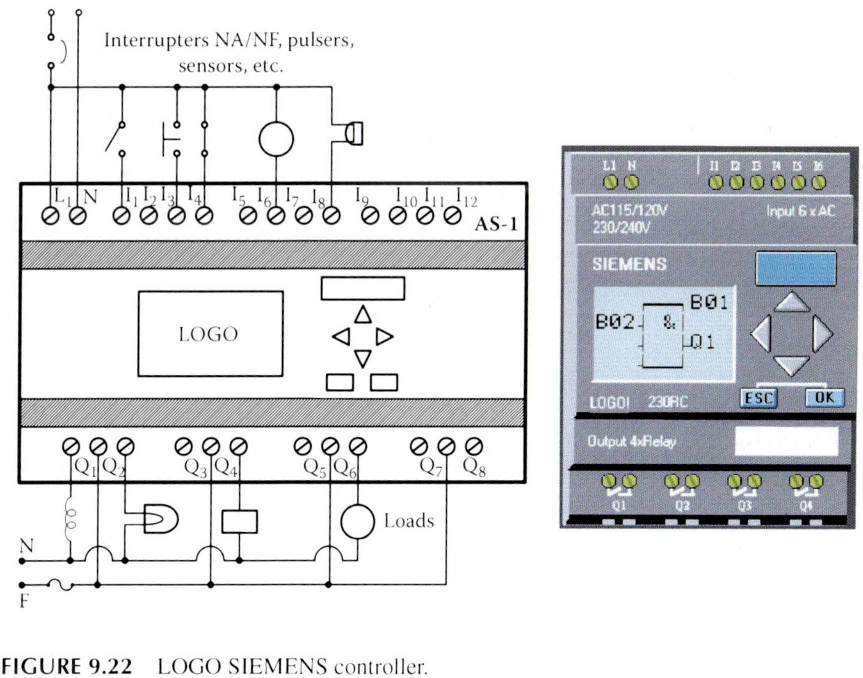

FIGURE 9.22 LOGO SIEMENS controller.

(Table 9.10) and specific (Table 9.11) functions for quick and simple programming, which can be commanded directly by a PLC or through a computer bus. Finally, it should be noted that the LOGO modules are affordable, and they have wide technical support and reference materials. In Table 9.12, the main PLC LOGO codes are reunited. There are a total of 30 blocks, a maximum of 7 blocks in series, and it is possible to download some commercial routine for CLP simulation for the following steps of project routine: written design, simulation, and implementation.

TABLE 9.10
Basic Functions of the PLC LOGO

Representation in Diagram Circuits	Representation in LOGO	Basic Function
 Series connection Closure contact	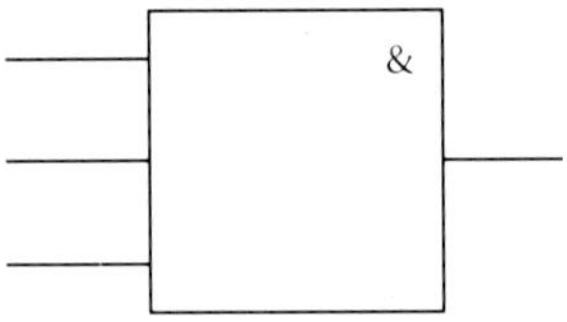	AND
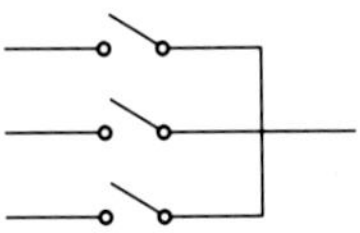 Parallel connection Closure contact	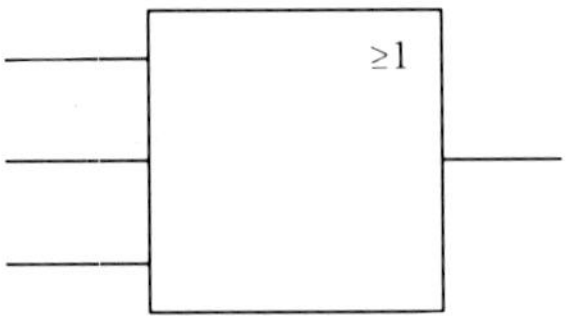	OR
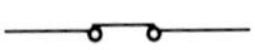 Inverter	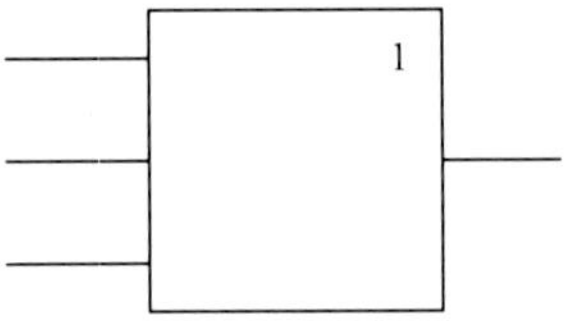	Inverter (NOT)
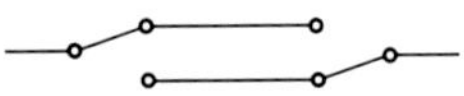 Dual alternator	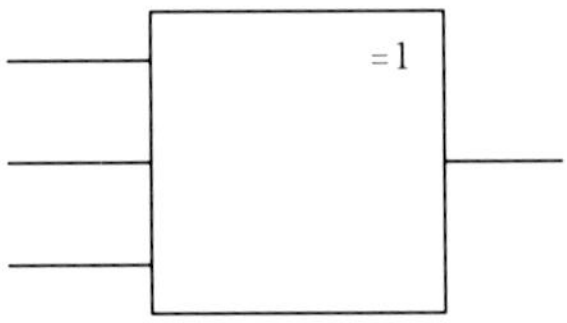	Exclusive OR (XOR)
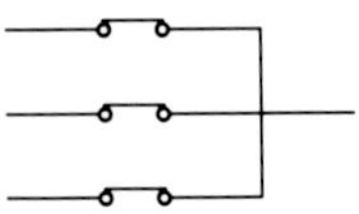 Parallel connection Opening contact	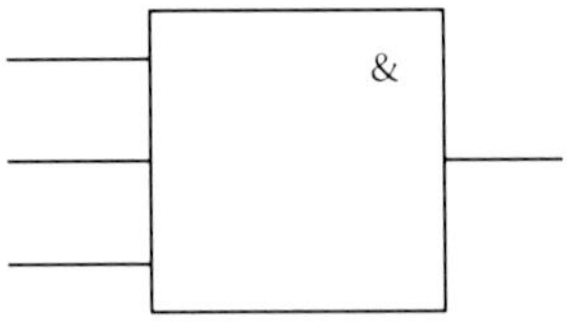	NAND
 Series connection Opening contact	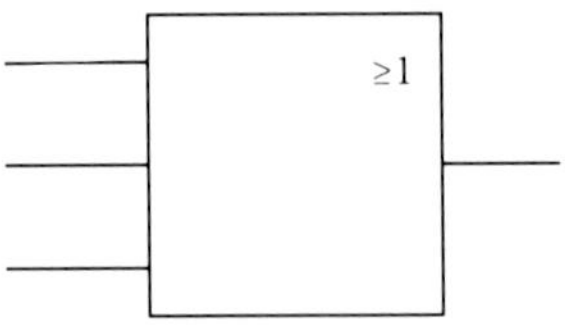	NOR

TABLE 9.11
Specific Functions of the PLC LOGO

Function	Representation in LOGO	Function	Representation in LOGO
Activation delay (trigger)	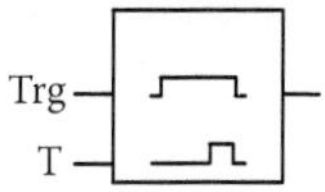	Memorable activation delay	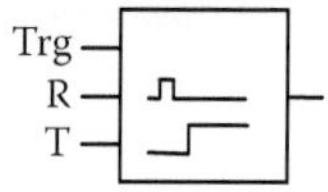
Deactivation delay	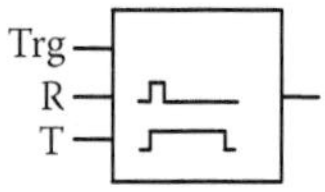	Up/down counter (dir = counting direction)	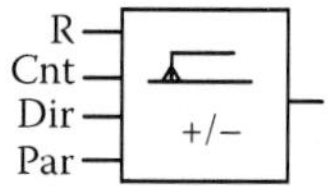
Impulse relay Pulse ≥ 50 ms	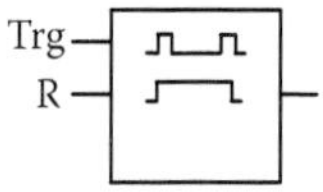	Counter of service hours (Ral = reset all)	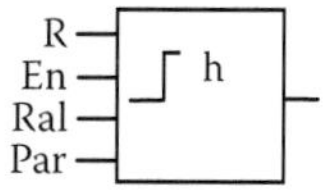
Timing clock Numbers 1, 2, 3	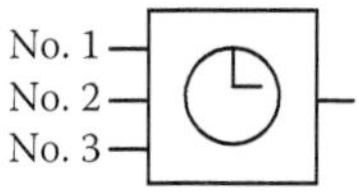	Dissipater relay/impulse output	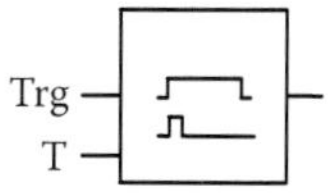
Self-retention relay (parameters)	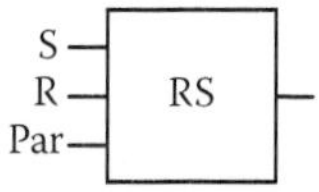	Limit value interrupter (max and min freq. meter)	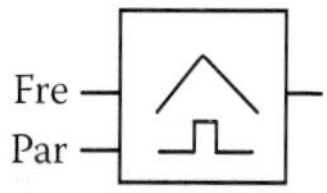
Impulse output when enabled	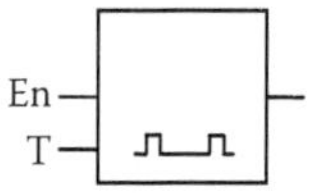		

TABLE 9.12
PLC LOGO Code Description

Code	Description	Code	Description
CO	Connector (I_1, I_2,…)	N/C	X
FS	Special function	Q_i	Output
GF	General function	I_i	Input
BN	Shared block		

EXERCISES

9.1 Mention the three main purposes of signal modulation in remote instrumentation and controls.

9.2 Draw the block diagram of a test system controlled by a computer using it to explain its operation.

9.3 Distributed control can be implemented with several computer modules communicating with one another and sometimes under a management layer. Typically, tasks are associated to computer modules as required by a parallel real-time controller. There are several ways to implement distributed control. Study books, Internet, technical references, papers, and reports and write a document describing how you can implement distributed control for smart grid applications.

REFERENCES

1. Horowitz, P. and Hill, W., *The Art of Electronics*, Cambridge University Press, London, England, 716pp., 2000.
2. Taub, H., *Digital Circuits and Microprocessors*, Publisher McGraw-Hill of Brazil Ltd, Rio de Janeiro, Brazil, 510pp., 1984.
3. Malvino, A. and Bates, D., *Electronic Principles*, 8th edn., WCB/McGraw-Hill, New York, 2015.
4. Simões, M.G. and Farret, F.A., *Modeling and Analysis with Induction Generators*, CRC Press, Boca Raton, FL, 2015.
5. Helfrick, A.D. and Jogging, W.D., *Modern Electronic Instrumentation and Techniques of Measurement*, Publisher Prentice-Hall of the Brazil-PHB, Englewood Cliffs, NJ, 324pp., 1994.
6. Catalog of B&B Electronics Mfg Co., Multi-interface Universal PCI Cards, Ottawa, IL, 2012.
7. Ipanaqué, W., Salazar, J., and Belupú, I., Implementation of an architecture of digital control in FPGA commanded from an embedded Java application, *2016 IEEE International Conference on Automatica (ICA-ACCA)*, Curicó, Chile, pp. 1–6, 2016, DOI: 10.1109/ICA-ACCA.2016.7778495.
8. Sozański, K., Signal-to-noise ratio in power electronic digital control circuits, *2016 Signal Processing: Algorithms, Architectures, Arrangements, and Applications (SPA)*, Poznan, Poland, pp. 162–171, 2016, DOI: 10.1109/SPA.2016.7763606.
9. Rajesh, D., Ravikumar, D., Bharadwaj, S.K., and Vastav, B.K.V., Design and control of digital DC drives in steel rolling mills, *IEEE International Conference on Inventive Computation Technologies (ICICT)*, Coimbatore, Tamilnadu, India, Vol. 3, pp. 1–5, 2016, DOI: 10.1109/INVENTIVE.2016.7830095.
10. Kanzian, M., Agostinelli, M. and Huemer, M., Modeling and simulation of digital control schemes for two-phase interleaved buck converters, *Austrochip Workshop on Microelectronics (Austrochip)*, Villach, Austria, pp. 7–12, 2016, DOI: 10.1109/Austrochip.2016.013.
11. Tektronix, *2016 Test and Measurement Solutions Catalog*, Tektronix, Portland, OR, 2016.
12. KIKUSUI Electronics Corp., *Operational Handbook*, Kikusui Electronics Corporation International Sales Department, Kawasaki, Japan, 2017.
13. AGILENT Technologies, Catalog of basic instruments, São Paulo-SP, Brazil, 2001.
14. Li, J., Gao, J., Yang, Y., Zhang, C., Li, W., and Zhang, X., Digital generator control unit for a variable frequency synchronous generator, *19th International Conference on Electrical Machines and Systems (ICEMS)*, IEEE Conference Publications, Chiba, Japan, pp. 1–6, 2016.
15. Bose, B.K., *Power Electronics and Variable Frequency Drives*, IEEE Press, New York, 640pp., 1997.
16. Kandray, D.E., *Programmable Automation Technologies—An Introduction to CNC, Robotics and PLCs*, Industrial Press, South Norwalk, CT, 2010, ISBN: 0831133465.

10 Software for Electric Power Instrumentation

10.1 INTRODUCTION

The previous chapters discussed how to represent the real world with digital signals for computer processing. Basically, the physical action and phenomena can be converted to voltage or current quantities through sensors and then translated and conditioned by the transducers for converting to digital levels by analog/digital converters. The main reason to convert an analog signal into a digital one is that digital processing allows programmability, and once real-world data is converted to digital data, a vast field of actions become possible, such as control, monitoring, and simulation.

Using proper software and adequate data acquisition systems, a computer can be turned into a universal instrument, as shown in Figure 10.1, where the real world outside of the computer is sensed and translated to digital signal, which could be handled through software programming and thereupon used to control external devices (e.g., lights, motors, power converters, and valves) through modulation techniques, such as pulse-width modulation (PWM).

Several software may be applied for such purpose; among them the LabVIEW, Arduino, and MATLAB® are underlined for power electronic application, especially for distributed generation systems. This chapter presents and describes these software applied to electrical power instrumentation.

10.2 LABVIEW DEVELOPMENT SYSTEM

LabVIEW (Laboratory Virtual Instrumentation Engineering Workbench) is a graphical control, test, and measurement application platform for measurement and automation that associates easy-to-use graphical development with the flexibility of a powerful programming language. LabVIEW was first launched in 1986, but has continued to evolve and extend into targets such as data acquisition, modular instruments, and embedded control and monitoring hardware. These packets include field-programmable gate arrays (FPGAs), sensors, microcontrollers, and other embedded devices.

The programs created in LabVIEW are called virtual instruments (VIs), since they are workstations endowed with a flexible software allowing modularity. Such graphical programs are based on the concept of data flow programming, which means that execution of a block diagram only occurs when all of its inputs are available. Then, the output data of the block are sent to all other connected blocks. The movement of data through the blocks determines the execution order of the VIs and functions. Data flow programming allows multiple operations to be performed in parallel, since its execution is determined by the flow of data and not by sequential lines of code [1].

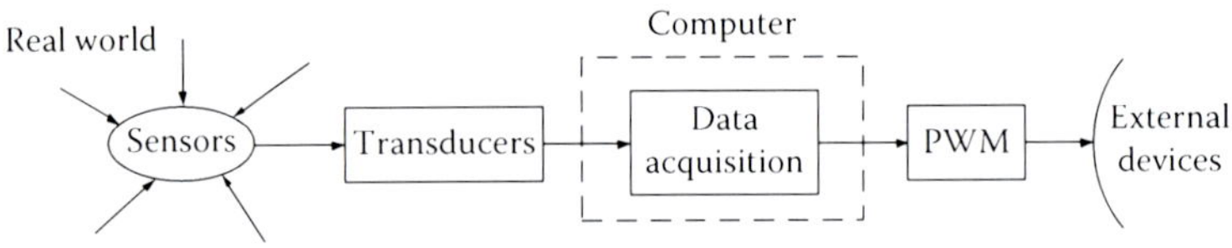

FIGURE 10.1 Universal instrument based on computer.

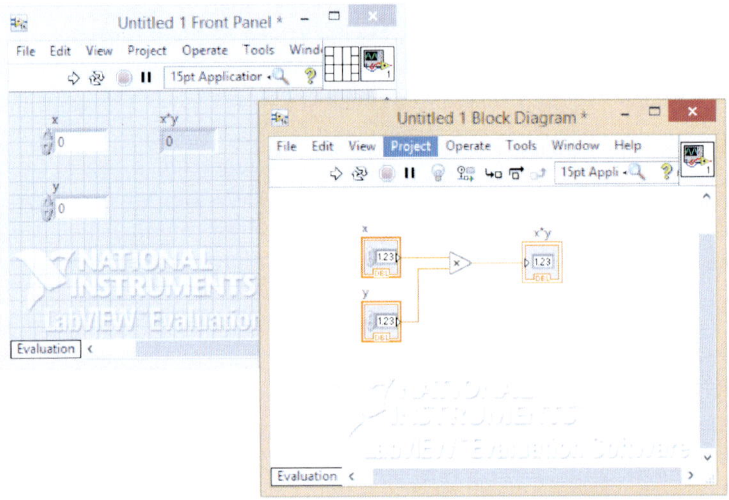

FIGURE 10.2 LabVIEW windows: front panel, block diagram, and icon connector.

A VI consists of three major components, which include a front panel (FP), a block diagram (BD), and an icon connector. The FP provides the user interface of a program, the BD incorporates its graphical code, while the icon connector defines inputs (controls) and outputs (indicators) of a VI. Figure 10.2 shows the three main components of a regular VI. When a VI is located within the BD of another VI, it is called a subVI. Both graphical programs VI and subVI are modular and can be run by themselves.

In the FP, one can add the number of inputs and outputs that the system requires. The basic elements inside the FP can be classified by controls and indicators. Knobs, pushbuttons, and dials are examples of controls, while graphs, LEDs (light indicators), and meters are examples of indicators. The general type of numerical data can be integers, floating, and complex numbers. Another type of data is the Boolean, useful in conditional systems (true or false), as well as strings, which are a sequence of ASCII characters that give a platform-independent format for information and data [2].

The BD consists of terminal icons, nodes, wires, and structures. Terminal icons are interfaces through which data are exchanged between an FP and a BD, and they correspond to controls or indicators that appear on the FP. Whenever a control or indicator is placed on an FP, a terminal icon gets added to the corresponding BD.

A node represents an object that has input and/or output connectors, and it performs a certain function. SubVIs and functions are examples of nodes. Wires establish the flow of data in the BD. Finally, the structures are used to control the data flow of a program, such as repetitions or conditional executions, e.g., *while* loop or *if-else* statement.

There is still the express VI, which is VI whose settings you can configure interactively through a dialog box. Express VIs appear on the block diagram as expandable nodes with icons surrounded by a blue field. The introduction of the express VIs allows designers faster and easier development of block diagrams for any type of data acquisition, analysis, and control application. Figure 10.3 shows the three main components of a VI with an express subVI (i.e., simulation signal).

Using loop structures, it is possible to repeat a sequence of programs or to enter the program conditions. A control loop structure is shown in Figure 10.3, where the

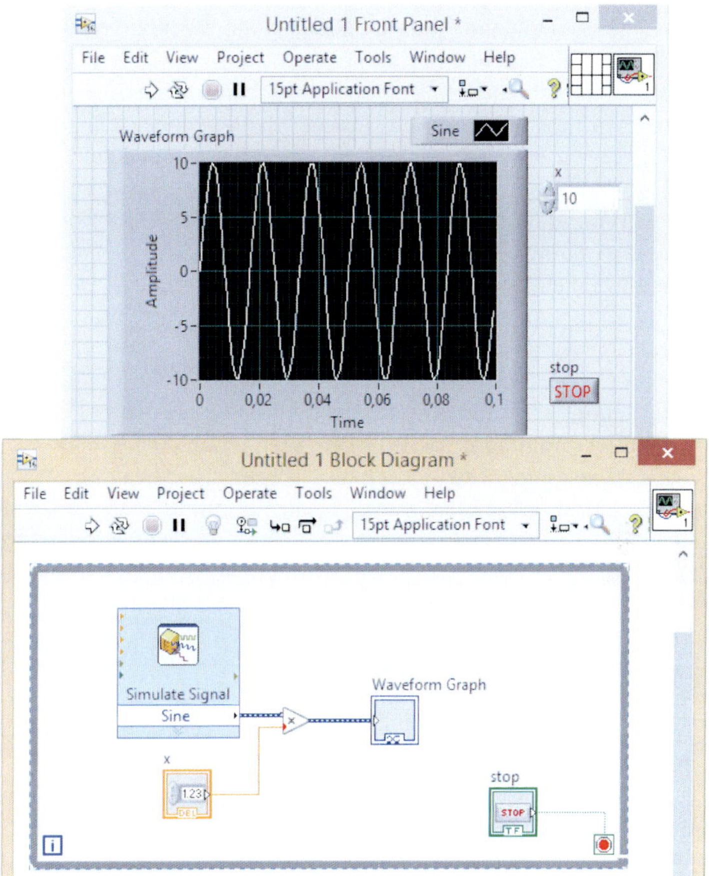

FIGURE 10.3 LabVIEW windows: front panel, block diagram, and icon connector using express VI.

while loop statement is interrupted by means of the pushbutton *stop* in the FP. It is also possible to analyze the outputs of the systems by means of a waveform chart or other block, by plotting the output data in the FP, as shown in Figure 10.3.

LabVIEW programming is widely used and many tutorials can be found in books, Internet, and courses. If one wants to get familiar with LabVIEW, to learn how to create a VI, how to use structures and subVIs, and how to build a system with regular VIs and express VIs, a good option is given in Reference 3.

10.2.1 Programming with LabVIEW

Besides LabVIEW software, the National Instruments (NI) Company provides modular hardware targets, which, combined with LabVIEW system, simplifies development and shorten time to commercial development. NI provides hardware platforms equipped with FPGA technology, which is reprogrammable silicon chip. All NI FPGA hardware products are built on a reconfigurable and modular I/O architecture, which features powerful floating-point processors. So, the programming oriented to FPGA targets is a wide field of application.

LabVIEW solves problems using a variety of programming models, such as data flow, single-cycle timed loop (SCTL), Textual Math, C code, ordinary differential equation (ODE), event-driven programming, and statechart module:

- Dataflow is the main LabVIEW programing model. It is based on the movement of data through the block diagrams, where the VI or function is just executed when all its input is available. Then, the output of block diagram is sent to all other connected blocks, determining the programming execution order.
- Single-cycle timed loop (SCTL) is a special use of the LabVIEW timed loop structure, whose all VIs and function are executed within one tick of the FPGA clock that has been chosen. A LabVIEW implementation involves three components to keep the data flow: the functions of interest, the synchronization, and the enabled chain. According to the SCTL model, the functions are performed within the cycle in a preset time and without the synchronization, as well as in an electrical circuit. Although it is supported for all targets, the execution of SCTL code is only effective for FPGA target, where the default selection is 40 MHz FPGA global clock.
- Textual Math is associated to LabVIEW through the LabVIEW MathScript RT Module allowing textual programming developed in MATLAB or GNU Octave software. Such environment compiles *.m* files from MATLAB platform.
- C code is based on C/C++ programming. LabVIEW is an open environment accommodating C and MATLAB codes as well as various applications such as ActiveX and DLLs (Dynamic Link Libraries). It is available also for FPGA target.
- Ordinary differential equation (ODE) is a specific solver and model loop rate to build a simulation model using the LabVIEW Control Design and Simulation Module based on differential equation.

- Interrupt-driven programming may be split into procedural-driven and event-driven program. The former executes a set of instructions in a specified sequence to perform a task. The structure and sequence of the program, not user actions, control the execution order of a procedural-driven applications. The program execution begins in main and then flows through method calls and control statements in a fairly predictable manner. The latter executes in an order determined by the user at run-time. In an event-driven program, the program first waits for events to occur, responds to those events, and then returns to waiting for the next event. How the program responds depends on the code written for that specific event. The order in which an event-driven program executes depends on which events occur and on the order in which those events occur. While the program waits for the next event, it frees up CPU resources that might be used to perform other processing tasks, and it simplifies the block diagram and minimizes CPU usage.
- Statechart programming algorithm is written in a flowchart-like manner using states, transitions, and events. It is available also for FPGA target.

Except for data flow, which is the general LabVIEW programming model, the abovementioned models most used for electrical power instrumentation are the SCTL for efficient FPGA programming with preset execution time, and the LabVIEW MathScript for textual programming through MATLAB or C code.

10.2.2 Virtual Instrument for Power Quality Analysis

The electrical power system is crucial for any country's industrial development, commercial use, and transport, among others. In the past years, energy consumption has steadily grown around the world, and all processes rely on high-quality electrical power for proper and efficient operation. Thus, the identification of power quality quantifiers, like power factor and total harmonic distortion, plays a very important rule in the modern power systems.

LabVIEW development platform can assist in this sort of application. Figure 10.4 shows the block diagram of the used elements for implementing a power quality monitoring instrument, as that described in Chapter 7. The main processing program is written using any solver in the BD of LabVIEW. The power quality quantifiers are displayed in the FP.

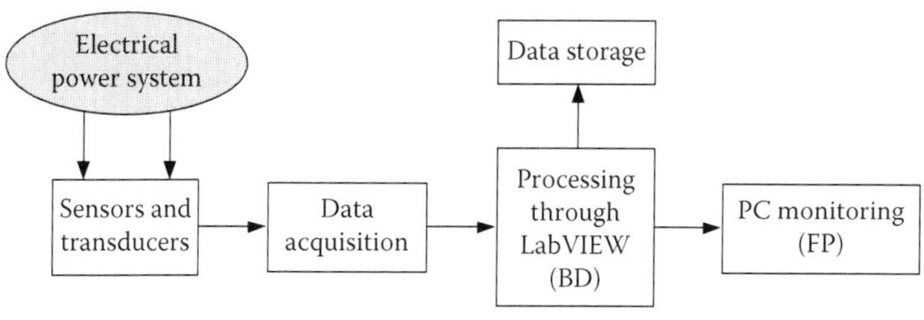

FIGURE 10.4 Main elements of power quality monitoring instrument.

The hardware of the power quality monitoring instrument may be devised by using National Instruments multifunctional data acquisition (DAQ) card [4]. An example is the DAQmx-PCI-6143-S board with 8 sampled analog input channels with a 16-bit resolution, up to 250 kS/s per channel. Two hardware boards for voltage and current signal sensing and conditioning must be devised using eight analog input channels, if applicable for three-phase four-wire power system. The current and voltage measurement signals can be obtained by using LEM transducers, e.g., four LA-55P for current and four LV-25P for voltage measurement, as shown in Figure 10.5.

In such an implementation, the computer is responsible for processing the acquired signal; thus, it is defined as an instrument based on computer. The main program is written inside the BD of LabVIEW, using the data flow programming model based on a producer/consumer architecture, as shown in Figure 10.6. The producer/consumer architecture is based on the master/slave pattern, and it is commonly used when acquiring multiple sets of data and sharing them between multiple loops running at different rates. The producer/consumer pattern's parallel loops are broken down into two categories: those that produce data and those that consume the data produced. Because queuing up (producing) these data is much faster than the actual processing (consuming), the producer/consumer design pattern is best suited for this application. Summarizing, the producer loop is decoupled from the consuming loop, and then it allows the produce loop to acquire queue additional data at the same time the

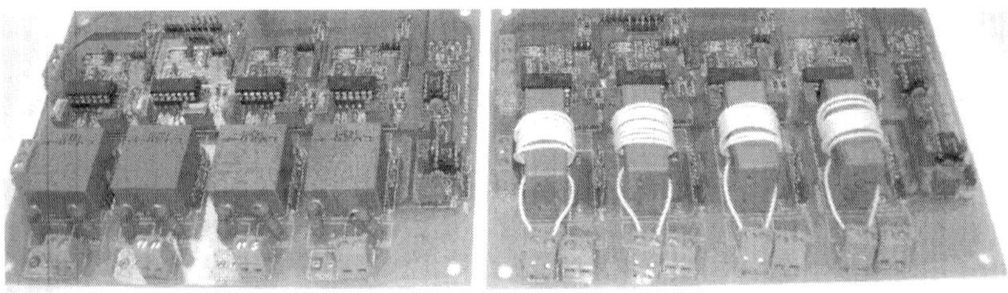

FIGURE 10.5 Two hardware boards for voltage and current signal (sensors and transducers).

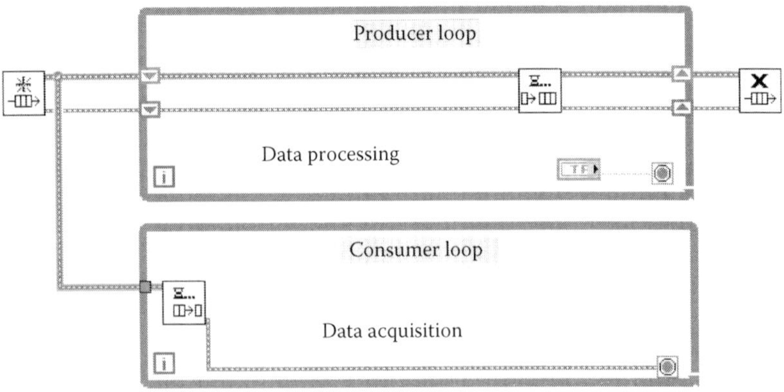

FIGURE 10.6 Producer/consumer architecture implemented in LabVIEW block diagram.

consuming loop is still processing old acquired data. If producer and consumer were conceivable at the same loop, the processing queue will not be able to add any additional data until the first piece of data has finished processing. This design pattern can also be used effectively when analyzing network communication.

In the basis of the previous description, the data acquisition occurs in the producer loop. It can be performed through the express subVI DAQ assistant from LabVIEW's function palette. This express subVI allows acquisition data configuration, such as output voltage range, custom scaling, task timing, and task triggering [5]. Figure 10.7 shows the subVI DAQ assistant gathering three voltage signals and four current signals, scaling these signals, and then displaying them to the front panel by means of waveform chart.

Besides data flow programming model, it is also possible to use Textual Math programming as C code. It facilitates writing mathematical equation, e.g., to calculate the active power. The call library function node block interprets *Dynamic Link Library* written in C code. When it is placed inside a *for* structure statement, as shown in Figure 10.8, it is processed every specified period of time, i.e., sampling period.

As a result of the devised virtual instrument for power quality, the user can analyze voltage, current, and instantaneous power waveforms, as shown in Figure 10.9. Moreover, implementing the power quality quantifiers, as those described in Chapter 6, the user can also analyze power quality factor parameters, like rms values, power terms, total harmonic distortion (THD), and voltage factors, as represented in Figure 10.10.

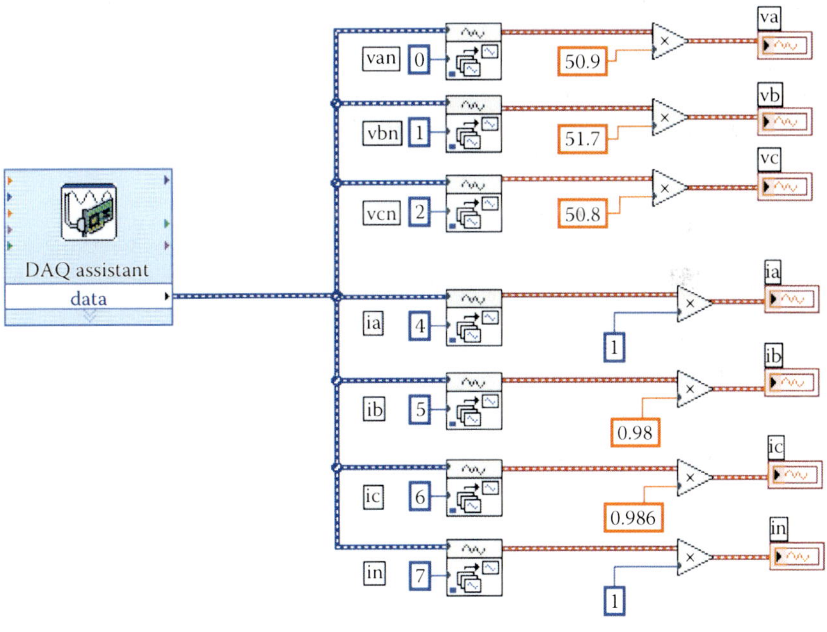

FIGURE 10.7 DAQ assistant block acquiring three voltage signals and four current signals.

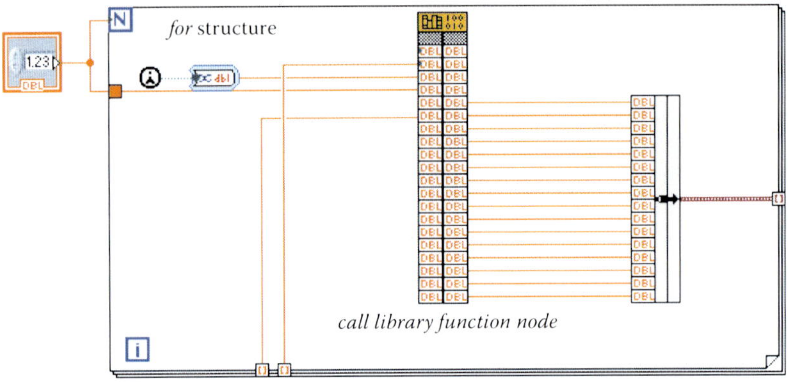

FIGURE 10.8 Combining textual C code programming with data flow programing model.

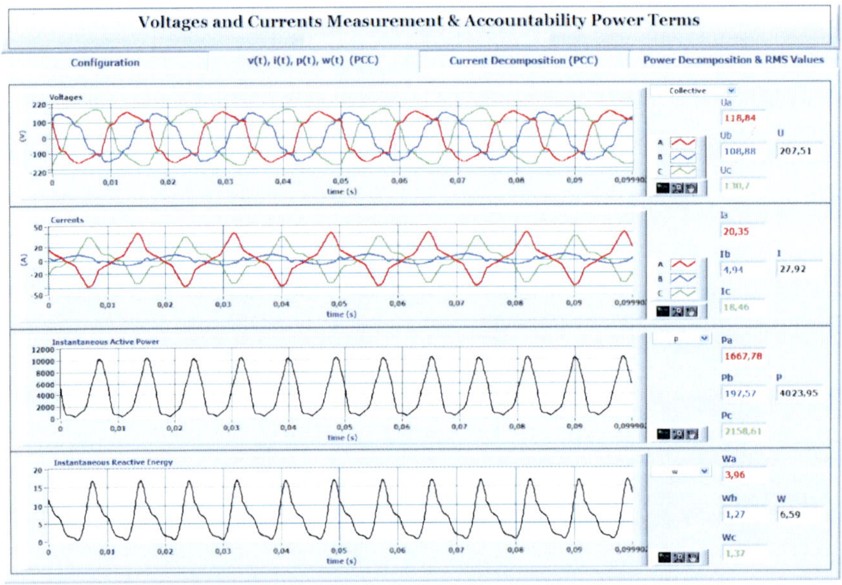

FIGURE 10.9 Voltage, current, and power waveforms displayed in the virtual instrument's FP of the power quality monitoring instrument.

10.2.3 Electric Power Instrumentation for Distributed Generation

Distributed generation is an approach that uses small-scale resources to generate electricity close to the consumers. DG resources often consist of renewable energy sources, e.g., photovoltaic and wind, and they offer several potential benefits, such as lower-cost electricity and higher power quality, reliability, and security than centralized power systems. In general, the DGs can be split into (1) power electronic

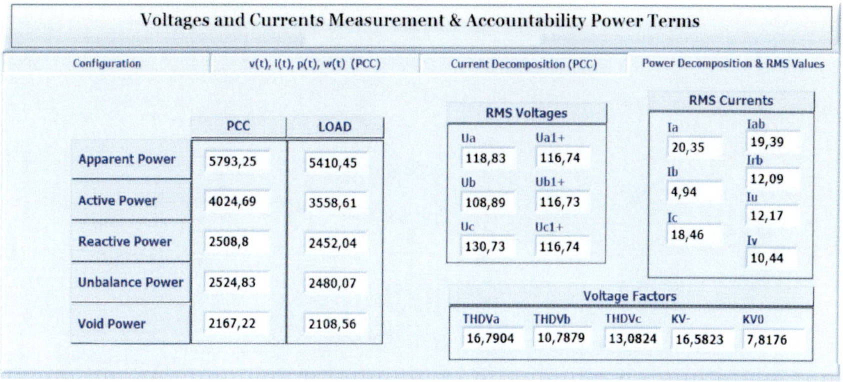

FIGURE 10.10 RMS voltage and current values, power terms, THD, and voltage factors displayed in the virtual instrument's FP of the power quality monitoring instrument.

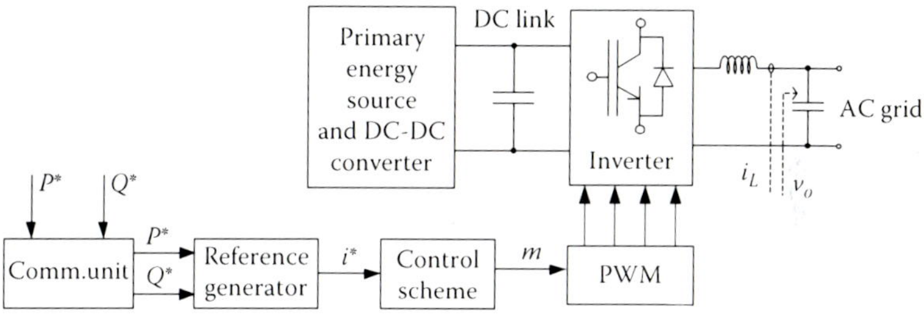

FIGURE 10.11 Typical distributed generation system structure.

converter (hardware and drive), (2) current and voltage control, (3) signal reference generator, and communication unit, as shown in Figure 10.11 and described in detail in Chapter 7.

Such systems can be easy to build through compact RIO platform or Single-Board RIO (sbRIO) with RIO Mezzanine card connector, as those shown in Figure 10.12. Both National platforms have digital I/O controllers, real-time processor, FPGA target and built-in peripherals like USB, RS232, CAN, and Ethernet. The possibility to program in both real-time processor and FPGA target allows two-level control implementation. A typical example is to implement the control loop scheme and PWM generation into the FPGA level (i.e., the faster processing), and the high level control, i.e., communication unit and the reference generator (i.e., lower dynamic), in the real-time processor.

The real-time processor and FPGA target can be programmed by means of controls and indicator blocks in the front panel of VI. The FPGA target can communicate with host computer through three different methods: (1) interactive FP communication, (2) programmatic FPGA interface communication, and (3) peer-to-peer streaming.

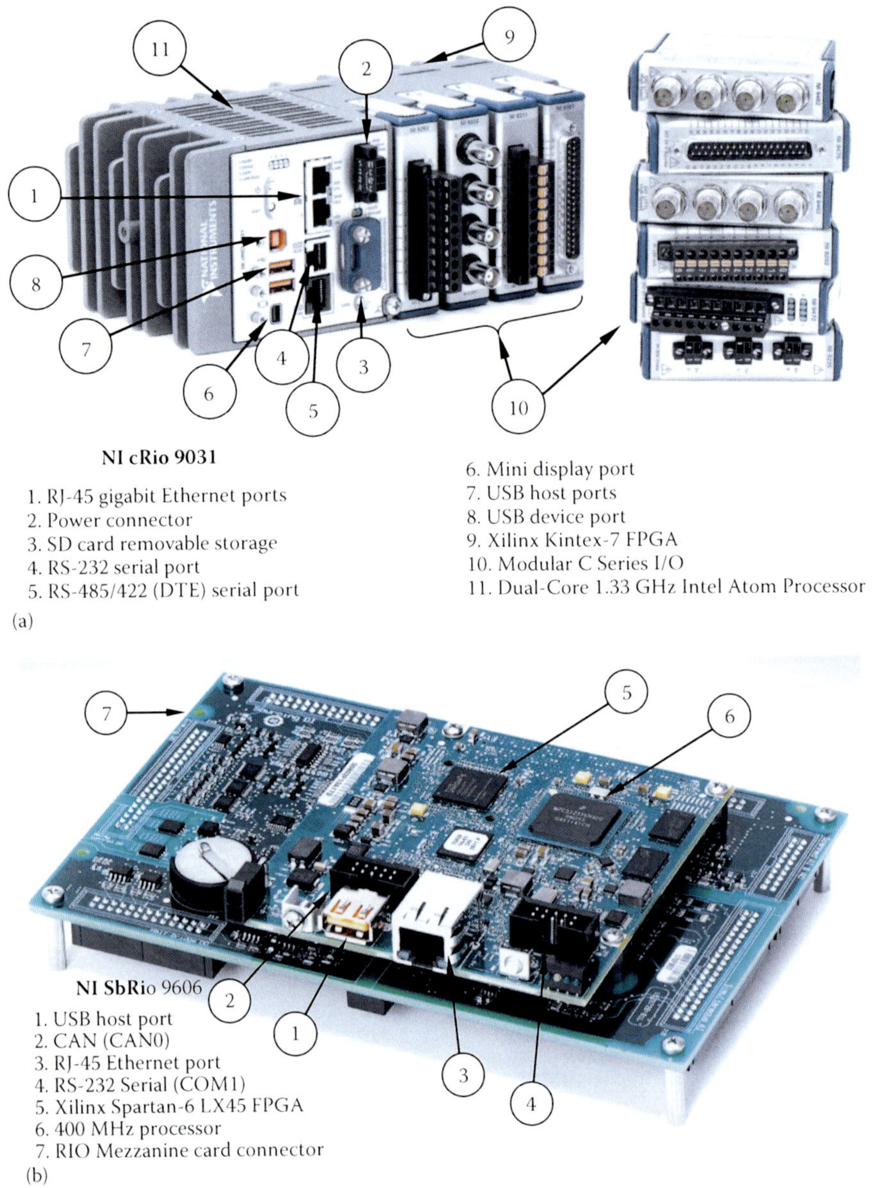

FIGURE 10.12 Reconfigurable embedded boards from National Instruments programmable through LabVIEW [6]: (a) compactRIO platform and (b) Single-Board RIO Mezzanine card connector.

The interactive FP communication is handled by a separate thread and is nondeterministic. Therefore, the parts of code that include the controls or indicators of the FP define the execution flow to the nondeterministic thread. In order to avoid that the management of communication can interfere with the programming execution flow, it is preferred not to use controls or indicators, rather use local or global variables. This method does not support communication using FIFOs or FPGA interrupts.

Programmatic FP communication is the most common programming method of communication with the FPGA target. It is useful when requesting a more robust form of communication, e.g., for reading or writing of data from the real-time processor with a regular frequency. The programming of this method of communication is made by means of dedicated interface functions, as shown in Figure 10.13a. Figure 10.13b shows the structure of programmatic FPGA communication built in the real-time processor. The *read/write control* block interfaces the real-time VI with the FPGA VI. A further advantage of this communication method is the minimization of the overhead associated with the transmission, as is the case in the field of communications for control operations. It is possible to use interactive FP communication and programmatic FPGA interface communication simultaneously.

NI peer-to-peer (P2P) streaming technology uses point-to-point transfers between multiple instruments without sending data through the host processor or memory. This enables devices in a system to share information without burdening other system resources. However, it requires two targets as well as a host computer.

A brief example of FPGA implementation for DG application is shown in Figures 10.14 and 10.15. In the former is built an oversampled PWM function with a switching frequency of 20 kHz. The *while* structure executes in 160 MHz, and the triangular carrier goes from −2000 to 2000 peak amplitude. The modulation variable is compared with the carrier generating the duty cycles, which pass through a dead time subVI and thereupon go to the digital output ports. Moreover, an interruption for synchronization between sampling and PWM is implemented.

Figure 10.15 shows the A/D synchronization and a proportional-integral controller. The flag signals (*set_v* and *set_pi*) generated from a flip-flop and created by

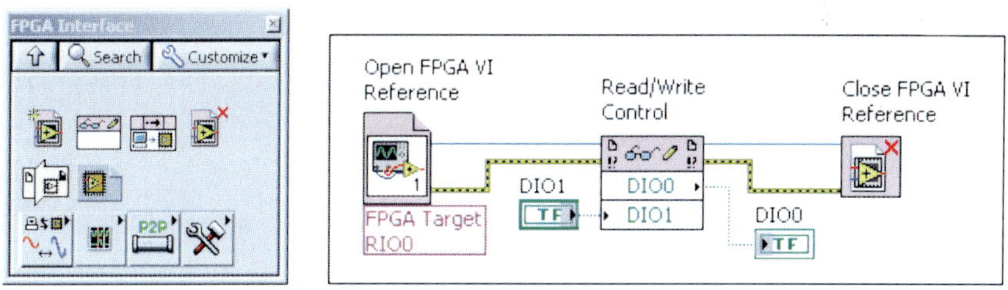

(a) (b)

FIGURE 10.13 Reconfigurable embedded boards from National Instruments programmable through LabVIEW: (a) FPGA interface palette and (b) support for programmatic FP communication.

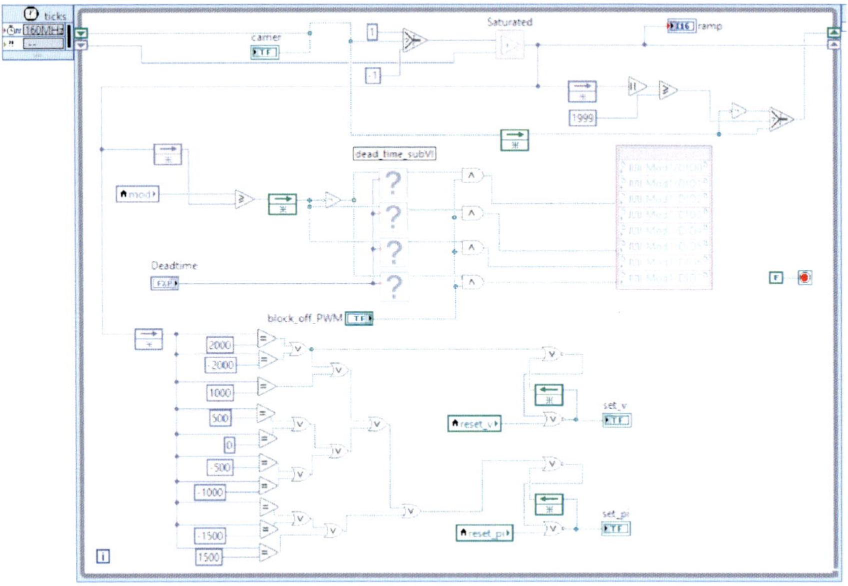

FIGURE 10.14 Oversampled PWM function.

synchronization algorithm in Figure 10.14 are then used to activate the analog/digital conversion, through a sequential structure block, the analog/digital conversion, synchronizing the sampling with the PWM carrier. In such program, the sampling frequency is 320 kHz. Also, in Figure 10.15, a PI controller is built for current control, in which its output (i.e., modulation) goes to the PWM function. The parameters K_{p_i},

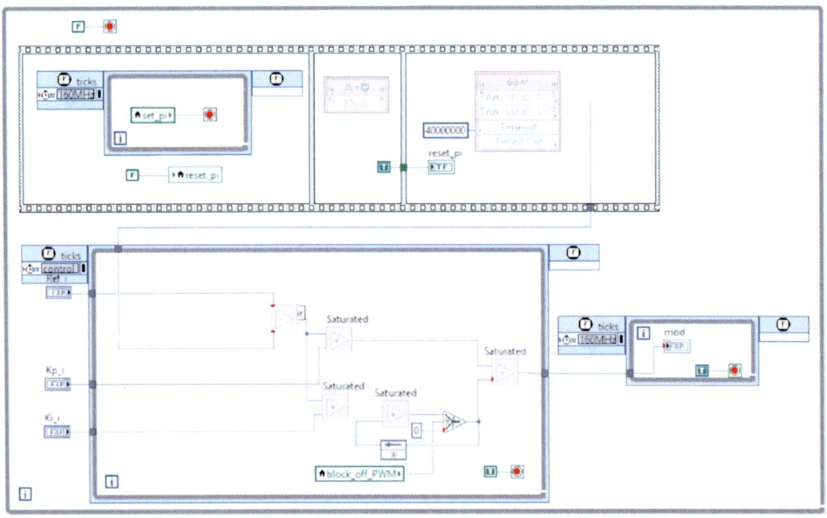

FIGURE 10.15 A/D conversion synchronized with PWM carrier and PI controller.

K_{i_i}, and $Ref_{\_i}$ are set by the user or come from the high-level program implemented in real-time VI. In Figures 10.13 and 10.14, there is a *block off PWM* flag responsible for disabling PWM modulation and reset the integral part of PI controller.

10.3 ARDUINO DEVELOPMENT SYSTEM

Arduino is an open-source platform based on easy-to-use hardware and software. It is intended for anyone making interactive projects. Arduino senses the environment by receiving inputs from many sensors and affects its surroundings by controlling lights, motors, and other actuators. One can tell Arduino what to do by writing code into the Arduino programming language and using the Arduino development environment. The source code for the Java environment is released under the GPL licenses and the C/C++ microcontroller libraries are under the LGPL licenses. Other important factor of Arduino platform is the pines compatibility among different versions of platforms. It facilitates the development of systems, which can be plug-in-play, and the interaction with other devices, even those from other manufactures. A typical example is the Arduino Shield designed with Ethernet and Bluetooth communication unit.

The most basic Arduino platform uses the Atmel AVR 8-bit microcontroller. Microcontrollers from AVR possess Harvard architecture, reduced instruction set computer (RISC), and has different memory and peripheral configurations. Moreover, the hardware supply and communication are performed through USB ports, not needing any extra power supply connection.

One of the simplest versions is the Arduino UNO, shown in Figure 10.16, which has Atmel ATMEGA328 microcontroller operating at 16 MHz, 32 kB of Flash, 2 kB of RAM, and 1 kB of EEPROM. The hardware interface has 23 pines, which may be

FIGURE 10.16 Arduino UNO platform. (From Arduino Community, Arduino UNO, https://www.arduino.cc/en/Main/ArduinoBoardUno, last accessed in December 2016 [9].)

set as I/O ports, or as dedicated function that executes its own instruction, e.g., serial communication, PWM, and external interruption.

Programs written for Arduino Software (IDE) are called sketches. These sketches are written in text editor and are saved with the file extension .ino. The editor has features for cutting/pasting and for searching/replacing text. The message area gives feedback while saving and exporting and also displays errors. The console displays text output by the IDE, including complete error messages and other information. The toolbar buttons allow one to verify and upload programs, create, open, and save sketches, and open the serial monitor. In general, Arduino programs can be divided in three main parts: *structure*, *values* (variables and constants), and *functions* [7,8].

Finally, it should be noted that these platforms are affordable, and it has technical support and reference materials. However, some applications may require access to specific resources, and, at this point, the application is limited to existing hardware and technology.

10.4 MATHWORKS MATLAB®/SIMULINK® DEVELOPMENT SYSTEM

MathWorks MATLAB is a tool that integrates mathematical functions, visualization, and a powerful language to provide a flexible environment for technical computing. MATLAB's open architecture makes it easy to explore data, create algorithms, and create custom tools that provide early insights and competitive advantages in product development and analysis projects. MATLAB is useful in a wide range of applications, including control systems design, digital signal processing, communications, test and measurement, image processing, and financial modeling [10].

Within MATLAB there is the Simulink® platform specific for electric and electronic power systems. Simulink is an interactive tool for modeling, simulating, and analyzing dynamic, multidomain systems. Simulink allows a user to accurately describe, simulate, evaluate, and refine a system's behavior through standard and custom block libraries. Simulink also integrates seamlessly with MathWorks MATLAB, providing access to an extensive range of analysis and design tools. Typical Simulink uses include control system, signal processing system, and communications system design and other simulation applications [10].

PLECS (piecewise linear electrical circuit simulation) is a Simulink add-on for the fast simulation of electrical and power electronic circuits. It allows the seamless integration of electrical circuits in the Simulink environment. Circuits are entered as netlists. They may consist of linear components (RLC), transformers, sources, measurements, and ideal switches. The use of ideal switches is advantageous for the simulation of switched networks such as power electronic systems. Running PLECS blockset requires MATLAB and Simulink on your computer [11].

MATLAB/Simulink has been widely used for hybrid programming in power electronic application. Hybrid programming is the association of textual and graphical programming, or the association of two different programming platforms, as, e.g., LabVIEW-MATLAB, PSIM-MATLAB, and Arduino-MATLAB. Such way of programming allows the designer to use the better of each platform and choose his/her preferred features of textual and graphical programming. Such association creates an easy-to-use code development, simulation, and DSP programming. This section describes some of the most useful hybrid program models using MATLAB/Simulink.

10.4.1 Interface with LabVIEW

The association between MATLAB and LabVIEW platforms combining textual and graphical models may save time and bring flexibility for designers. There are many options for importing or exporting file from MATLAB environment to LabVIEW. MATLAB codes can be used into LabVIEW environment through the LabVIEW MathScript RT Module. MathScript is an add-on module for high-level textual programming, with syntax similar to MATLAB. It is also possible to create custom-made *.m* file. The integration of MathScript function and LabVIEW can be performed by MathScript Node. The MATLAB Script Node is a structure that allows MATLAB textual programming, and it can be found on the Scripts & Formulas VIs palette. It allows defining inputs and outputs on the MathScript Node border to specify the data to be transferred from the graphical LabVIEW to the textual MathScript code. Figure 10.17 shows the Scripts & Formulas palette and an example of MATLAB code implemented in LabVIEW. It requires MATLAB license and LabVIEW installed on the same machine.

Users can also use directly the LabVIEW MathScript Window for an interactive interface in which *.m* file scripts can be loaded, saved, developed, and executed. In this window, one can write commands one-by-one for quick calculations, script debugging, or learning. It is necessary to install LabVIEW and LabVIEW MathScript RT Module.

A third option is to save the MATLAB code in ASCII format, using the following code [12]:

```
>>SAVE filename X -ascii -double -tabs
```

This command generates a file named *filename*, which contains the X data in ASCII format properly tabulated. Importing this file into LabVIEW is done using the *Read From Spreadsheet File* VI located on the File I/O VIs and Functions palette.

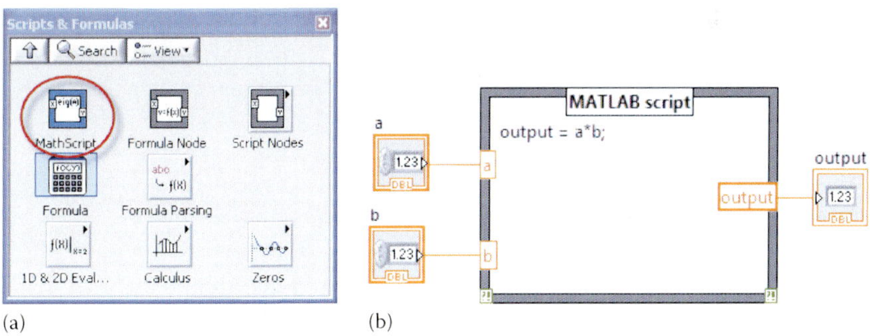

FIGURE 10.17 MATLAB®/LabVIEW association by means of MATLAB script: (a) Scripts & Formulas palette and (b) MATLAB script structure.

If the goal is to import a file generated by LabVIEW into MATLAB, simply use the code below in the MATLAB command window:

```
>>LOAD filename
```

where *filename* is the name of the LabVIEW file that is required to be opened. It is important to note that it is necessary to update the directory where the file is saved. The "load" command transforms the data file into a variable that will be saved in the MATLAB workspace, whose name will be the same as the file. Double-clicking on the variable will bring up a window displaying all the data contained in the variable.

10.4.2 Interface with Arduino

Arduino has an easy integration with sensors and many built-in features for motors and analogic and digital control. It makes the board versatile and suitable for small projects and learning purposes. However, to perform plotting graphs of sensor data, dealing with advanced math, signal processing, or control routines on Arduino itself is quite labor-intensive and time-consuming.

Through the association of MATLAB/Simulink and Arduino, one can benefit from both platforms: the easy interface Arduino has with sensors and actuators, and MATLAB that facilitates advanced analysis, plotting features, and programming language. MATLAB has two main approaches to address this integration [13]:

1. MATLAB support package for Arduino
2. Simulink support package for Arduino

The benefits of using MATLAB for Arduino programming include the following:

- Possibility of read/write sensor data interactively without the necessity to compile the code first. Read and write sensor data interactively without waiting for code to be compiled.
- With the help of a huge variety of prebuilt functions, it is possible to analyze the sensor data. There are functions for signal processing, machine learning, and mathematical modeling, among others.
- Visualize data promptly using MATLAB's vast array of plot types.

The benefits of using Simulink for Arduino programming include the following:

- Automatic code generation can be used to run on the Arduino on your own developed algorithms and also simulate them.
- Incorporate signal processing, control design, state logic, and other advanced math and engineering routines in the hardware projects.
- While the algorithm runs on the Arduino, it is possible to optimize parameters and tune interactively.

10.4.2.1 MATLAB® Support Package for Arduino

This package supports tasks of reading, writing, and analyzing data from Arduino sensors. The programs written in MATLAB are able to read and write data to Arduino and its peripherals, such as motor drives, I2C (e.g., real-time clocks, digital potentiometers, temperature sensors, digital compasses, memory chips, FM radio circuits, I/O expanders, LCD controllers, and amplifiers), and SPI devices (e.g., shift registers, sensors, and SD cards). MATLAB is a high-level interpreted language, making programming easier than with other languages, and, even more, it supports the handy capability of seeing results from I/O instructions immediately.

Connected to a computer running MATLAB, Arduino can benefit from the thousands of built-in math, engineering, signal processing, machine learning, mathematical modeling, vast plotting types functions, and more, allowing quick analysis and visualization of data collected from Arduino, and all processing is done on the computer running MATLAB.

10.4.2.2 Simulink® Support Package for Arduino

This package supports user to develop algorithms that run stand-alone on Arduino. Users can benefit from the Simulink block diagram environment for modeling dynamic systems and developing algorithms that may include all MATLAB's features.

The package adds new blocks to Simulink's library for configuring Arduino's sensor devices and acquires and writes data on Arduino and its outputs and peripheral devices. Before deploy to Arduino, user is able to modify parameters during run-time execution and simulate its effects. Once the algorithm development is completed, the consolidated program can be downloaded to Arduino, for deployment and stand-alone operation.

Simulink integrations package allows user to focus, for instance, on the modeling process rather than on writing low-level code. Specifically, Simulink has libraries that allow user to use variables undeclared, and it structures the code in the way a typical Arduino compiler requires, saving time and making things easier, especially for beginners. In the end, many different mathematical blocks and interesting visualizations are available to choose from with Simulink and MATLAB, and it provides the tools to make simulations.

10.4.2.2.1 Brief MATLAB®-Arduino Tutorial

First, make sure MATLAB and Simulink support packages for Arduino are installed. You can get the latter on the navigator to the Add-Ons menu and click Get Hardware Support Packages [14].

Once the packages are installed, connect the Arduino board to the PC, and then type the following command in MATLAB command window:

```
>> a = arduino()
```

"a" is the MATLAB Arduino object.

If you have more than one Arduino connected to your PC, it is necessary to specify the board type you are communicating with:

```
>> a = arduino('com3', 'uno')
```

such as "com3" is the port, and "uno" is the board type.

MATLAB will attempt to communicate with the Arduino board. If it succeeds, MATLAB will display the properties of the board such as the port on which your board is connected, the model of Arduino board, and available pins and libraries.

Once connected, there are several functions and commands that can be used. Below, there are some example codes to illustrate these commands:

```
%-- connect to the board
a = arduino('COM9')

%-- specify pin mode
a.pinMode(4,'input');
a.pinMode(13,'output');
%-- digital i/o
a.digitalRead(4)              % read pin 4
a.digitalWrite(13,0)          % write 0 to pin 13

%-- analog i/o
a.analogRead(5)               % read analog pin 5
a.analogWrite(9, 155)         % write 155 to analog pin 9

%-- serial port
a.serial                      % get serial port
a.flush;                      % flushes PC's input buffer
a.roundTrip(42)               % sends 42 to the arduino and back

%-- servos
a.servoAttach(9);             % attach servo on pin #9
a.servoWrite(9,100);          % rotates servo on pin #9 to 100
                              degrees
val=a.servoRead(9);           % reads angle from servo on pin #9
a.servoDetach(9);             % detach servo from pin #9

%-- encoders
a.encoderAttach(0,3,2)        % attach encoder #0 on pins 3
                              (pin A) and 2 (pin B)
a.encoderRead(0)              % read position
a.encoderReset(0)             % reset encoder 0
a.encoderStatus;              % get status of all three encoders
a.encoderDebounce(0,12)       % sets denounce delay to 12
                              (~1.2ms)
a.encoderDetach(0);           % detach encoder #0

%-- close session
delete(a)
```

Regarding the Simulink, the Simulink support package for Arduino provides a block library with several functions, as shown in Figure 10.18.

10.4.3 Interface with PSIM

The PSIM is a high-performance simulation software, mainly used for converters and power conditioners, electric machine drives, renewable energy systems, electric vehicles, and many other applications in power electronics. Among its main

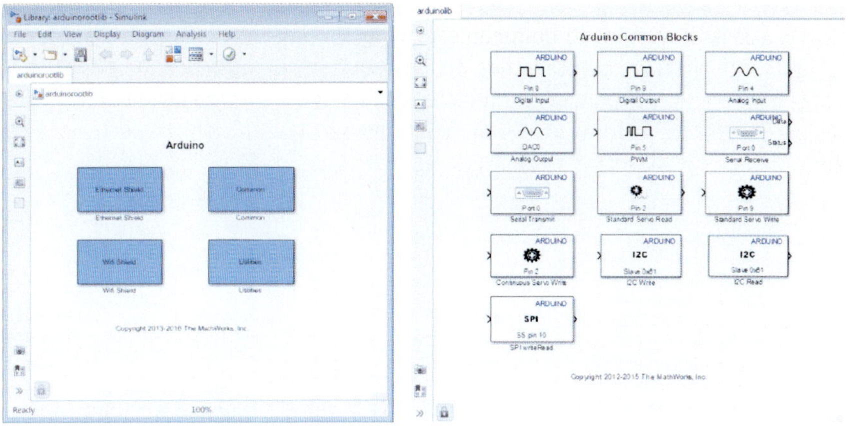

FIGURE 10.18 Simulink® support package for Arduino platform. (From Open block library for Arduino hardware, https://www.mathworks.com/help/supportpkg/arduino/ug/open-block-library-for-arduino-hardware.html, last accessed in December 2016 [15].)

characteristics, it gives the possibility of integration with the MATLAB/Simulink platform, through the SimCoupler module.

MATLAB/Simulink is widely used for simulation of control modules by having a variety of powerful toolboxes. On the contrary, PSIM software is widely used for power module simulation, because it has a simple graphical interface to use, even counting with almost negligible simulation time. Therefore, in order to implement and control an electronic power circuit, such association of two platforms is appealing because it brings together the best of both tools in a single hybrid simulation.

This association is performed through the SimCoupler, which allows creating an interface between MATLAB/Simulink and PSIM. Part of the circuit, usually the power circuit part, is implemented in the PSIM and the control scheme in Simulink. This association allows a significant improvement in simulation time, benefiting the user of the PSIM with all the variety of Simulink's toolboxes.

An easy-to-understand tutorial can be found in Reference 16 for Simulink-PSIM association. Following such tutorial, an average current control mode for buck converter can be devised. In this example, the power electronic circuit is implemented in PSIM, while the control scheme plus PWM generation is built into Simulink. Figure 10.19 shows the implementation through SimCoupler module.

10.4.4 Programming DSP through MATLAB®/Simulink®

Digital signal processing gained applicability in past and in recent years. This phenomenon is related to the development of digital signal processors (DSPs) and the development of advanced techniques for digital signal analysis. Another important two point are the versatility and tolerance to noise of the digital systems, whereas the analog systems are susceptible to interference due to working with signals of small

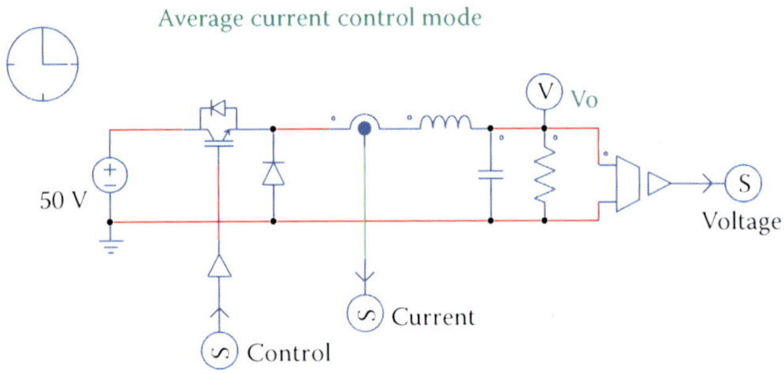

(a)

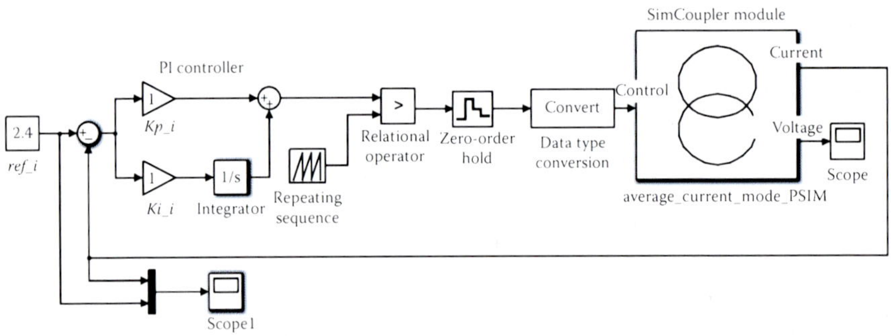

(b)

FIGURE 10.19 Average current control mode for buck converter by means of PSIM Tutorial—how to use SimCoupler module: (a) power circuit implement in PSIM and (b) control scheme devised in MATLAB®.

amplitude, besides deteriorating with time. Digital system only works with two logical levels, which reduces their susceptibility to noise [17].

MATLAB (Matrix Laboratory) is a high-level software created to work with matrix calculations, and, over time, it has undergone changes and gained functionalities. These features are often added through the so-called Toolbox; for digital signal processing applications, MATLAB has the DSP System Toolbox. The DSP System Toolbox has features to accelerate the development, simulation, and analysis of signal processing systems in MATLAB and Simulink [10].

A great advantage of using Simulink is that since the circuit structure resembles the design that will be implemented, mapping tools are capable of transcribing such structure into other programming languages used to program the DSP. The DSP application is designed with the use of block diagrams and is implemented in real time. It allows creating a high-level program of relatively easy implementation that allows fast corrections and changes. Therefore, there is no need for a specialized training in any particular programming language.

With the DSP System Toolbox, one can develop and analyze digital filters, as well as enabling the generation of C/C++ codes that can be used in embedded systems. Such procedures accelerate the development of projects using DSP, enable functions tests, and help in the development of code in MATLAB language, which is widely known by engineers.

In order to start programming the DSP through MATLAB/Simulink, it is necessary to install the Embedded Target for DSP to integrate with Texas Instruments eXpressDS tools. Once the supported development board is successfully installed, to start developing the desired application, start MATLAB, and in the command prompt, type c200lib. A new Simulink blockset will open that which contains libraries, including blocks used for C2000 devices. The process to build the model is exactly the same as used for any other Simulink models. However, it is important to select blocks only from the following libraries: appropriate Target Preferences library block; from the fitting libraries in the c2000lib block library; from Real-Time Workshop; from Fixed-Point Blockset; from discrete time blocks from Simulink; from any other blockset that is necessary, although it must operate in the discrete time domain.

The next step is to set the simulation parameters. Open the model and select Simulation Parameters located at the Simulink option. In the dialog box, choose Real-Time Workshop. It is important to remember to specify the correspondent version of the target system file and template makefile. Click browse and choose the appropriate one from the list of targets. A GRT (generic real-time) target is configured to generate a model code, for a real-time system, that would be executed on your workstation. An ERT (embedded real-time) target is the target used to generate a model code for an independent, embedded real-time system. A Real-Time Workshop Embedded Coder is necessary for the use of this option. It is also mandatory to specify discrete time by choosing fixed-step and discrete in the solver panel in the simulation parameters.

Finally, it is possible to create a real-time executable and download it into de TI development board just by clicking build on the Real-Time Workshop pane. A C code will be automatically generated by the Real-Time Workshop, which adjusts the I/O device drives as commanded by the hardware blocks used on the Simulink block diagram.

All the abovementioned platform development systems are highly used for students, researchers, and professors in undergraduate and graduate courses. This chapter aimed at presenting an overall description of LabVIEW, Arduino, and MATLAB, highlighting the main features and characteristics of each software platform for electrical power instrumentation. Many other materials including books, tutorial, reports, PowerPoint material, papers, etc., can be found in order to assist the first users.

EXERCISES

10.1 Implement in LabVIEW a VI to generate sinusoidal signals with first, third, fifth, and seventh frequency order. Plot the corresponding waveform in a waveform chart as shown in Figure 10.3.

10.2 Implement in LabVIEW a subVI using data flow programming model to calculate the total harmonic distortion. Test and evaluate such subVI using the previous signal generator.

10.3 Implement in LabVIEW a subVI using textual programming model to calculate the total harmonic distortion. Use the MATLAB script structure to develop such programming, as shown in Figure 10.17. Test and evaluate such subVI using the signal generator of Problem 10.1.

10.4 Develop Problems 10.1 and 10.3 in MATLAB/Simulink.

10.5 Implement the signal generator of Problem 10.1 in PSIM, using graphical programming model.

10.6 Perform a hybrid programming Simulink-PSIM to evaluate the MATLAB THD programming through PSIM signal generator using the SimCoupler module as shown in Figure 10.19.

REFERENCES

1. Kehtarnavaz, N. and Mahotra, S., *Digital Signal Processing Laboratory: LabVIEW-Based FPGA Implementation*, 1st edn., BrownWalker Press, Boca Raton, FL, 2010.
2. Ponce-Cruz, P. and Ramírez-Figueroa, F.D., *Intelligent Control Systems with LabVIEW*, 1st edn., Springer-Verlag London Limited, Mexico city, Mexico, 2010.
3. Kehtarnavaz, N. and Kim, N., *Digital Signal Processing System-Level Design Using LabVIEW*, Elsevier's Science & Technology Rights Department, Oxford, UK, 2005.
4. National Instruments Co., Data acquisition (DAQ), http://www.ni.com/data-acquisition/, last accessed in May 2017.
5. National Instruments Co., NI-DAQmx express VI tutorial, http://www.ni.com/tutorial/2744/en/, last accessed in March 2017.
6. National Instruments Co., LabVIEW hardware, http://www.ni.com/en-us/shop.html, last accessed in March 2017.
7. Arduino Community, Getting started with Arduino, http://www.arduino.org/, last accessed in January 2017.
8. Arduino Community, Tutorials on Arduino project, https://www.arduino.cc/, last accessed in March 2017.
9. Arduino Community, Arduino UNO, https://www.arduino.cc/en/Main/ArduinoBoardUno, last accessed in December 2016.
10. MathWorks, MatWorks&MATLAB, https://www.mathworks.com/, last accessed in December 2016.
11. MATLAB central, Power electronics in Simulink using PLECS, https://www.mathworks.com/MATLABcentral/newsreader/view_thread/19355, last accessed in December 2016.
12. National Instruments Co., How do I transfer data between the MathWorks, http://digital.ni.com/public.nsf/allkb/2F8ED0F588E06BE1862565A90066E9BA, last accessed in December 2016.
13. Arduino Community, Overview of MATLAB/Simulink with Arduino, https://learn.adafruit.com/how-to-use-MATLAB-and-simulink-with-arduino/overview, last accessed in December 2016.
14. Arduino Community, Legacy MATLAB and Simulink support for Arduino, http://www.mathworks.com/MATLABcentral/fileexchange/32374-legacy-MATLAB-and-simulink-support-for-arduino, last accessed in December 2016.
15. Arduino Community, Open block library for Arduino hardware, https://www.mathworks.com/help/supportpkg/arduino/ug/open-block-library-for-arduino-hardware.html, last accessed in December 2016.
16. UTFPR - Universidade Tecnológica Federal do Paraná, PSIM tutorial: How to use the SimCoupler module, http://paginapessoal.utfpr.edu.br/waltersanchez/PROGRAMAS/Tutorial%20-%20Simcoupler%20Module.pdf/at_download/file, last accessed in December 2016.
17. Isen, F.W., *DSP for MATLAB and LabVIEW I: Fundamentals of Discrete Signal Processing*, 1st edn., Morgan & Claypool Publishers, San Rafael, CA, ISBN-13: 978-1598298901, 2008.

11 Introduction to Smart Grid Systems

Luciane N. Canha, Alzenira R. Abaide, and Daniel P. Bernardon

11.1 INTRODUCTION

Smart Grid is characterized by a series of integrated technologies, methodologies, and procedures for planning and operation of electrical systems, being favored by the increase of resources related to information, communication, and engineering technologies. Another feature is the diversity of application areas, e.g., energy technologies, transportation efficiencies, utility transformations, and building innovations. In this context, electronic instrumentation is a key part, since the active management of power systems is performed by complex monitoring and control instrumentation and networks.

The electrical systems can be considered one of the largest and most complex dynamic systems built by humanity. These systems are responsible for the interaction of generation, transmission, and distribution of electrical power to comply with reliability and quality standards with minimal cost and environmental impact. This complex interconnection of information, communication, and engineering technologies, as demanded by smart grids, can be supplied by the several instrumentation techniques discussed in the previous chapters. In particular, this chapter presents an overview of the technologies and applications for Smart Grid.

According to Beyond Petroleum [1], the use of energy grows about 5% every year in the world. This growth rate makes sure that alternative sources are sought to comply with this demand with reliability and quality by exploring the technological innovation. Therefore, the electrical systems in the world will have significant changes in the coming years because of the integration of information and communication technology (ICT) infrastructures. Therefore, the electrical systems must also be prepared for the advent of electric vehicles and the large penetration of distributed generation sources, especially the renewable ones. This new concept of the network will transform the ordinary electrical systems into an intelligent system: Smart Grids [2]. This chapter introduces principles established by the main worldwide organizations involved in smart grids and their interconnections with power systems.

The concept of Smart Grid is quite comprehensive, but there is no unique definition. The IEEE P2030 Project (IEEE Smart Grid Roadmap [3]) has some conceptual ideas and formulations. One of them defines a Smart Grid as follows: "In a generic and holistic definition, the Smart Electric Grid or Smart Power (Smart Grid) is a complex end to end system that is composed of multiple subpower systems interconnected and interrelated to each other through multiple protocols that contain multiple layers of technologies (energy, ICT and control/automation)." It means that the electrical network uses advanced technology to monitor and manage in real time the transportation of energy, through the power flow and a two-way information between the power supply and the end customer, integrating and enabling actions by all users connected to this system, providing a sustainable, economical, and safe energy efficiently. Besides that, this system provides customers with information about their consumption, rates, taxes, and product and service quality in real time.

The introduction of Smart Grids will produce a marked convergence among the infrastructure of generation, transmission, and distribution of energy, digital communications, and data processing [4]. The latter will work as an Internet Equipment connecting the so-called IEDs (*Intelligent Electronic Devices*) and exchanging information and control actions among different segments of the grid. This convergence of technologies will require the development of new control, automation, and optimization methods for the electrical system, with a strong trend to use techniques for a distributed problem-solution based on multiagents.

Some of the common characteristics assigned to Smart Grids are as follows [5]:

- *Self-healing*: Online evaluation and precontrol of system failures, automatic fault, and restoring
- *Interactivity*: The ability to execute interactions between the distribution network and the consumer to get the power flow information and two-way interaction flow of capital
- *Robust*: Better response time in cases of natural human disasters, protection, and information security
- *Optimization*: Improvement of the asset utilization, reduction of operating and maintenance costs, and reduction of losses
- *Compatibility*: Large generators, alternative energy sources, distributed generation, etc.
- *Integration*: Improvement of processes, integration of information, and standardization/management refinements

Since the concept of Smart Grid is quite broad, their development goals also change according to the strategy and grid codes of each country.

Regarding the interests of the power utility, in general, the following benefits are associated with Smart Grid, consumers, and regulating agents [6]:

- Improvement of the service quality provided by the utilities
- Reduction of CO_2 emission
- Promotion of energy efficiency
- Integration of renewable sources and grid, increasing the participation of the distributed generation
- Preparation for the advent of the electric vehicle
- Increase of the system reliability
- Ability of self-healing
- Online monitoring and control
- Smart meters with two-way communication

- Reduction of the operating costs
- Improvement of the network expansion planning
- Increase of the asset management
- Promotion of the technology industry

A Smart Grid consists primarily of four parts, all of them permeated by ITC:

- *Home Area Network* (HAM): Home networks include smart appliances, self-generation, electric vehicles, and smart meters that work as gateways that interact with the user and are related to the proposal of moderate consumption and efficient use of energy. They are critical infrastructures for the development of Smart Grids. Their main function is enabling the consumer and the system to establish contact with the load as well as the consumer participation in the system, who can even provide energy to the grid. The smart energy meter acts as a sensor and establishes the information exchange between network and load, allowing functions such as the deployment of hourly rates, monitoring power quality, remote switching (disconnection/connection), and prepayment.
- *Advanced Distribution Operation* (ADO): Full automation of all control devices and operating functions of the distribution network with the aim of smart grid self-correction and efficiency. As examples, there are the automatic controls of switches, voltage, and reactive flow, as those described in Chapter 7.
- *Advanced Transmission Operation* (ATO): Focus on the management of energy exchanges and reduction of blackout risks, providing improved information exchange among subsystems, and quick response to disasters and dynamic disturbances. Examples include the integration of synchronized phasor measurements, high-temperature superconductor cables, flexible AC transmission (FACTS), advanced protection relays, and high-voltage DC transmission (HVDC).
- *Advanced Asset Management* (AAM): A tool for asset system management, which is divided into four layers: consumer, logic, application services, and system services. AAM is responsible for the equipment control, ECS failure detection, maintenance, and spare part stocks, among others.

11.2 DISTRIBUTION SYSTEM AUTOMATION

A high degree of automation is expected in a Smart Grid. The network infrastructure must support data management of electronic meters (MDMs), monitoring and control of network status (overload, reactive power control, etc.), load and distributed generation management, and charging of plug-in hybrid electric vehicles (PHEVs), among other features. Information systems should communicate with each other at different implementation levels, such as power regulation strategies, billing, maintenance management, consumer and network databases, and geographic information systems (GIS). Interoperability between standards and communication protocols plays a critical role in the advancement of Smart Grids. An important reference guide is published by the National Institute of Standards and Technology (NIST) [7].

At the operation center, the supervisory system (SCADA) performs the interface between the technical team (operator, planner, supervisor, etc.) and the network devices. At the distribution network, some automation features may include automatic adjustment of protection devices, automatic regulation of voltage levels, control of capacitor banks and transformer's taps, control of distributed generation, and self-reconfiguration, in addition to the automatic management of loads and consumption measurements [8]. Furthermore, the automation requires remote-controlled equipment (switches, reclosers, circuit breakers, etc.), digital controllers, and Intelligent Electronic Devices (IEDs).

11.3 ADVANCED METERING INFRASTRUCTURE

According to Reference 9, the modernization of power distribution networks has taken place in several countries since the 1980s, especially with the replacement of electromechanical meters by the electronic ones. In order to replace them, it is a necessary, but not sufficient, condition for smart metering . Smart metering is often regarded as the smart grid itself, but it is only the beginning to achieve the overall concept (see Figure 11.1) [10].

The following perceptions can be focused:

- AMR (*Automated Meter Reading*), which is a concept related to a one-way communication with the Measurement Control Center (MCC) for the process of power generation bill, aiming greater accuracy in measurements and cost-saving costs to users.
- *A smart meter* is an electronic meter (or digital), whose capacity is far beyond the measurement of energy consumption for electrical power applications. This equipment is capable of recording data at configurable time intervals, and it allows bidirectional communications with the MCC. To reach the overall concept of Smart Grid, the *smart meter* must also allow the integration of smart appliances.

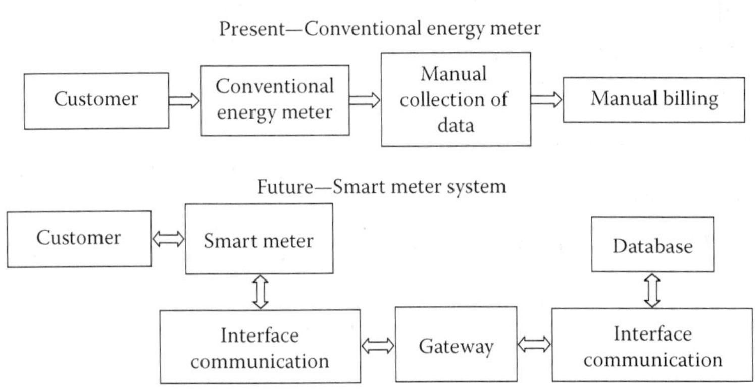

FIGURE 11.1 Current and future metering methodology.

- AMM (*Advanced Meter Management*) is related to a technical management platform of *smart meters* arranged in communication networks, being capable of reading load profiles in time intervals of less than one hour. This can be thought of as any management action of the meter as grid asset. Its basic functions are (i) device parameter management, (ii) group management to allow control of groups of devices, as configuration and firmware upgrade, and (iii) communication platform management, ensuring reliable communication between meters and MCC and reporting network status, communication performance, and special situations. It is important to notice that the AMM does not save the data collected by the meters (even temporarily)—it just transmits them to MDM.
- MDM (*Meter Data Management*) processes and manages data generated by the meters, including information beyond just energy consumption, such as power factor and quality indexes. It aims to improve processes such as billing, operational efficiency, customer services, forecasting of power demand, Distribution Management System (DMS), fraud management, demand management, and others. The issue is not just on how to collect remote data (AMM function) but also on how to manage them for further information. Its basic functions are (i) working as a data repository of records, events, and alarms and (ii) processing and analyzing data from the meters, wherever the validation and correction of inconsistent data are applied and transforming elementary load profiles into useful information to the utility.
- AMI (*Advanced Metering Infrastructure*): Some authors use the term AMI as a synonym for smart metering, covering the concepts of AMM and MDM. In practical terms, it is related to the means of communication infrastructure that are required to allow the functionalities of smart metering.

The automation of equipment and how to decide when to consume energy regarding the set of customer preferences can reduce the peak load, which has a large impact on the power generation costs, and postpone the need for construction of new power plants [11,12].

Advanced Metering Infrastructure (AMI) involves all elements necessary for communication between consumers and utilities or suppliers. The communication is bidirectional and allows the dealer to inform the consumer, e.g., the real-time value of the energy [13]. One of its structure is the Smart Meter, which allows real-time measurement and reception of power utility commands.

Smart meters can communicate and execute commands by the remote or local control as the most advanced way to monitor and control electronic devices installed at the client site. It is responsible for collecting system diagnostic information plus communicating with other smart meters. It can still limit the maximum power consumption and remotely disconnect or reconnect the power supply [14,15]. It warns outages of energy and alerts frauds.

Conventional energy meters read and record consumption continuously, and the person responsible for reading needs to go to the consumer and check the accumulated consumption displayed in the meter interface. With smart meters, data reading is performed instantly and sent to the central control, and also displayed on the meter itself, to keep the consumers alert and in control of their consumption. In addition to this, the cuts and reclose can be done remotely, and in the case of faults, the exact problem of the fault can be located quickly.

11.4 SMART HOME

Any home device using electricity can be plugged in the home network and receive a command under the smart home concept. Whether it is possible to execute that command by voice, remote control, tablet, or smartphone, the home reacts. Most applications are related to lighting, home security, energy efficiency, entertainment, and thermostat regulation.

Much of this is due to the jaw-dropping success of smartphones and tablet computers. These ultra-portable computers have constant Internet connections, meaning they can be configured to control a myriad of other online devices. It has led toward the new definition of Internet of Things.

The Internet of Things refers to the objects and products that are interconnected and identifiable through digital networks. This web-like sprawl of products is getting bigger and better every day.

A smart home could change people's life. Not only can it free up time in the day that could be better spent elsewhere, but it can help regulate the people's routine. Most smart home newbies start with lighting because it is the easiest and least risky modification one could make. Smart lighting serves two functions: automated contextual lighting (e.g., by the time of day) and energy efficiency (e.g., turning off unnecessary lights).

11.5 INFORMATION AND COMMUNICATION TECHNOLOGIES

The network to be used in a Smart Grid must allow two-way communication between consumers and the distribution system operator (DSO) in power system generation and distribution. Also, it is an important highlight that communication requires a high level of availability and reliability, since the information transmitted by the communication network may have a high level of criticality. In this aspect, various types of technologies may be used to exchange information [16,17].

In these terms, the alteration or falsification of information can cause major damage to the system or interrupt important services going on. Thus, about transmitted information, the following criteria need to be considered:

- Confidentiality relates to information privacy. Thus, only the emitter and receiver could have access to it.
- Authenticity: Refers to the identity of the message sender, i.e., the receiver makes sure that the information received has not been sent by an imposter.
- Data integrity: It guarantees that the information has not undergone any tampering, i.e., the information that has reached the receiver is exactly what was sent by the sender.
- To provide communication between active devices in a Smart Grid, it is necessary to use network protocols.

With this purpose, the power system may perform monitoring and management through a specific software, such as the SCADA system (Supervisory Control and Data Acquisition) [17].

Supervisory systems can perform readings from sensors through communication networks and accomplish it by analyzing these data. These data can be used by other applications with the purpose to provide a specific functionality or service. Some applications are reconfiguring the power system, AMI, and protection. However, the application accuracy of a Smart Grid depends directly on the quality of the data obtained by SCADA systems. Therefore, one should ensure that the measured data are coming from reliable sources. Therefore, it can be observed how security issues such as authenticity, integrity, and confidentiality are essential. To allow that, it is indispensable to permit only communications with the SCADA system from legitimate devices in the network. As can be seen, these are common problems in the context of computer networks and information technology [18]. However, a Smart Grid acquires even greater importance given the vital dependence the society has toward the energy use. In this regard, several efforts are being made to make the communication network and the systems used for implementation of Smart Grids reliable. Standards such as those present in IEC 61850 specify issues relating the automation of the system control of power substations [19]. There are also other standards that take directly into consideration the security issues involved in these applications, such as the IEC 62351 family of standards [20]. However, each SCADA system provides these standards by implementing security in a specific way.

11.5.1 Standard 61850

IEC 61850 is a standard regulating communications within a power substation environment. It is a standard that has gained global acceptance by both suppliers and customers, which, in this case, is represented by power utilities. This standard sets strict rules interoperability between independent manufacturer devices, thus providing protection, monitoring, control, and automation [19]. Figure 11.2 shows the main communication services defined by standard IEC 61850 and also the MODBUS protocol [19,21,22]. These services are provided in application layers where different communication modes are used, such as MMS, GOOSE, and TCP/IP [19].

The standard IEC 61850 consists of 10 parts, each one dealing with a specific topic, which broadly addresses the topic proposed for substation automation systems. The IEC 61850 series provides not only data transfer, but it is a complete structure that separates the application and communication systems through the use of an abstract interface [19].

11.5.2 Standard 62351

The standard IEC 62351 establishes the information security standards for control operations in power system. The standard IEC 62351 is currently divided into 11 parts, each one dealing with a specific topic [20]. IEC 62351 provides different methods to ensure different communication types in the standard IEC 61850 [19].

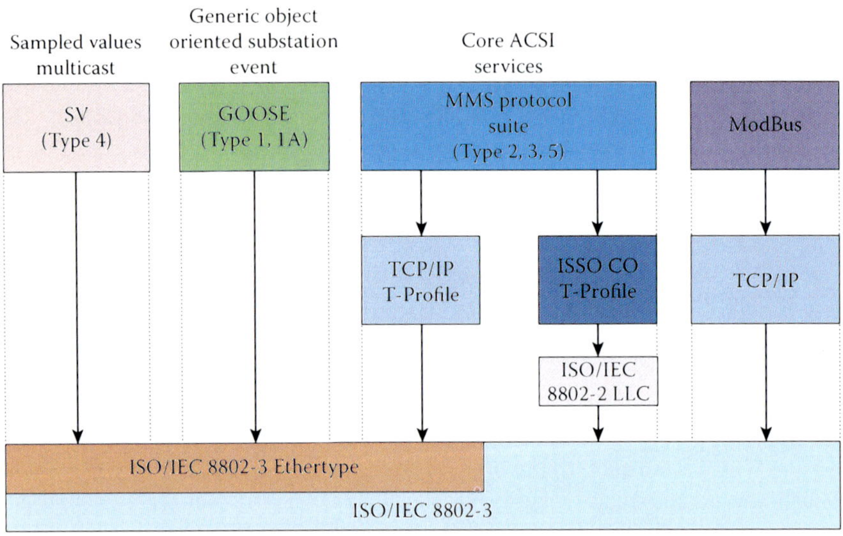

FIGURE 11.2 Adapted from IEC 61850 communication and services.

11.5.3 Attacks on SCADA Systems

The SCADA systems are part of electrical infrastructures all over the world, both critical and noncritical. Nowadays, the communication networks are converging to widely used and open standard protocols, like the IP-based networks. However, the implications of a deliberate attack on SCADA systems would be serious. Because it may jeopardize the confidentiality, integrity, and availability of the systems involved [23,24], security must be a fundamental rule at its implementation.

The security routine in IP networks is an intense field of study, given multiple vulnerabilities that can be exploited with even greater importance than on the Internet. Therefore, the security vulnerabilities are very important issues in an SCADA environment. Based on that, some of the main types of vulnerabilities are considered [23,24]:

1. IP spoofing, which replaces computer's IP, with a false address, to try to access any service or resource connected to the network. Thus, it is possible to attack devices without fear of being tracked, because the address sent to the recipients is false.
2. Spying and tampering network packets to find critical information, or tampering them to lead the system to an inconsistency state.
3. Denial of Service attacks (DoS) and Distributed Denial of Service attacks (DDoS). The DoS is used to flood a specific resource with several requests to overload it, making it unavailable. In this way, legitimated requests cannot be treated due to the overload caused by unnecessary requests. There is also the DDoS, where the attacks are carried out by several simultaneous elements, making it even more effective and more difficult to avoid attacks. In the context of Smart Grids, an attack accomplished in an SCADA system can make

data obsolete or even inaccessible to applications using data provided by the SCADA. Thus, applications available on the Smart Grid, including protection, may no longer work, causing serious consequences to the power system.

A common scenario of vulnerability is to obtain data through the SCADA system over an IP network, with monitoring devices. In this scenario, the described vulnerabilities can be easily identified through the use of tools analyzing the data traffic in communication networks. The use of masking techniques, such as IP spoofing, where an attacker changes its address to look like a legitimate device, is widely treated in the literature [24]. Network sniffers like Wireshark or TCPDump can be easily used to analyze network traffic. With this analysis, critical information can be intercepted and tampered if the messages do not use data encryption. Using the Man-in-the-Middle Attack, the data can easily be intercepted and tampered. The information read by a sensor, such as smart meter, could be modified. This can lead the system to an inconsistent state. Likewise, a DoS or DDoS attack can easily be performed, making several requests, apparently legitimate, thus overloading the SCADA system and making it unavailable. With this purpose, some packet insertion tools, such as HPING3, could be used.

Based on the vulnerabilities and security challenges presented above, several efforts must be made to bring possible solutions to mitigate vulnerabilities. It must pay special attention to delay tolerance of messages specified by standard IEC 61850 [19]. Once the use of digital signature or encryption requires great processing power and memory, they are still a challenge for smart grid designers.

11.6 ELECTRIC VEHICLES (EV) IN SMART GRID SYSTEMS

A large portion of the energy consumed across the world is destined to the transport sector (Figure 11.3), as the majority of vehicles used in this segment are powered by fossil fuels, and so it becomes responsible for much of the pollutant emissions in the atmosphere, seconded only by the industrial sector. Thus, the need for insertion of a more efficient and cleaner technology becomes necessary, e.g., by replacing fossil fuels with other sources such as diesel oil alternatives like electrical power. Currently, the idea of the hybrid car has been popularized, which replaces petrol automobiles with electric vehicles. The residential and transport sectors represent 27% of the global energy consumption in 2012 [25].

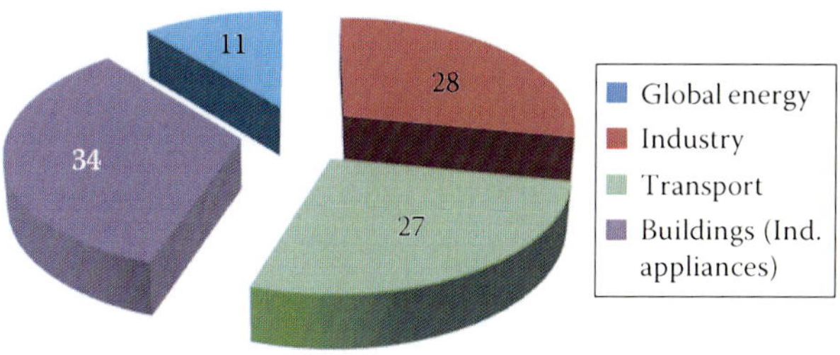

FIGURE 11.3 Percentage of global energy demand by sector (2012).

Electric vehicle (EV)/plug-in electric vehicle (PEV) is a general term used to describe any car that uses a power source to drive an electric motor for its propulsion purposes. The first one is called the VE plug-in, where a battery is used to supply energy to an electrical motor charged from an electrical power outlet. It has the same characteristics of a conventional HEV, having an electric motor and an additional internal combustion engine.

The second one is the traditional hybrid car, where the battery is charged exclusively by a primary combustion engine or by regenerating the braking energy. The option of charging batteries from a power grid makes the plug-in model operate with a reduced amount of fossil fuel. The majority of the plug-in cars are privately owned.

At the end of 2014, there were approximately 665k EVs registered in the world, representing 0.08% of the total privately owned cars. The United States is the leading country, representing 39%, followed by Japan holding 16%, and China with 12%.

For a wide use of plug-in EVs, the available infrastructure has to be reviewed, being represented by charging stations or planning adjustment of the power system load availability, since they require power grid connection to charge the battery. Therefore, it is one more load connected to the electrical power system [26].

Moreover, the batteries used in EVs and HEVs can be used to supply energy to the grid when the battery-stored energy is available, introducing the concept of the vehicle to grid (V2G) [27]. This technology can be applied with better results after the smart grid implementation when the communication systems will be fully integrated into the electrical power system. In the EVs case, the smart grid allows coordinating the power flux, indicating the number of hours the vehicle can charge the batteries or supply energy to the network.

11.7 A SMART OPERATION EXAMPLE

In this section, an application for automatic restoration of a power supply in distribution systems using remote-controlled switches is presented. This example uses the Smart Grid concepts for self-healing.

The main features of the proposed system include the following:

- A methodology for automatic restoration of the power supply in the distribution network
- A multicriteria decision-making method, AHP (Analytic Hierarchy Process [28]) applied to restoration of the power supply
- A computer analysis integrated with SCADA of remote-controlled switches to allow performing an automatic restoration in real time

11.7.1 Methodology for Automatic Restoration of a Power Supply

The logic for a power restoration is presented herein considering the hypothetical example of the simplified distribution network illustrated in Figure 11.4. The normally closed (NC) and normally open (NO) switches in Figure 11.4 are remotely controlled.

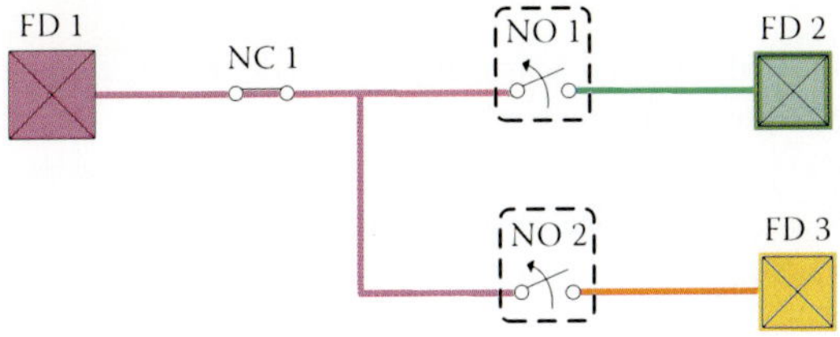

FIGURE 11.4 Example of distribution network. FD, feeder; NC, normally closed switch; NO, normally open switch.

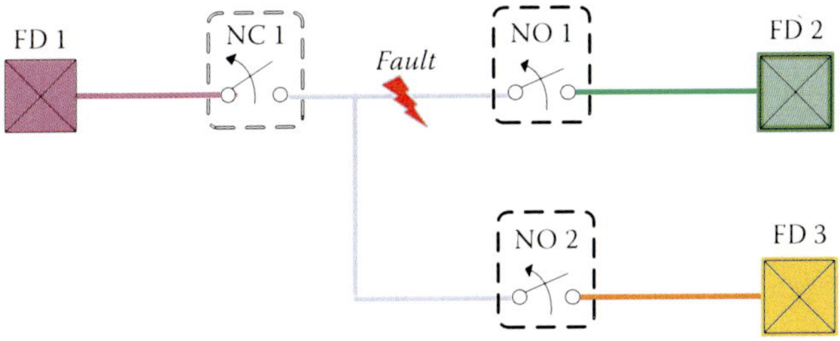

FIGURE 11.5 Fault downstream of NC-1.

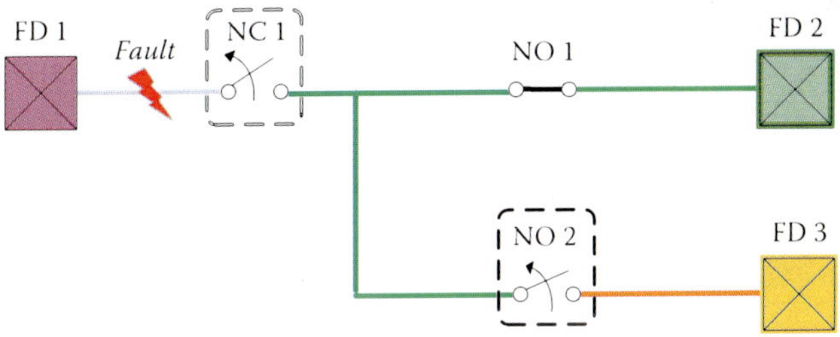

FIGURE 11.6 Fault upstream of NC-1, e.g., transferring the consumers to FD-2.

Assuming that an outage has occurred in feeder FD-1, the procedure for electrical power restoration is as follows [29]:

- Fault downstream of an NC-1 switch: In the event of this fault, the current values of short-circuiting will be flagged online in the SCADA system. So, it is assumed that the failure occurred downstream of an NC-1 switch; then, this switch is operated automatically to isolate the fault (Figure 11.5).
- Fault upstream of the NC-1 switch: In the event of this fault, the current values of short-circuiting will not be flagged in the SCADA system. So, it is assumed that the failure occurred upstream of the NC-1 switch, automatically operating the remote-controlled switches to open NC-1 and to close NO-1 or NO-2 to transfer consumers downstream of NC-1 to another feeder (Figure 11.6).

The technical and operational feasibility of load transfers using remote-controlled switches is verified by computer simulations. If there is more than one option of load transfer (e.g., to FD-2 or FD-3), the best option will be chosen considering the defined objective functions and constraints by a multicriteria decision-making algorithm. After this analysis, the developed tool automatically sends the necessary commands to the equipment. Moreover, the automatic operation of remote-controlled switches is carried out only after having tried to restart the protection devices, i.e., they are done only in case of a permanent fault, after a maximum of 3 minutes needed to complete the computer simulations and switch maneuvers since the instant of fault identification.

11.7.2 Objective Functions and Constraints

After a contingency, the challenge is to decide which one is the best option to restore the energy, depending on objective functions and constraints. This is a multiple criterion decision-making problem since various types of objective functions can be considered.

The most common objectives are the maximization of restored consumers and restored energy; however, generally, it is not possible to optimize the grid for both objectives simultaneously. Furthermore, it is also important to ensure the distribution system reliability through continuity indicators. The basic parameters are the SAIDI (System Average Interruption Duration Index) and the SAIFI (System Average Interruption Frequency Index) [30]. In this approach, some expected values are based on the probability of failure in the system.

The constraints to restore the energy system are the maximum loading of the electrical elements, protection settings, and the allowable voltage drop in the primary network. Typically, the last two restrictions are the hard ones. In contrast, a percentage of overloading the network elements is acceptable in a temporary situation, assuming that the fault can be fixed in a couple of hours.

The following objective functions and constraints were defined in the analysis of load transfers in case of contingencies.

11.7.2.1 Objective Functions

First level:

- Maximization of the number of restored consumers
- Maximization of the amount of restored energy

Second level:

- Minimization of the expected SAIFI
- Minimization of the energy losses

11.7.2.2 Constraints

- Current magnitude of each element must lie within its permissible limits:

$$|I_i| \le I_{i\max} \tag{11.1}$$

- Current magnitude of each protection equipment must lie within its permissible limits:

$$|I_i| \le I_{jprot} \tag{11.2}$$

- Voltage magnitude of each node must lie within its permissible ranges:

$$V_{j\min} \le V_j \le V_{j\max} \tag{11.3}$$

where

SAIFI is the value of the System Average Interruption Frequency (failures/year)
I_i is the current at branch i (A)
$I_{i\max}$ is the maximum current accepted through branch i (A)
I_{jprot} is the pickup current threshold of the protection device j (A)
V_j is the voltage magnitude at node j (kV)
$V_{j\min}$ is the minimum voltage magnitude accepted at node j (kV)
$V_{j\max}$ is the maximum voltage magnitude accepted at node j (kV)

The second level of the objective functions is only performed if the solutions of the first level are equal.

Verification of the objective functions and constraints is made by calculation of the load flow [31] for the various alternatives in real time. The expected SAIFI is obtained by applying the classical equations of reliability during the load flow calculating process [32].

11.7.3 Multicriteria Decision-Making Method

Identifying the best option to restore the energy is not simple since two objective functions are used in each level. For example, one particular option may have the largest number of consumers to be transferred and the other one the largest amount of energy to be transferred. To solve this, the Analytic Hierarchy Process (AHP) method is usually chosen because of its efficiency in handling quantitative and qualitative criteria for the problem resolution. The first step of the AHP is to clearly state the goal and recognize the alternatives that could lead to it. Since there are often many criteria considered important in a decision-making, the next step in AHP is to develop a criteria hierarchy with most general criteria at the top of it. Each top-level criterion is then examined to check if it can be decomposed into subcriteria. The next step is to determine the relative importance of each criterion against all the other criteria associated with it, i.e., it establishes weights for each criterion. The final step is to compare each alternative against all others on each criterion on the bottom of the hierarchy. The result will be a ranking of the alternatives complying with the stated goal according to the defined hierarchy of the criteria and their weights [33].

In the proposed approach, the main criterion is to choose the best option for automatic restoration of power supply, and the subcriteria are the proposed objective functions. The alternatives are the options for load transfers.

An example of the AHP algorithm is defined as follows:

1. The setup of the hierarchy model.
2. Construction of a judgment matrix. The element values in the judgment matrix reflect the user's knowledge about the relative importance of every pair of factors. As shown in Table 11.1, the AHP creates an intensity scale of importance to transform these linguistic terms into numerical intensity values.

   Assuming $C_1, C_2, \ldots, C_n$ to be the set of objective functions, the quantified judgments on pairs of objectives are then represented by an n-by-n matrix:

$$\mathbf{M} = \begin{array}{c} C_1 \\ C_2 \\ \vdots \\ C_n \end{array} \begin{bmatrix} 1 & a_{12} & \cdots & a_{1n} \\ 1/a_{12} & 1 & \cdots & a_{2n} \\ \vdots & \vdots & \ddots & \vdots \\ 1/a_{1n} & 1/a_{2n} & \cdots & 1 \end{bmatrix} \quad (11.4)$$
$$\begin{array}{cccc} C_1 & C_2 & \cdots & C_n \end{array}$$

   where n is the number of objective functions and the entries the following rules define j $(i, j = 1, 2, \ldots, n)$:

   - If $a_{i,j} = \alpha$, then $a_{j,i} = 1/\alpha$, where α is an intensity value determined by the operators, as shown in Table 11.1.
   - If C_i is judged to be of equal relative importance as C_j, then $a_{i,j} = 1$, and $a_{j,i} = 1$; in particular, $a_{i,i} = 1$ for all i.

TABLE 11.1
Scale of Importance for Comparison of Criteria

Intensity of Importance	Definition
1	Equal importance
3	Weak importance of one over another
5	Essential or strong importance
7	Very strong or demonstrated importance
9	Absolute importance
2, 4, 6, 8	Intermediate values between adjacent scale values

Source: Saaty, T.L., *The Analytic Hierarchy Process: Planning, Priority Setting, Resource Allocation*, McGraw-Hill, New York, 1980.

3. Calculate the maximal eigenvalue and the corresponding eigenvector of the judgment matrix **M**. The weighting vector containing weight values for all objectives is then determined by normalizing this eigenvector. The form of the weighting vector is as follows:

$$W = \begin{bmatrix} w_1 \\ w_2 \\ \vdots \\ w_n \end{bmatrix} \tag{11.5}$$

4. Perform a hierarchy ranking and consistency checking of the results. To check the effectiveness of the corresponding judgment matrix, an index of consistency ratio (CR) is calculated as follows [34]:

$$CR = \frac{\left(\frac{\lambda_{max} - n}{n-1}\right)}{RI} \tag{11.6}$$

where
λ_{max} is the largest eigenvalue of matrix **M**
RI is the random index

A table with the order of the matrix and RI value can be found in Reference 31. In general, a consistency ratio of 0.10 or less is considered acceptable.

11.7.4 Application of AHP Method

The AHP method was implemented in the methodology described above, which is detailed considering the example of distribution network illustrated in Figure 11.7, where the normally closed (NC) switches and the normally open (TS) tie switches are remotely controlled. The main goal is to define the best option to restore the

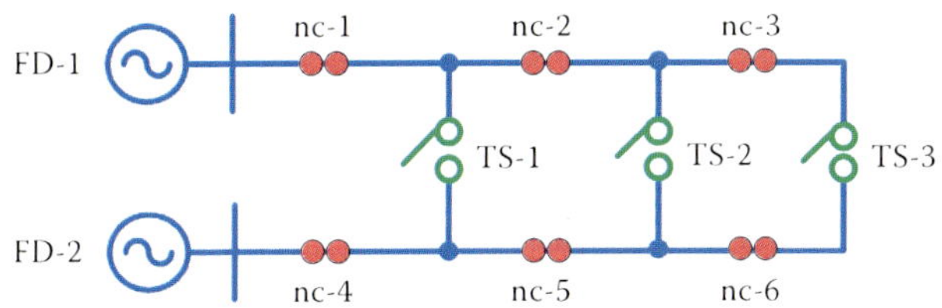

FIGURE 11.7 Distribution network.

energy, considering that an outage has occurred in feeder FD-1 due to a fault between NC-1 and NC-2 switches.

Upstream consumers of NC-1:

- Automatic opening NC-1 and reclosing FD-1

Downstream consumers of NC-2:

- Option (a): Automatic opening NC-2 and automatic closing TS-2 to transfer downstream consumers from NC-2 to feeder FD-2
- Option (b): Automatic opening NC-2 and automatic closing TS-3 to transfer downstream consumers from NC-2 to feeder FD-2

According to the method AHP, they were defined weights of the objective functions:

First level:

$$W_1 = \begin{bmatrix} W_a \\ W_b \end{bmatrix} = \begin{bmatrix} 0.64 \\ 0.36 \end{bmatrix} \tag{11.7}$$

Second level:

$$W_2 = \begin{bmatrix} W_c \\ W_d \end{bmatrix} = \begin{bmatrix} 0.61 \\ 0.39 \end{bmatrix} \tag{11.8}$$

where

W_1 is the weight vector of the first level
W_a is the weight of the number of restored consumers
W_b is the weight of the amount of restored energy
W_2 is the weight vector of the second level
W_c is the weight of the expected SAIFI
W_d is the weight of the energy losses

Considering that the two options are equal to the first level, Tables 11.2 and 11.3 show the result options for the second level. In this case, there is no restriction violation.

TABLE 11.2
Results of the Analysis for Second Level

Options	Expected SAIFI (Failures/Year)	Energy Losses (kWh)
1. Open NC-2 and Close TS-2	8.03	390.00
2. Open NC-2 and Close TS-3	7.94	420.00
Base Selected	7.94	390.00

TABLE 11.3
Normalized Values of Table 11.2

Options	Expected SAIFI	Energy Losses
1. Open NC-2 and Close TS-2	0.99	1.00
2. Open NC-2 and Close TS-3	1.00	0.93

The results using the AHP method were obtained by Equation 11.9:

$$\begin{bmatrix} Op1 \\ Op2 \end{bmatrix} = \begin{bmatrix} 0.99 & 1.00 \\ 1.00 & 0.93 \end{bmatrix} \cdot \begin{bmatrix} 0.61 \\ 0.39 \end{bmatrix} = \begin{bmatrix} 0.99 \\ 0.97 \end{bmatrix} \tag{11.9}$$

According to the method, the option "1" is considered the best solution. Thus, the system performs the commands to perform the load transfer, open NC-2, and close TS-2.

REFERENCES

1. Beyond Petroleum, BP statistical review of world energy, London, U.K., June 2011, http://www.bp.com/content/dam/bp-country/de_de/PDFs/brochures/statistical_review_of_world_energy_full_report_2011.pdf, last visited in May 2017.
2. International Energy Agency, Impact of smart grid technologies on peak load to 2050, OECD/IEA, Paris, France, 2011, https://www.iea.org/publications/freepublications/publication/smart_grid_peak_load.pdf, last visited in September 2016.
3. Annaswamy, A., Smart grid research: Control systems—IEEE vision for smart grid control: 2030 and beyond roadmap, DOI: 10.1109/IEEESTD.2013.6648362, http://ieeexplore.ieee.org/document/6648362/, last visited in September 2016.
4. Brena, R.F., Handlin, C.W., and Angulo, P., A smart grid electricity market with multiagents, smart appliances and combinatorial auctions, *2015 IEEE First International Smart Cities Conference (ISC2)*, Guadalajara, Mexico, October 25–28, 2015.
5. Xue-song, Z., Li-qiang, C., and You-jie, M., Research on smart grid technology, *2010 International Conference on Computer Application and System Modeling (ICCASM)*, Taiyuan, China, Vol. 3, pp. V3-599–V3-603, October 22–24, 2010.
6. EPRI—Electric Power Research Institute; Mossé, A., Smart Grids: Challenges and Reality (Redes inteligentes: Desafios e Realidades), *Smart Grid Forum*, São Paulo, Brazil, 2009.
7. National Institute of Standards and Technology, NIST framework and roadmap for smart grid interoperability standards, Release 2.0. NIST Special Publication 1108R2, Washington, DC, 2012.
8. Bernardon, D.P., Mello, A.P.C., and Pfitscher, L.L., Chapter 2: Real-time reconfiguration of distribution network with distributed generation, in (ed.) Jian, K., *Real-Time Systems*, InTech, Rijeka, Croatia, pp. 8–28, 2016, ISBN 978-953-51-2397-2.
9. International Energy Agency, Technology roadmap smart grids, OECD/IEA, Paris, France, 2011, www.iea.org/publications/freepublications/publication/smartgrids_roadmap.pdf, last visited in August 2016.
10. Depuru, S., Wang, L., Devabhaktuni, V., and Gudi, N., Smart meters for power grid—Challenges, issues, advantages and status, *Renewable and Sustainable Energy Reviews*, 15(6), 2736–2742, August 2011.

11. Bouhafs, F., Mackay, M., and Merabti, M., Links to the future: Communication requirements and challenges in the smart grid, *IEEE Power & Energy Magazine*, 10(1), 27–28, 2012.
12. Vojdani, A., Smart integration, *IEEE Power & Energy Magazine*, 6(6), 71–79, November 2008.
13. Hart, D.G., Using AMI to realize the smart grid, *Proceedings of IEEE Power & Energy Society General Meeting—Conversion and Delivery of Electrical Energy*, Pittsburgh, PA, 2008.
14. Zheng, J., Lin, L., and Gao, D.W., Smart meters in smart grid: An overview, *IEEE Green Technologies Conference*, Denver, CO, 2013.
15. SCP Clearinghouse—Sustainable Consumption and Production, http://www.scpclearinghouse.org/c/17-energy-efficiency.html, last visited in December 2013.
16. Wang, X. and Yi, P., Security framework for wireless communications in the smart distribution grid, *IEEE Transactions on Smart Grid*, 2(4), 809–818, 2011.
17. Ekanayake, J. et al., *Smart Grid: Technology and Applications*, John Wiley & Sons, Hoboken, NJ, 2012.
18. Ghansah, I., Smart grid cyber security potential threats, vulnerabilities and risks, PIER Energy-Related Environmental Research Program, CEC-500-2012-047, California Energy Commission, Sacramento, CA, pp. 1–93, 2012.
19. IEC 61850 International Standard, Communication networks and systems in substations—Part 8-1: Specific Communication Service Mapping (SCSM)-Mappings to MMS (ISO 9506-1 and ISO 9506-2) and ISO/IEC 8802-3, First edition, 2004–05.
20. IEC TS 62351 Technical Specification, Power systems management and associated information exchange—Data and communications security—Part 3: Communication network and system security—Profiles including TCP/IP, First edition 2007–06.
21. Internet Engineering Task Force. RFC 2026. MODBUS Application Protocol, 2002.
22. Makhija, J., Comparison of protocols used in remote monitoring: DNP 3.0, IEC 870-5-101 & MODBUS, No. 03307905, pp. 1–19, 2003.
23. Nicholson, A., Webber, S., Dyer, S., Patel, T., and Janicke, H., SCADA security in the light of Cyber-Warfare, *Journal of Computers and Security Archive*, Elsevier, 31(4), 418–436, 2012.
24. Aloul, F., Al-Ali, A.R., Al-Dalky, R., Al-Mardini, M., and El-Hajj, W., Smart grid security: Threats, vulnerabilities, and solutions, *International Journal of Smart Grid Clean Energy Smart*, 1(1), 1–6, 2012.
25. SCP Clearinghouse, Sustainable consumption and production, http://www.scpclearinghouse.org/c/17-energy-efficiency.html, last visited in December 2013.
26. Kempton, W. and Tomić, J., Vehicle-to-grid power fundamentals: Calculating capacity and net revenue, *Journal of Power Sources*, 144(1), 268–279, 2005.
27. Global-EV-Outlook-2015, International Energy Agency (iea). Global EV Outlook 2016. https://www.iea.org/publications/freepublications/publication/Global_EV_Outlook_2016.pdf, last visited in August 2016.
28. Saaty, T.L., *The Analytic Hierarchy Process: Planning, Priority Setting, Resource Allocation*, McGraw-Hill, New York, 1980.
29. Bernardon, D.P., Sperandio, M., Garcia, V.J., Pfitscher, L.L., and Reck, W., Chapter 3: Automatic restoration of power supply in distribution systems by computer-aided technologies, in (ed.) Kongoli, F., *Automation*, InTech, Rijeka, Croatia, pp. 45–60, 2012, ISBN 978-953-51-0685-2.
30. Brown, R.E., *Electric Power Distribution Reliability*, 2nd edn., CRC Press, New York, 2009, ISBN: 978-0-8493-7567-5.
31. Kersting, W.H. and Mendive, D.L., An application of ladder network theory to the solution of three-phase radial load-flow problems, *IEEE Power Engineering Society Winter Meeting*, A76(044-8), 1–6, 1976.

32. Tsai, L., Network reconfiguration to enhance reliability of electric distribution systems, *Electric Power Systems Research*, 27, 135–140, 1993.
33. Baricevic,T., Mihalek, E., Tunjic,A., and Ugarkovic, K., AHP method in prioritizing investments in transition of MV network to 20 kV, *CIRED 2009—20th International Conference on Electricity Distribution*, Prague, Czech Republic, pp. 1–4, June 2009.
34. Saaty, T.L. and Tran, L.T., On the invalidity of fuzzifying numerical judgments in the Analytic Hierarchy Process, *Mathematical and Computer Modelling*, 46, 962–975, 2007.

Index

A

Adaptive neuro-fuzzy inference system (ANFIS), 187
ADC, *see* Digital-analog converter
Advanced Asset Management (AAM), 259
Advanced Distribution Operation (ADO), 259
Advanced Metering Infrastructure (AMI), 260–262
Advanced Meter Management (AMM), 261
Advanced Transmission Operation (ATO), 259
Alternating current (AC) voltmeters
- advantages, 100
- average/effective sinusoidal signals, 101
- environmental conditions, 100
- high-value input resistance R_{in}, 100–101
- signal rectification, 100–101
- sinusoidal full-wave voltage, 101–102
- voltage-to-current conversion, 100–101

AMR, *see* Automated Meter Reading
Analog signal processing
- AGC, 49–50
- crystal oscillators, 47–49
- phase-shift amplifier, 39–41
- relaxation oscillator, 45–47
- Schmitt trigger comparator, 41–42
- signal linearization, 38–39
- VCO/VFC, 45
- Wien oscillator, 42–45

Analog voltmeter, 103–104
Analytic Hierarchy Process (AHP) method
- application, 271–273
- example, 270–271
- quantitative and qualitative criteria, 269

ANFIS, *see* Adaptive neuro-fuzzy inference system
Application-specific integrated circuits (ASIC) technology, 213
Artificial neural networks (ANNs)
- characteristics, 179
- DGSs, 182–183
- industrial systems, 181–182
- renewable energy, 181

ATO, *see* Advanced Transmission Operation
Automated Meter Reading (AMR), 260
Automatic gain control (AGC), 49–50

B

Bipolar transistor (BJT), 97–98
Block diagram (BD), 235–237
Brushless DC motor (BLDC) resolvers, 91–92
Buffer amplifier, 14

C

Capacitive isolation, 32–33
Capacitive sensors
- cylindrical sensor, 71
- displacement, 70
- insertion sensor, 70
- planar capacitors, 69
- rotation sensor, 71

C code, 238
Center of Excellence in Energy and Power Systems (CEESP), 208–209
Charged coupled device, 82
Chopper amplifier, 99–100
Class B push-pull amplifier, 20
Common-mode rejection ratio (CMRR), 3, 6, 85
Computer-based systems
- components, 214
- data transmission lines, 225–226
- frequency counter, 217–218
- input drivers, 217–218
- output drivers, 217–219
- parallel DIPI IEEE488, 215–217
- RS485 network biasing, 226–227
- serial interface IEEE485, 219–220
- tests, 214–215
- USB
  - connections, 221–222
  - enumeration, 223–224
  - examples, 222
  - features, 224
  - hub, 222–223
  - ZIP drives, 221–222

Consistency ratio (CR), 271
Continuous wavelet transform (CWT), 150
Current-controlled oscillator (ICO), 125

D

Data acquisition
- boards, 206–207
- DSP
  - architecture, 203–204
  - CPU data flow and processing, 204
  - data bus, 204–205
  - operations and programmed decisions, 205
  - peripherals, 205–206

example, 208–210
microcontroller, 207–208
Dataflow, 238
DC values, *see* Direct current values
Decade counting units (DCUs), 105–106
Delta-sigma circuit, 53
Derivative amplifier, 12–13
Device under test (DUT), 130, 215
DG, *see* Distributed generation
DGSs, *see* Distributed generation systems
Differential amplifier
A_c calculation, 5
CMRR, 6
current mirror circuit, 6–7
A_d calculation, 4
error equation
bipolar amplifiers, 19–20
drift and offset values, 17–18
Thevenin resistance, 9, 17–18
voltage and current, 18–19
gains, 2–3
parameters, 7–8
programmable gain, 28–30
standard instrumentation, 27–29
voltage feedback, 25–27
Differential instrumentation amplifier, 29
Differential pressure sensors, 88–89
Digital-analog converter (ADC), 150
Digital control
computer, 228
interface functions, 227
output bit status, 227–228
PLCs
application and operation principles, 229
architecture, 228–229
FBD, 229
industrial and automation processes, 230
input/output technology, 230–231
ladder programming language, 229
LOGO SIEMENS controller, 231–233
SFC, 229–230
structured text, 230
Digital network analyzers, 113–115
Digital signal processors (DSPs)
data acquisition
architecture, 203–204
CPU data flow and processing, 204
data bus, 204–205
operations and programmed decisions, 205
peripherals, 205–206
MATLAB, 253–255
signal processors, 210–211
Digital-to-analog (D/A) converter, 29
Digital voltmeters (DVMs)
characteristics, 110
continuous balance voltmeter, 108–109
dual-slope, 106–108
ramp voltmeter, 105
single-slope meter, 105–106
successive approximation type, 109–110
Direct current (DC) values
electronic DC voltmeter, 97–98
electronic voltmeter-ammeter, 98
output voltages, 74
Distributed generation systems (DGSs)
ANNs, 182–183
automonitoring distribution generator, 158–159
conventional rotating generators, 159
current-based, 160
fuzzy logic, 182–183
power quality monitoring, 158–159
PV power system, 157–158
smart inverter, 158
structure of, 159–160
voltage source-based, 160–161
Distributed generators (DGs)
automonitoring inverter
active power filter, 159, 170–171
electronic power side, 160, 171
signal reference, 160, 171
simulation results, 171–173
unity power factor, 168, 171
voltage and current, 159, 171
conventional rotating generators, 159
current-based, 160
FPGA, 243, 245
power quality
harmonic circulation, 164, 166–167
in-phase term, 163, 166
issues, 166
power factor, 168–169
smart inverter, 158, 166
THD, 167–168
P2P streaming technology, 245
PV-based inverter instrumentation
current source-based, 160, 169
maximum power, 158, 169
power reference, 163, 169
simulation results, 169–170
structure of, 160, 169
unity power factor, 168–169
PWM function, 245–247
resources, 242–243
RIO platform, 243
sbRIO, 243–244
signal reference
active and reactive power, 162
equivalent circuit, 162
indirect current command, 161
in-phase term, 163
nonfundamental frequency terms, 164–166
nonlinear expression, 161–162
quadrature terms, 163–164

smart inverter instrumentation
indirect current command, 161, 169
signal reference, 163, 169
simulation results, 170–172
structure of, 160, 169
voltage source-based, 160–161
Drain source (DS) voltage, 49
DSPs, *see* Digital signal processors
DUT, *see* Device under test
DVMs, *see* Digital voltmeters

E

Elastic sensors, 87
Electric vehicle (EV), 265–266
Electronic multimeter, 103–104
Emulator, 119, 123, 131
Equalization loop network, 82

F

Fast Fourier transform (FFT), 150, 152, 197–199
Feedback inverter amplifier, 9–10
Field-effect transistor (FET), 49, 97–98
Field-programmable gate arrays (FPGAs), 213–214, 243, 245
Finite impulse response (FIR), 150
Frequency synthesizer
direct method, 127
indirect method, 123–124
PLLs
coordinate conversion, 126
frequency divider, 125
internal VCO, 125
local and internal signal, 124
operating principle, 124–125
product of functions, 126
VZC detector, 125
Front panel (FP), 236
Function block diagram (FBD), 229
Fuzzy logic
adaptive control, 184
advantages, 180
characteristics, 179
classical control, 183
decoupled *d-q* equations, 184
DGSs, 182–183
disadvantage, 180
discrete event systems, 184
features, 180
function approximation, 197–198
learning/probabilistic reasoning, 180
mathematical approach, 184
neuro-fuzzy techniques, 180–181
nonlinear methods, 183
optimal control, 183
optimization
artificial intelligence, 191, 193
asymmetric triangular membership functions, 191–192
decision variables, 190
FLC-1, FLC-2, and FLC-3, 196–197
flux programming, 190–191
optimal response, 189–190
torque-speed curves, 193
wind energy system, 193–196
PID controllers, 183
PI-like control loops, 185–186
robust control, 183
rule matrix and membership functions, 187
speed control algorithm, 184, 187–188
state-space methods, 183
stochastic control, 184
universe of discourse, 184–186
variables, 184–186

G

Gain-bandwidth product (GBP), 7–8
Geiger-Müller tube (GM) tube, 83–84

H

Hall effect device (HED), 79–81
Harmonic analysis
characteristics, 136
Fourier series, 136–137
harmonic distortion
analog frequency analyzers, 147
feed-forward filter, 149
heterodyne analyzer, 148–149
network parameters, 139–143
power factor, 144–147
RF analyzers, 147
spectrum analyzer, 147
tuned filter, 147–148
switched current signal, 137–138
three-phase systems, 139
train of blocks, 137–138
High-pass (HP) filter, 128
Home Area Network (HAM), 259

I

Induction motor (IM), 208–209
Inductive sensors
infinite cylindrical inductor, 72–73
LVDT, 73–74
output voltages, 74
parameters, 74–75
relative magnetic permeability, 72
reluctance, 72
VDI, 73

Infinite impulse response (IIR), 150
Instrumentation amplifier, 30
  analog signal processing
    AGC, 49–50
    crystal oscillators, 47–49
    phase-shift amplifier, 39–41
    relaxation oscillator, 45–47
    Schmitt trigger comparator, 41–42
    signal linearization, 38–39
    VCO/VFC, 45
    Wien oscillator, 42–45
  capacitive isolation, 32–33
  circuit boards, 57–58
  differential amplifier
    A_c calculation, 5
    CMRR, 6
    current mirror circuit, 6–7
    A_d calculation, 4
    error equation, 9, 17–20
    gains, 2–3
    parameters, 7–8
    programmable gain, 28–30
    standard instrumentation, 27–29
    voltage feedback, 25–27
  large-signal push-pull amplifiers, 20
  magnetic isolation, 30–31
  operational amplifier
    applications, 1
    differential signal block, 2
    digital and analog processing, 9–17
    gain variations, 1
  optical isolation, 31–32
  signal conditioning
    analog-to-digital conversion, 53–57
    attenuator, 49
    digital-to-analog conversion, 51–53
    impedance matching, 50
    NTSC, 50–51
    pulsed signal, 50
  V-to-I
    differential input with grounded load, 37–38
    differential input with ungrounded load, 36
    single input with grounded load, 34–36
    single input with ungrounded load, 34
Intermediate frequency (IF), 147
Interrupt-driven programming, 239

L

Laboratory Virtual Instrumentation Engineering Workbench (LabVIEW)
  distributed generation
    FPGA, 243, 245
    P2P streaming technology, 245
    PWM function, 245–247
    resources, 242–243
    RIO platform, 243
    sbRIO, 243–244
  programming models, 238–239
  virtual instruments
    block diagram, 235–237
    front panel, 236
    icon connector, 236–238
    power quality analysis, 239–242
Large-signal push-pull amplifiers, 20
Least significant (LSB), 109
Light-dependent resistors (LDRs), 66
Linear variable differential transformer (LVDT), 73–74
Low-pass (LP) filter, 128

M

Magnetic isolation, 30–31
MATLAB
  Arduino, 250–253
  DSPs, 253–255
  LabVIEW, 249–250
  PLECS, 248
  PSIM, 252–254
  Simulink® platform, 248–249
Metal-oxide-semiconductor FETs (MOSFETs), 99–100
Meter Data Management (MDM), 261
Model reference adaptive controller (MRAC), 188–189
Modulated by amplitude (MA), 129
Modulated by frequency (MF), 129
Most significant bit (MSB), 55, 109

N

National Instruments (NI) Company, 238
Negative effect of Thomson, 76
Neural network (NN)
  characteristics, 179
  DGSs, 182–183
  function approximation
    applications, 198
    backpropagation, 198–199
    speed vector control, 198–199
    vector control, 199–200
  heuristics and learning, 184
  induction motor speed, 184–185
  industrial systems, 181–182
  nonlinear system, 188–189
  optimal condition, 190
  renewable energy, 181
  wind turbine, 194–195
Neural network inverse model, 188–189
Noninverting amplifier, 13

Normally closed (NC) switches, 271–272
Normally open (TS) tie switches, 271–272
Nuclear magnetic resonance (NMR), 87

O

Ohmmeters
  changing resistance R_{fs}, 111–113
  current and voltage measurements, 110–111
  full-scale adjustment R_{sc}, 112
  improved configuration, 112
1-bit converter, 53
Operational amplifier
  applications, 1
  differential signal block, 2
  digital and analog processing
    buffer amplifier, 14
    derivative amplifier, 12–13
    full-wave precision rectifier, 15–16
    high gains, 16–17
    integration amplifier, 11–12
    inverting amplifier, 9–10
    level detectors/comparators, 14–15
    noninverting amplifier, 13
    precision diodes, 15–16
    summing amplifier, 10–11
  gain variations, 1
Optical isolation, 31–32
Ordinary differential equation (ODE), 238

P

Peer-to-peer (P2P) streaming technology, 245
Phase-locked loops (PLLs), 94
  coordinate conversion, 126
  frequency divider, 125
  internal VCO, 125
  local and internal signal, 124
  operating principle, 124–125
  product of functions, 126
  VZC detector, 125
Photocouplers, 31–32
Photovoltaic (PV) power system
  current source-based, 160, 169
  DGSs, 157–158
  maximum power, 158, 169
  power reference, 163, 169
  simulation results, 169–170
  structure of, 160, 169
  unity power factor, 168–169
Photovoltaic sensors, 78–79
Piecewise linear electrical circuit simulation (PLECS), 248
Piezoelectric sensors, 78
π-type attenuator, 120–121
Platinum, 63–64
Plug-in electric vehicle (PEV), 266
Pneumatic sensors, 87–88
Potentiometer sensors, 66
Power electronic simulator (PSIM), 252–254
Power sources and adapters, 115–116
Probability density function (PDF), 197
Programmable logic controllers (PLCs)
  application and operation principles, 229
  architecture, 228–229
  FBD, 229
  industrial and automation processes, 230
  input/output technology, 230–231
  ladder programming language, 229
  LOGO SIEMENS controller, 231–233
  SFC, 229–230
  structured text, 230
Programmable logic devices (PLDs), 214
Proportional-integral-derivative (PID) controllers, 183
Pulse amplitude modulation (PAM), 31
Pulse density modulation (PDM), 31
Pulse frequency modulation (PFM), 31
Pulse width modulation (PWM), 31, 245–247

R

Radio frequency (RF) analyzers, 147
Reduced instruction set computer (RISC), 247
Resistive sensors
  effect of temperature
    noninvertible input resistor, 68–69
    platinum, 63–64
    potentiometer, 66
    semiconductors, 64
    strain gauge, 64–65, 68
  physical effects, 63
  resistance changes, 63
  temperature compensation
    load cell, 67–68
    pressure measurement, 67–68
    Wheatstone bridge, 66–67
RIO platform, 243
Rotary shaft sensors, 89–90

S

SCTL, *see* Single-cycle timed loop
Semiconductors, 64–65
Sensors
  active electric sensors
    blood sensor alcohol, 86
    GM tube, 83–85
    HED, 79–81
    image intensifiers, 82
    NMR, 87
    photomultipliers, 81–82
    photovoltaic, 78–79
    piezoelectric, 78

        probes, 84–85
        SQUID, 87
        thermocouples, 75–78
    examples, 61–62
    instrumentation chain, 61–62
    mechanical-based sensors
        differential pressure, 88–89
        elastic, 87
        pneumatic, 87–88
        rotary shaft, 89–90
        shaft encoder, 90–95
        turbine, 89
    passive electric sensors
        capacitive sensors, 69–71
        inductive, 72–75
        resistance, 63–69
Sequential function chart (SFC), 229–230
Shaft encoder
    feedback movement sensors, 91–92
    magnetic, 90
    optical sensors, 90–91
    PLL, 94
    ratiometric tracking conversion, 94
    resolver-converters, 92–93, 95
    resolvers, 91
    rotor position/speed, 93–94
Signal generators and simulators
    attenuator
        power reduction, 119–120
        types, 120–121
    audio generators, 131
    frequency division, 128
    frequency synthesizer
        direct method, 127
        indirect method, 123–124
        PLLs, 124–126
    function generators, 131
    MF, 129
    pulse generators and rectangular waves, 131
    sweep generator, 129–130
    waveforms, 122–123
Signal processors
    ASIC technology, 213
    DSPs, 210–211
    FPGA, 213–214
    parallel processing, 211–212
    PLDs, 214
    transputers, 212–213
Silicon-intensified target (SIT), 82
Simulink® platform, 248–249, 253–255
Single-Board RIO (sbRIO), 243–244
Single-cycle timed loop (SCTL), 238
Single slope meter, 105–106
Sketches, 248
Smart Grid
    AAM, 259
    ADO, 259
    AHP method, 271–273
    AMI, 260–262
    ATO, 259
    benefits, 258–259
    characteristics, 258
    constraints, 268–269
    decision-making method, 269–271
    distribution network, 259–260
    electric vehicles, 265–266
    home networks, 259, 262
    objective functions, 268
    power restoration, 266–268
    two-way communication
        IEC 61850, 263–264
        SCADA system, 263–265
        standard IEC 62351, 263
        transmitted information, 262
Smart inverter
    indirect current command, 161, 169
    power quality, 158, 166
    signal reference, 163, 169
    simulation results, 170–172
    structure of, 160, 169
Smart meter, 260
Software
    Arduino, 247–248
    LabVIEW
        distributed generation, 242–247
        power quality analysis, 239–242
        programming models, 238–239
        virtual instruments, 235–238
    MATLAB
        DSPs, 253–255
        interface, 249–253
        PLECS, 248
        Simulink, 248–249
    universal instrument, 235–236
Spectral analysis
    electromagnetic noise, 149
    Fourier transform, 150
    frequency domain, 150–151
    frequency scanning
        dynamic range, 153
        good-quality, 152
        Nyquist-Shannon theorem, 151–152
        real time, 150–151
        resolution limitations, 152
        scanning problem, 152
        virtual time, 150–151
    mathematical analysis, 153–155
    scaling coefficients, 150
    wavelet transform, 150
Statechart programming algorithm, 239
Strain gauges, 65
Successive-approximation-register (SAR), 109
Summing amplifier, 10–11

Superconducting quantum interference device (SQUID), 87
SuperSpeed (SS), 221

T

Textual Math, 238
Thermistors, 64–65
Thermocouples
  equivalent EMF, 76–77
  Peltier effect, 75
  reference temperature, 76–77
  Seebeck effect, 76–77
  Thompson effect, 75–76
  types, 77–78
Total harmonic distortion (THD), 167–168
  active and reactive powers, 141
  active/average power, *P*, 143
  apparent power, 140
  with DC component, 140
  without DC component, 140
  instantaneous active power, p_a, 142–143
  instantaneous nonactive power, p_{na}, 143
  phase difference, 141–142
  power factor, 144–147
  signal distortion analyzers
    analog frequency analyzers, 147
    feed-forward filter, 149
    heterodyne analyzer, 148–149
    RF analyzers, 147
    spectrum, 147
    tuned filter, 147–148
  voltage/current, 139–140
Transducers, *see* Sensors
Transmitted datum (TD), 225
Turbine sensors, 89

U

Uninterruptible power supply (UPS), 160–161
Unity gain frequency, 7–8
Universal Serial Bus (USB)
  connections, 221–222
  enumeration, 223–224
  examples, 222
  features, 224
  hub, 222–223
  ZIP drives, 221–222

V

Variable differential inductor (VDI), 73
Virtual instruments (VIs)
  block diagram, 235–237
  front panel, 236
  icon connector, 236–238
  power quality analysis
    elements, 239
    hardware, 240
    parameters, 241, 243
    producer/consumer architecture, 240–241
    subVI DAQ assistant, 241
    Textual Math programming, 241–242
    voltage, current, and power waveforms, 241–242
Voltage-controlled oscillator (VCO), 45, 124
Voltage-to-current conversion (V-TO-I)
  differential input with grounded load, 37–38
  differential input with ungrounded load, 36
  single input with grounded load, 34–36
  single input with ungrounded load, 34
Voltage-to-frequency conversion (V/f), 53
Voltage-to-frequency converter (VFC), 45
Voltage zero crossings (VZC) detector, 125
Volt-ohm-milliamp (VOM), 112–113

W

Wave distortion meter, 148–149

Z

Zener diode feedback, 14–15